Climate Change Management

Series editor
W. Leal Filho, Hamburg, Germany

More information about this series at http://www.springer.com/series/8740

Walter Leal Filho
Editor

Innovation in Climate Change Adaptation

Springer

Editor
Walter Leal Filho
Hamburg University of Applied Sciences
Faculty of Life Sciences
Hamburg
Germany

ISSN 1610-2010 ISSN 1610-2002 (electronic)
Climate Change Management
ISBN 978-3-319-25812-6 ISBN 978-3-319-25814-0 (eBook)
DOI 10.1007/978-3-319-25814-0

Library of Congress Control Number: 2016930165

Springer Cham Heidelberg New York Dordrecht London

Printed on acid-free paper

Springer International Publishing AG Switzerland is part of Springer Science+Business Media (www.springer.com)

Preface

It is widely believed that climate change is one of the most important challenges of modern times. Its origins are global in scope, in the sense that greenhouse emissions, coupled with unsustainable models of development—most of which based on the depletion of natural resources and ecosystems—lead to disruptions in eco-geological cycles and to substantial changes in weather conditions. These, combined with socio-economic problems such as poverty and limited access to technologies, are leading to losses in agriculture production, reductions in water availability, sea level rise and damages to properties due to extreme events, among others.

There are currently hundreds of projects being undertaken around the world, focusing on matters related to climate change. They include both, schemes focusing on mitigation as well as on adaptation. Global expenditures on climate change are well in excess of US$ 5 billion per year, a substantial proportion of which to fund projects in developing countries. Most of these schemes focus on capital projects on infrastructure, as well as schemes in the field of agriculture, water/rainwater or on research aimed at making fossil energy technologies cleaner and less harmful to the people. This includes photovoltaics, post-combustion capture of $CO2$ from engines or power plants, and direct removal of greenhouse gases from the atmosphere, among others. A number of projects have therefore a technological focus, also based on the premise of agencies such as the Organisation for Economic Cooperation and Development (OECD), which believe that technological change is one of the keys to ensuring that climate change can be addressed without compromising economic growth.

But apart from technology-based initiatives, there is a perceived need to find new, innovative ways to pursue climate change adaptation. We need, in other words, more innovation in climate change, especially in respect of adaptation. The problem is that even though innovation on climate change adaptation has proven successful in a number of ways, and that there are many initiatives which show how effective it can be, there are very few publications which have focused on this topic.

This book fills in the gap. It gathers a set of 25 chapters from authors from various continents, which show how innovation in processes, in planning, use of resources or use of existing technologies, can help to foster climate change adaptation.

Bearing in mind the broad field of innovation on climate change, this book is divided into two parts. Part I, with chapters 1–11, deals with a wide range of innovation issues related to climate change adaptation, in different contexts, such as planning, reforms, technology and transformative processes. It also considers the environmental, social and economic elements which are associated with the implementation of some innovative approaches to climate change adaptation here documented, with examples from various parts of the world.

Part II, with chapters 13–25, focuses on innovation in sectorial approaches to adaptation, such as in agriculture and city planning. It also includes the health sector, an area which is often overlooked during discussions on climate change adaptation plans. Moreover, changes in organisations are also outlined in one of the chapters, a pre-condition for successful institutional climate change adaptation plans. Similarly to Part I, the second part of the book also contains examples and case studies from a wide range of countries and contexts.

Among other issues, one of the main lessons from this book is that innovative approaches to climate change adaptation also need to consider the acceptability issue, which varies substantially among countries and social groups and is influenced by various social and cultural factors. We need a greater willingness to adapt and to be creative in finding unconventional ways to use innovation to the advantage of climate change adaptation. Indeed, if innovation on climate change adaptation is to become more common and more widely practised, knowledge and communication gaps between citizens and policy makers also need to be reduced.

I want to thank all authors for sharing their know-how, as well as for the support they provided in producing this book. I hope this publication will provide a valuable support to international climate change adaptation efforts and will foster the use of innovative approaches round the world.

Hamburg, Germany
Winter 2015/2016

Walter Leal Filho

Contents

Part I
Innovation in Planning, Reforms, Technology and Transformative Processes

Chapter 1
Innovative Approaches to Climate Change Adaptation

Walter Leal Filho

Abstract The process of climate change adaptation is characterised by a great deal of complexity. Its successful implementation may only be achieved by a combination of a wide range of approaches, methods and processes. Climate change adaptation also needs innovation. Based on the perceived need to explore the links between climate change adaptation and innovation, this paper defines how innovation can support climate change adaptation, and suggest a variety of approaches which may help to realise its potential.

Keywords Climate change • Innovation • Adaptation • Urban environment

Introduction

Thanks partly to the wide body of research and data deriving from many studies on the origins and consequences of climate change, the theme as a whole and its impacts in particular, has become very prominent. Indeed, as shown in Table 1.1, there has been a noticeable increase in the levels of emphasis and awareness about climate change over the past 25 years, which allows one to conclude that the topic has evolved considerably since 1990. Indeed, the current level of international interest afforded to this topic and the high levels of expenditures on climate change mitigation and adaptation allows one to conclude that climate change has become one of the most prominent scientific themes of modern times.

In addition to the increases in temperatures, which most people associate with climate change, some of the major challenges seen in this field include sea level rise, increases in the frequency of extreme events such as intense rainstorms, floods,

W. Leal Filho (✉)
School of Science and the Environment, Manchester Metropolitan University, Chester Street, Manchester M1 5GD, UK

Faculty of Life Sciences, Hamburg University of Applied Sciences, Ulmentliet 20, 21033 Hamburg, Germany
e-mail: walter.leal@haw-hamburg.de

W. Leal Filho (ed.), *Innovation in Climate Change Adaptation*, Climate Change Management, DOI 10.1007/978-3-319-25814-0_1

Table 1.1 Increases in the emphasis given to climate change: 1990–2015

Items	Stand in 1990	Stand in 2015
Level of government interest	Low	High
Level of government commitment to climate change goals	Low (UNFCCC was approved at UNCED in Rio in 1992)	High
Levels of expenditures on climate change projects	Low, little statistics available	High, many statistics available
Level of public interest	Very low	High
Media coverage	Sporadic	Intensive
Engagement from developing countries	Low	Very high

droughts, and heat waves (Local Governments for Sustainability et al. 2013; Jabareen 2013; European Commission 2013c), among others. Impacts may also be seen in areas as varied as ecosystems and biodiversity, agriculture and forestry, water resources, marine resources and fisheries, which have been the subject of much research. In addition, they also affect at different levels of intensity areas such as:

(a) Tourism
(b) Energy supply
(c) The built environment and infrastructure
(d) Human health
(e) Land management
(f) Regional planning and
(g) Insurance services (ETAP 2014)

It is here appropriate to remind that despite the wide range of affected areas, most people tend to associate climate change predominantly with increases in temperature. An example of this trend is the fact that a study performed in a number of European cities reported that 81 % of surveyed cities have experienced periods of very hot weather or heat waves, 78 %—flooding from heavy rainfall, 69 %—storms (Perks 2013).

In some occasions, climate change may lead to semi-permanent or permanent damages to economic activities and livelihood (such as in agriculture). In other areas—such as tourism—climate change may also yield positive impacts, since moderate temperature increases in temperate countries may help in the development of more tourism activities. Figure 1.1 illustrates the interrelations between these two (i.e. positive and negative) examples of climate change impacts from an economic standpoint.

Over and above the direct impacts of climate change, many communities face its indirect effects such as increased strains on materials and equipment, higher peak electricity loads (and their timing), voltage fluctuations, transport disruptions, increased need for emergency management, diminished air quality, and sub-optimal performance of key infrastructure (IPCC 2014; NPCC 2009, Wardekker et al. 2003 in Jabareen 2013; Perks 2013). Therefore, they are required to be more resilient and prepared to address the threats head-on, otherwise many

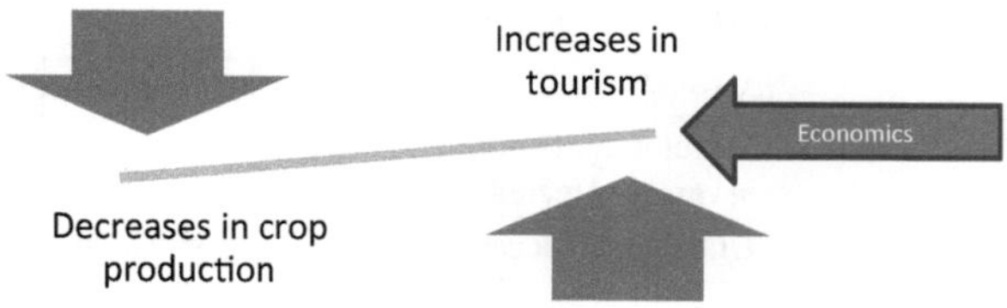

Fig. 1.1 Interrelations of climate change impacts

communities—especially in the developing world—are more and more likely to become even more vulnerable to risks (UNISDR 2010 in Jabareen 2013).

The most common approaches to address climate change impacts are mitigation and adaptation, whereas adaptation measures are considered as an indispensable complement to climate mitigation strategies (European Commission 2013b).

Adaptation may be defined as an adjustment in natural or human systems—in both urban and rural areas—in response to actual or expected climatic stimuli or their effects. This adjustment is needed in order to moderate (or alleviate) the potential negative consequences of climate change, or exploits beneficial opportunities (such as the example provided in the tourism sector). Climate change adaptation increases the ability of individuals, groups, or organizations to cope with current and future changes (Perks 2013; European Commission and European Environmental Agency 2015; Kazmierczak and Carter 2010). Adaptation approaches may include, for instance, the development or adoption of a technology and/or capacity building in form of improved risk management or knowledge enhancement (West and Gawith 2005 in Tompkins and Eakin 2012).

The development of an adaptation strategy that defines the main stages required to be fully implemented, is one of the initial steps in pursuing it. It is important to note that adaptation strategies are influenced by a variety of factors, which are illustrated in Table 1.2.

Other factors could be added to the list summarised in Table 1.1, whereas a deeper look at specific issues such as the limited emphasis to research on climate change adaptation in the countries which needs it most, is also needed. Also, the importance of well prepared climate change adaptation strategies cannot be emphasised often enough. Paradoxically, despite their relevance and the key role they can play, there are either non-existing or—when they do—not properly implemented.

One of the examples of an adaptation strategy for a whole region is a document produced by the European Commission, namely the "EU Strategy on Adaptation to Climate Change". It aims to make Europe more climate-resilient (European Commission 2015). In particular, the document highlights the importance of implementing adaptation measures at city level (European Commission 2013c). Among the main actions of the EU Adaptation Strategy are:

- Bridging the knowledge gap,
- Facilitating the climate-proofing of the Common Agricultural Policy (CAP), and

Table 1.2 Some of the factors which influence adaptation strategies

Factors	Variable	Consequences
Geographical location	Developing countries more severely affected	More intensive climate change impacts
Access to financing	Unbalanced access	Some countries better able to finance adaptation than others
Access to technologies	Unbalanced access	Some countries better able to use innovation than others
Government climate change policies	When absent, lack of targets or a basis for long-term action	No guidance as to what needs to be done and where
Public awareness about climate change and its impacts	Lack of awareness perpetuates some practices	Increased vulnerability (e.g. buildings in river catchment area)
Knowledge basis	When low, limited capacity to adapt	Greater vulnerability and lower resilience

- Ensuring more resilient infrastructure and promoting insurance and other financial products for resilient investment and business decisions (European Commission 2013d).

But despite the progresses seen in the past, many needs regarding the development and the implementation of climate change adaptation remain today. Therefore, innovative ways to look at the problem and at its various variables, are needed. The subsequent parts of this paper will thus look at how innovation can assist climate change adaptation, focusing on some key areas.

Innovation in Climate Change Adaptation

Innovation can be defined as a process via which specific developments (or changes) in ways to handle a process, an issue or a problem, may take place. Innovation processes differ from others since they are continuous and tend to evolve. They sometimes—but not always—are connected with technologies. Indeed, much innovation is process-based.

In the particular case of climate change adaptation, innovation may have a technical/technological and a non-technical dimension. Technological innovation is on the one hand expensive and requires investments many countries are not promptly willing to make, but it could lower the cost of achieving environmental objectives, on the other. In cases where the estimated costs of reducing greenhouse gas emissions are influenced by the technological trajectory of a country's economy, technological innovation can offer the chance to leap frog from a given problem (for instance, dependence on coal fired power stations), going straight to its technical solution (e.g. use of hydroenergy to generate energy and to reduce emissions).

Due to their closeness, technological and non-technological adaptation options should be seen in a combined matter, both being therefore the emphasis of this chapter. The subsequent description will be divided into three different categories, all of which focusing on infra-structure: **soft non-structural or 'social' infrastructure approaches, grey and green infrastructures.**

Soft Non-structural or 'Social' Infrastructure Approach

The soft non-structural or 'social' infrastructure approach corresponds to the design and implementation of procedures, and employing, inter alia, land-use controls, information dissemination and economic incentives, to:

1. Reduce vulnerability
2. Encourage adaptive behaviour and
3. Avoid maladaptation

Some of these measures can facilitate the implementation of grey or green approaches (Local Governments for Sustainability et al. 2013; ACT—Adapting to Climate change in Time 2014).

Among the measures being applied at present, mentioned can be made to awareness raising campaigns. One example is the campaign. "The Netherlands Live with Water" (Kazmierczak and Carter 2010). Here the innovation is seen in respect of the active engagement of the relevant stakeholders (e.g. public, business), coupled with gathering data and information and defining baselines, which can be used for the development of climate change action plans, policies and standards.

In terms of policies and standards, **Eurocodes** can be mentioned as an example. It is a set of European Standards (EN) for the structural design of buildings and civil engineering works, produced by the European Committee for Standardisation (CEN) to be used in the European Union. They provide the requirements for mechanical strength, stability and safety as basis for design and engineering contract specifications (European Commission 2013a). Taking into account the damages climate can lead to both property and infra-structure, such standards can be very helpful. And so can tools such as

Insurance. It support adaptive practices by helping to manage climate change risks, providing incentives for climate risk prevention and disseminating information on climate change risks and risk prevention measures (European Commission 2013a). Moreover, **building codes and zoning** could directly regulate housing and indirectly influence housing markets through transportation and infrastructure planning and investment (World Bank 2011), also bearing in mind possible influences of climate change.

One of the examples of soft approaches implementation is development of proactive **heat alert system** in Toronto, Canada. The Heat-Health Alert system relies on computer modelling of various weather factors, including apparent temperature (a measure of human discomfort due to combined heat and humidity), cloud cover, wind direction and speed, and air mass. The model predicts when the probability of excess mortality due to certain oppressive air masses rises above expected thresholds. When the conditions are such that it could potentially rise above 65 %, the Toronto Medical Officer of Health issues a Heat Alert. When the probability rises above 90 %, an Extreme-Heat Alert is issued (Mehdi et al. 2006).

Generally speaking, the adaptation of new and existing infrastructures to a changing climate could be implemented in two main ways: **grey and green.**

Grey Infrastructure

The grey infrastructure approach corresponds to physical interventions or construction measures using engineering services to make buildings and infrastructure more compatible with the social and economic needs (and the well-being) of society, being more capable of withstanding extreme events and to avoid infrastructure lock-ins that may provide little to no adaptive capabilities in the future (Local Governments for Sustainability et al. 2013; ACT—Adapting to Climate change in Time 2014).

This issue is a matter of great concern, because of the large share of existing physical infrastructure that has been planned and built without any consideration to climate change or to projected climate impacts (Local Governments for Sustainability et al. 2013).

Engineering measures such as additional cooling circuits for power plants or design standards for distribution poles can significantly increase the robustness and reliability of installations (European Commission 2013a). **Building insulation** allows energy-saving for heating and maintains lower temperatures in hot periods (Perks 2013).

The growth of urban population and its impacts on habitats and ecosystems require considering of alternative in planning physical form of a city. Experts suggest a number of criteria of evaluation, which are helpful in evaluating plans from the perspective of eco-form:

- **Compactness**: Use of urban land more efficiently by increasing the density of development and activity (Jenks 2000 in Jabareen 2013). Compact urban space can minimize the need to transport energy, materials, products, and people (Elkin, McLaren and Hillman 1991 in Jabareen 2013).
- **Mixed land uses**: The diversity of functional land uses, such as residential, commercial, industrial, institutional, and transportation leads to decrease of the travel distance between activities, encouraging walking and cycling, and reducing the need for car travel (Parker 1994, Alberti 2000, Van and Senior 2000, Thorne and Filmer-Sankey 2003 in Jabareen 2013).
- **Passive solar design**: Orientation, layout, landscaping, building design, urban materials, surface finish, vegetation, and bodies of water facilitate the optimum use of solar gain and microclimatic conditions and reduces the need for the

heating and cooling of buildings by means of conventional energy sources (Owens, 1992, Thomas 2003, Yanns 1998 in Jabareen 2013).

- **The retrofit of homes and buildings**, such as addition of light-coloured/white roofs, highly reflective surfaces or materials, or sun shading to alleviate heat, water storage space and smart ventilation, could reduce energy demand for cooling or heating (World Bank 2011).
- **Pervious surfaces** help to make transportation more resilient, by decreasing both ponding and runoff during rainstorms.

Research has shown that pervious pavements can lead to a reduced need for road salt application on streets in the winter by as much as 75 %, as well as a reduction in road noise of 10 dB (CNT 2010 and Schwartz 2010 in Foster et al. 2011 in World Bank 2011).

For instance, the **City of Toronto** in the frame of their adaptation strategy expands the implementation of sustainable urban drainage systems including permeable pavements (Kazmierczak and Carter 2010).

Green/Green–Blue Infrastructure and Ecosystem Services

Green infrastructure or **Green blue infrastructure** contributes to the increase of ecosystems resilience and reduction of biodiversity loss and degradation of ecosystem, and restoration of water cycles. At the same time, green infrastructure uses the functions and services provided by the ecosystems to achieve a more cost effective and sometimes more feasible adaptation solution than grey infrastructure (Local Governments for Sustainability et al. 2013; ACT—Adapting to Climate change in Time 2014).

Green and blue infrastructure is interconnected networks of open spaces and natural areas, such as forests, extensive grasslands, rivers, wetlands, parks, gardens, green walls, and roofs, greenways, water streams and canals. Such infrastructure enables ecosystem services like flood protection, temperature regulation, filtering of air and providing recreation areas (Local Governments for Sustainability et al. 2013).

Natural vegetation provides a carbon sink, lowering temperatures in the city, reduces flooding risk, and improves water quality (Center for Neighborhood Technology 2011 in World Bank 2011; EEA 2012 in Perks 2013). Planting real trees makes a genuine impact. Planting 100 m^2 of trees in a city can help reduce the temperature by 1 °C. Green surfaces are 20 °C cooler than artificial (ETAP 2014).

For instance, the **city of Bologna, Italy**, has developed a public-private partnership to promote urban forestry, with the intention of sequestering carbon dioxide, mitigating the urban heat island effect, and reducing air pollution (ICLEI 2011b in World Bank 2011).

In addition, the city of **Rio de Janeiro** in **Brazil**, has the world's largest urban forest (Tijuca Forest), which, with 3.953 ha, occupies around 3.5 % of the city's territory. The forest, which has the status of a national park, encompasses 328 animal and 1619 plant species, and is known to help to regulate the climatic balance of the city.

In **Taiwan**, a strategic wetland conservation greenway along the west coast provides valuable habitat and diverse ecosystem services, including flood protection (Hsieh et al. 2004 in World Bank 2011).

Approximately 780,000 ha (20 %) of Austrian forests are classified as 'protection' forests ("Bann- und Schutzwälder"). The primary benefits of these mountainous forests are protection from natural disasters (such as avalanches, mudslides, rock falls), biodiversity enhancement, recreation and tourism (European Commission 2013a).

Another measure is the **green roofs** approach (e.g. planted trees and shrubs, or grass on rooftops), which creates shade, reflects heat and hence help to reduce the impacts of heat in cities. Non-vegetated roofs can exceed temperatures of 50 °C, while vegetated roofs remain at 25 °C. The green roofs also help cool the urban core, as well as reduce the impact of heavy precipitation, beautify the area, improve air quality, and reduce energy costs (Mehdi et al. 2006). The plants and the soil on green roofs capture 50–80 % of annual rain, improve air quality, increase vegetation and reduce urban heat islands (European Commission 2014). Green roods provide habitats to support wildlife. Moreover, building owners benefit as green roofs last twice as long as conventional roofs, insulate the buildings, reduce infrastructure costs and increase property value (ETAP 2014).

One of the cities where these approaches are successfully applied is **Copenhagen**. According to the city mandatory green roof policy, all roofs with a pitch under 30° are to be landscaped, it also includes refurbishment of older roofs (Ansel and Appl 2011; Livingroofs.org 2014). In 2014 **Rotterdam** has over 185,000 m^2 of Green Roofs, by the year 2030 the city plans to achieve 800,000 m^2 (Geisler et al. 2014).

The greening of facades of buildings is also known to help to reduce temperatures or humidity changes, naturally cools down temperatures during the summer, and helps to keep warmth in winter. It also isolates from noise, provides clean and oxygenated air, stores humidity and protects from extreme weather events. Green facades are a natural habitat for fauna and flora, have a positive impact on the local microclimate and are actively contributing to the conservation of the environment and nature and to reducing operational costs in the long run (European Commission 2013a).

The **City of Toronto** in its adaptation strategy emphasises actions focusing on use of green and blue spaces, such as increase of tree cover, provision of green roofs, or actions aiming at improved plant health. The city increases a number of storm water retention ponds, creates and preserves green spaces in low-lying areas for flood management and increase shoreline buffers to protect against increased runoff from more intense storms (Kazmierczak and Carter 2010).

The city of **Stuttgart** in Germany has been planning to exploit the role of natural wind patterns and dense vegetation in reducing problems of overheating and air pollution. A Climate Atlas was developed for the Stuttgart region, presenting the distribution of temperature and cold air flows according to the city's topography and land use. Based on this information, a number of planning and zoning regulations are recommended which aim to preserve open space and increase the presence of vegetation in densely built-up areas (Kazmierczak and Carter 2010).

The city of **Almada** in Portugal, has undertaken efforts to protect the site "Fonte da Telha" which faces the impacts of demographic pressure and climate change, and is prone to sea flooding and heavy impacts from storms. As adaptation measures, the municipality decided to reconstitute the dune system and replant local plant species that will conserve and enhance biodiversity and allow environmental conservation (Local Governments for Sustainability et al. 2013).

Further examples of innovation in climate change adaptation are now provided in the areas of water management and food security.

Water Management

Innovation approaches for climate change adaptation in water management are mainly based on sophisticated technological instruments and processes, among which mention can be made to desalination of sea water, or the improvement of water-use efficiency by recycling water or physical improvements (e.g. pipe retrofits) (IPCC 2014; Danilenko 2010, ADB 2006 in World Bank 2011).

To protect vulnerable areas, municipalities can adapt public areas in the way that they will store the rainwater, for instance by using intelligent street profiles, another possibility, construction of underground water storage in combination with other techniques is useful in places with limited space (Vertalingen and Fox 2013).

In **Spain**, many cities apply advanced technologies to produce water through desalination, and recycling and reusing water through its drinking-water and advances wastewater treatment plants to tackle problems related to water scarcity and drought and climate change to increase water availability up to the level requires in 2007 (ETAP 2014).

Copenhagen improved its rain-water management by developing a separation of sewage and rain water drainage, innovations in sustainable drainage systems and clearing road water run off by removing oil and heavy metals (ETAP 2014).

Among the adaptation 'water' projects implemented by **Rotterdam are** 5000 m^3 of additional water storage space in Tjalklaan, the water square at Bellamyplein and the water square Kleinpolderplein, along with 10,000 m^3 of underground water storage under Museumpark car park and Eendragtspolder rowing course. This is combined with 4 million m^3 of a water storage facility with recreation area and (top) sports facility (Cities Climate Leadership Group (C40) and Connecting Delta Cities (CDC) 2015).

An individual household can apply adaptation approaches on a small scale, by insulating their houses, installing double glazing (Posthumus et al. 2008, Johnson and Priest 2008, Howgate and Kenyon 2009, Erdlenbruch et al. 2009 in Tompkins and Eakin 2012), by paving their gardens, or by covering them in wooden decking instead of traditional grassy lawns (Tompkins and Eakin 2012). They also can adopt a **rainwater harvesting system**, to reduce pressure on national water infrastructure, or shared groundwater resources. Or, an individual may choose to invest their own time to clear a public drain outside their own home to minimise flood risk to their own property (Reidsma et al. 2010).

As an example of the strategic value and viability of rainwater harvesting systems, mention can be made to the project Baltic Flows, funded as part of the European Commission's seventh Framework Programme.

The Baltic Flows project shall lay the foundation for development of new capacities and policies for effectively monitoring and managing the quality and quantities of rainwater in the Baltic Sea Region. The project focuses on streams, rivers and cities in Baltic Sea catchment areas, not on the sea itself. The strategies, knowledge and expertise created during the project can be exploited elsewhere in the European Union and in other global regions. The project will support the development of research-driven clusters in each region; enhanced capacities in diffuse load monitoring and urban stormwater management will lead to new business opportunities in the global market for water monitoring and management know-how and solutions.

In addition, the project AFRHINET (ACP-EU Technology-Transfer Network on Rainwater Harvesting Irrigation Management for Sustainable Dryland Agriculture, Food Security and Poverty Alleviation in sub-Saharan Africa) offers another example of an innovative climate change adaptation project on rainwater harvesting, with a focus on irrigation for agriculture. It is funded as part of the European Union's ACP Programme. Through the implementation of integrated theoretical and practical capacity-building, and the development of technology-transfer and demonstration projects in the field of rainwater harvesting irrigation (RWHI), the knowledge and use of RWHI management for small-scale irrigation will be enhanced in rural dryland areas of sub-Saharan Africa. In addition, through the development of research and technology-transfer centres, and a transnational network, a platform for co-operation and the exchange of experience in RWHI management will be created. The project network comprises micro-enterprises, non-governmental and public actors, academic/scientific institutions, and rural dryland local communities, especially farmers, women and youth groups.

Food Security: Urban Farming/Agriculture

Urban areas typically produce very little of their own food and rely on very often long-distance transportation food supplies. **Urban farming** commonly referred to as urban agriculture, is one of the ways to reduce 'food-miles' and increase food security. The approach includes family food gardens, school gardens, and garden plots, which are located throughout the city (Specht et al. 2015). It provides new green spaces, control of runoff and provision of shade that offsets the heat of the concrete city (Dimitri et al. 2015).

Cities can also choose to set aside municipal land for urban agriculture through zoning and regulation. By locating city-owned plots near poor areas and giving priority over the plots to those most in need, city governments can also help to build the resilience of vulnerable populations (Henriques 2009 in World Bank 2011).

Apart from the reduction of distances in transporting food, other benefits of urban agriculture include carbon sequestration, potentially reduced urban heat island effect, improved physical and mental health, improved aesthetics, community building, employment opportunities, improved local land prices, provision of habitat for wildlife and waste recycling (Brown and Jameton 2000, Slater 2001, Twiss et al. 2003, Hynes and Howe 2004, Pearson et al. 2010 in Mok et al. 2014).

According to official statistics, 3000 ha (3 % of **Berlin**'s area) fall under the official land use code of an allotment garden ("Schrebergarten") (Specht et al. 2015).

Conclusions

There is a wide acceptance of the fact that climate change is no longer merely a theoretical scenario. It is already in progress, and its effects are experienced by communities in many locations, across the world. In the coming decades, despite all efforts and many achievements in climate mitigation, the challenges involved with adaptation to climate change will grow (European Commission 2013b). By means of innovative adaptation measures, the adverse impacts of climate change on natural, social, and economic systems may be reduced—or even avoided in some occasions—thus minimising damages and costs. But as climate change progresses, the opportunities for successful adaptation may shrink and the costs associated with adaptation may will increase. Because of the long life-spans of new construction projects, buildings, and infrastructure, it is important to take future climate change into account now, in the planning and development stages (European Commission 2013b).

New technologies and innovative approaches could play a more significant role in the adaptation process, by improving resource efficiency, reducing costs and improving aesthetics, among others. And since many of the technological and infrastructure knowledge already exists, it is a question of putting the available experiences into practice. Better transnational cooperation, and the joint execution of projects may be useful in allowing experiences to be exchanged and in ensuring that tested and proven experiences—as well as best practice—may be more widely disseminated.

References

Adapting to Climate change in Time (ACT) (2014) Planning for adaptation to climate change. Guidelines for municipalities. Available at: http://www.actlife.eu/medias/306-guidelines versionefinale20.pdf

Ansel W, Appl R (2011) Green roof policies – an international review of current practices and future trends. Nürtingen, Germany. Available at: http://www.igra-world.com/images/news_and_events/IGRA-Green-Roof-Policies.pdf
Cities Climate Leadership Group (C40) and Connecting Delta Cities (CDC) (2015) Rotterdam climate change adaptation strategy. Available at: http://www.deltacities.com/cities/rotterdam/climate-change-adaptation. Accessed 25 Apr 2015
Dimitri C, Oberholtzer L, Pressman A (2015) The promises of farming in the city: introduction to the urban agriculture themed issue. Renew Agric Food Syst 30(1):1–2, Available at: http://journals.cambridge.org/action/displayAbstract?fromPage=online&aid=9530308&fileId=S174217051400043X
ETAP (2014) Adapting to climate change through eco-innovation. Available at: http://ec.europa.eu/environment/archives/ecoinnovation2009/2nd_forum/pdf/report_copenhagen.pdf
European Commission (2013a) Adapting infrastructure to climate change. Available at: http://ec.europa.eu/clima/policies/adaptation/what/docs/swd_2013_137_en.pdf
European Commission (2013b) Guidelines on developing adaptation strategies. Available at: http://ec.europa.eu/clima/policies/adaptation/what/docs/swd_2013_134_en.pdf
European Commission (2013c) Launch event: EU strategy on adaptation to climate change. Available at: http://ec.europa.eu/clima/events/articles/0069_en.htm. Accessed 24 Apr 2015
European Commission (2013d) The EU Strategy on adaptation to climate change. Available at: http://ec.europa.eu/clima/publications/docs/eu_strategy_en.pdf
European Commission (2014) Copenhagen is European Green Capital 2014. Available at: http://ec.europa.eu/environment/europeangreencapital/wp-content/uploads/2012/07/ENV-13-004_Copenhagen_EN_final_webres.pdf
European Commission (2015) What is the EU doing? Adaptation to climate change. Available at: http://ec.europa.eu/clima/policies/adaptation/what/index_en.htm. Accessed 25 Apr 2015
European Commission and European Environmental Agency (2015) Glossary. Available at: http://climate-adapt.eea.europa.eu/glossary#linkAdaptation. Accessed 25 Apr 2015
Geisler L, Lange K, Schoor E (2014) Water governance assessment of the green roof policy in Rotterdam. Available at: http://ucwosl.rebo.uu.nl/wp-content/uploads/2014/10/Group-Paper-Law-Green-roofs-Geisler-Lange-Schoor.pdf
IPCC (2014) Climate Change (2014) Synthesis Report. Contribution of Working Groups I, II and III to the 5th assessment report of the intergovernmental panel on climate change. IPCC, Geneva
Jabareen Y (2013) Planning the resilient city: concepts and strategies for coping with climate change and environmental risk. Cities 31:220–229. doi:10.1016/j.cities.2012.05.004
Kazmierczak A, Carter J (2010) Adaptation to climate change using green and blue infrastructure. A database of case studies. Available at: http://www.grabs-eu.org/membersArea/files/Database_Final_no_hyperlinks.pdf
Livingroofs.org (2014) Copenhagen and its green roof ambitions. Copenhagen's Green Roof Ambitions. Available at: http://livingroofs.org/copenhagen-green-roofs. Accessed 25 Apr 2014
Local Governments for Sustainability, European Secretariat (ICLEI) and CEPS (Centre for European Policy) (2013) Climate change adaptation: empowerment of local and regional authorities, with a focus on their involvement in monitoring and policy design. Available at: http://cor.europa.eu/en/documentation/studies/Documents/climate-change-adaptation.pdf
Mehdi B, Mrena C, Douglas A (2006) Adapting to climate change an introduction for canadian municipalities. Available at: http://www2.gnb.ca/content/dam/gnb/Departments/env/pdf/Climate-Climatiques/AdaptingClimateChangeIntroduction.pdf
Mok HF et al (2014) Strawberry fields forever? Urban agriculture in developed countries: a review. Agron Sustain Dev 34(1):21–43. Available at: http://download-v2.springer.com/static/pdf/881/art:10.1007/s13593-013-0156-7.pdf?token2=exp=1430166529~acl=/static/pdf/881/art%3A10.1007%2Fs13593-013-0156-7.pdf*~hmac=a7dbd6c9792c2caf4a3b2b41f90ce9b752f713b63efa088ffa3ceb55cfa7388d

Perks J (2013) Adaptation strategies for European cities. Available at: http://climate-adapt.eea.europa.eu/documents/18/11155975/Adaptation_Strategies_for_European_Cities_Final_Report.pdf
Reidsma P et al (2010) Adaptation to climate change and climate variability in European agriculture: the importance of farm level responses. Eur J Agron 32(1):91–102
Specht K et al (2015) Zero-acreage farming in the city of Berlin: an aggregated stakeholder perspective on potential benefits and challenges. Sustainability 7(4):4511–4523, Available at: http://www.mdpi.com/2071-1050/7/4/4511/htm
Tompkins EL, Eakin H (2012) Managing private and public adaptation to climate change. Glob Environ Chang 22(1):3–11. doi:10.1016/j.gloenvcha.2011.09.010
Vertalingen F, Fox R (2013) Rotterdam climate change adaptation strategy. Rotterdam, Netherlands. Available at: http://www.deltacities.com/documents/20121210_RAS_EN_lr_versie_4.pdf. Accessed 25 Apr 2015
World Bank (2011) Guide to climate change adaptation in cities. Available at: http://www.ppiaf.org/sites/ppiaf.org/files/publication/Urban_Handbook_Final.pdf

Chapter 2
Adaptation Planning Process and Government Adaptation Architecture Support Regional Action on Climate Change in New South Wales, Australia

Brent Jacobs, Christopher Lee, Storm Watson, Suzanne Dunford, and Aaron Coutts-Smith

Abstract This paper reports progress of the Government of New South Wales (NSW), Australia, in implementing climate adaptation responses through the establishment of an effective adaptation architecture and incorporation of the elements of best practice adaptation policy development. Ideally, adaptation policy development should be grounded in practice; support adaptation processes that reduce social and environmental vulnerability; account for short-term variations and longer-term changes in climate; recognise the importance of scale from the local to the global; be assessed in the context of human development; and, employ participatory processes throughout its formulation and implementation.

At the centre of the NSW Government's approach, Enabling Regional Adaptation (ERA) is an on-going, multi-region, stakeholder-led process designed to inform local and regional adaptation planning and action. ERA consists of several phases that include: integrated assessment of vulnerability at regional scale (climate and socio-economic profiling, impact pathways development, adaptive capacity assessment and identification of collective actions); development of strategic adaptation pathways, change models and process benchmarking; and, place-based dialogue on transformational adaptation with local stakeholders. ERA is supported by an adaptation architecture that includes: regional capacity building, enhancement of social capital, knowledge dissemination, research partnerships and dedicated funding. Since 2010, the project has engaged 720 regional decision-makers through

B. Jacobs (✉)
Institute for Sustainable Futures, University of Technology Sydney, PO Box 123, Broadway, NSW 2007, Australia
e-mail: brent.jacobs@uts.edu.au

C. Lee • S. Watson • S. Dunford • A. Coutts-Smith
Impacts and Adaptation, NSW Office of Environment and Heritage, PO Box A290, Sydney South, NSW 1232, Australia
e-mail: christopher.lee@environment.nsw.gov.au

W. Leal Filho (ed.), *Innovation in Climate Change Adaptation*, Climate Change Management, DOI 10.1007/978-3-319-25814-0_2

33 participatory workshops and assessed adaptation in five NSW planning regions covering 75 % of the State's population and 64 % of Local Government Areas.

Keywords Climate change • Regional adaptation • Government policy • Best practice

Introduction

In NSW, Australia the State Government is implementing an on-going, multi-region, stakeholder-led process designed to inform local and regional adaptation planning. The process, named Enabling Regional Adaptation (ERA), seeks to incorporate the elements of best practice in policy development and establish an adaptation architecture that is robust, inclusive and participatory across policy silos and tiers of government.

NSW occupies a land area exceeding 800,000 km^2 with a coast line of about 2000 km in length (Department of Water Resources 1994). While considered to have a temperate climate, it varies considerably depending on proximity to the coast and mountains (EPA 1997). NSW contains a diverse range of landscapes from a narrow coastal plain in the east, [where up to 80 % of the State's population of approximately 7.4 million people live, primarily in the global city of Sydney, ABS (2015)], to a sparsely-populated, broad riverine plain in the west. Its vegetation types range from rainforests in the north east to alpine forests in the south east and semi-arid rangelands in the west (EPA 1997). The State's size makes a summary of future climate difficult although, broadly, increases in temperature and changes in rainfall amount, intensity and seasonality are expected (OEH 2015). These climatic changes, often experienced through the impacts of extreme climate events (wild fires, riverine flooding, droughts), are expected to drive adaptation in the State's human settlements and the natural environments, built infrastructure, government services and economic activity that support them (Jacobs and Boronyak-Vasco 2014; Jacobs et al. 2014a, b). A regional approach to adaptation planning for climate change in NSW is essential to account for the diversity of contexts in which government operates (Fig. 2.1).

Government policy on climate change adaptation sits at the intersection of strategic planning by a range of actors (local government, businesses, state and national agencies) on diverse issues (land use, economic development, natural resource sustainability) and autonomous adaptation by individuals, households and communities (Jacobs et al. 2014b). Primarily, government's role is to provide the correct legal, regulatory and socio-economic environment to support autonomous adaptation (Fankhauser et al. 1999). However, this role should be performed flexibly so that it does not constrain adaptive responses but still ensures that planetary boundaries are not breached and the socially vulnerable are protected (Leach et al. 2012). In addition, the likelihood that adaptation pathways leading to

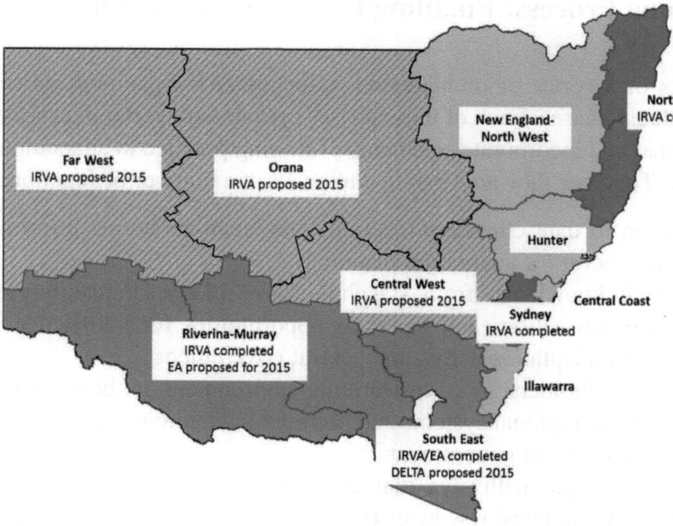

Fig. 2.1 Map illustrating the geographical extent of the ERA process. *Source*: authors

transformation may be essential for survival should be embraced, which is often an uncomfortable space for government (Pelling 2011; Leach et al. 2010).

Brooks and Adger (2005) described the attributes of ideal practice in adaptation policy development. Their assessment included policy development that is grounded in practice; supports adaptation processes that reduce social and environmental vulnerability; accounts for short-term variations and longer-term changes in climate; recognises the importance of scale from the local to the global; is assessed in the context of human development; and, employs participatory processes throughout its formulation and implementation. Furthermore, Smith et al. (2009) described the essential components of a government's adaptation architecture as consisting of: political leadership and institutional organisation; stakeholder involvement; climate change information and appropriate use of decision analysis techniques; explicit consideration of barriers to adaptation; funding for adaptation; technology development and its diffusion; and adaptation research.

This paper will describe the components of ERA that inform policy and practice, the concordance of these components with leading practice, and the comprehensive adaptation architecture established to support regional adaptation in NSW.

The Planning Process: Enabling Regional Adaptation (ERA)

ERA consists of a series of multi-region, stakeholder-led planning processes in three inter-related phases. Each of these phases operates at overlapping scales from local to regional and incorporates elements of leading practice in adaptation policy development. Together they are designed to perform a number of functions:

- To present up to date climate projections and socio-economic information at spatial scales useful for adaptation planning.
- To elicit tacit knowledge from stakeholders of likely climate impacts and current adaptation actions and to create opportunities for knowledge sharing among regional adaptation actors and central government.
- To engage stakeholders in a co-learning environment to better understand the complexity of regional systems and promote interactions across government policy silos and tiers of government.
- To develop strategic pathways that allow government to consider planned transitions away from business-as-usual and toward necessary system transformation that is both desirable and more resilient to on-going climate change.

The three phases of ERA are: Integrated Regional Vulnerability Assessment (IRVA), Enabling Adaptation (EA) and Dialogue to Enhance Local Transformative Adaptation (DELTA). Regional boundaries are set according to those of the NSW Government Department of Planning, which consists of 12 regions each of which is currently at a different stage in the process (Fig. 2.1). ERA is most advanced in the South East Region because it has been used to pilot each process before extension to other regions of NSW.

(a) **Integrated Regional Vulnerability Assessment (IRVA)**. The IRVA process (Jacobs et al. 2014b) draws on contemporary practices in vulnerability assessment to identify exposure and sensitivity of region scale social, economic and biophysical systems to the impacts of climate change, the direct and flow-on effects of these impacts on government service delivery and the adaptive capacity of local government administrations and state government agencies to continue to service the community. Separate sector-based assessments are conducted and the findings then integrated to develop a set of key regional vulnerabilities, which emphasises the potential for maladaptive outcomes through unilateral action in policy silos. The IRVA:

    - Employs a systems thinking approach that acknowledges communities exist within human–natural (or social-ecological) systems, and encourages a plural-conditional approach to adaptation policy development (Stirling 2010)
    - Makes extensive use of participatory engagement in which stakeholders co-create an understanding of vulnerability through their knowledge (often tacit) of the region (Fig. 2.2)

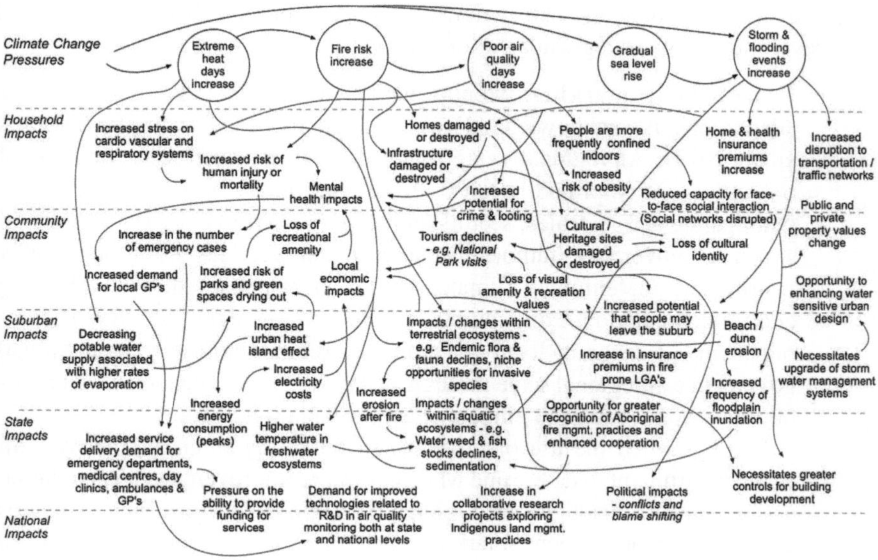

Fig. 2.2 Impact pathways generated through a participatory workshop with the government service providers in the natural and cultural assets sector of the Greater Sydney Region in the Sydney IRVA. *Source*: authors

- Focuses on developing an understanding of the constraints to adaptation, and on identifying opportunities for building adaptive capacity so communities can deal better with climate shocks regardless of their nature or timing,
- Identifies opportunities for collective action that operate across scales, sectors and disciplines, and encourages regional decision makers to seek collaborative projects to address complex vulnerability drivers, or apply regional capacities; and
- Supports qualitative analysis wherever possible with quantitative information (regional climate projections and socioeconomic trends), which acknowledges that societal interactions are complex and contradictory in nature, and not amenable to external, expert-led, reductionist approaches to problem analysis.

(b) **Enabling Adaptation (EA)**. EA aims to catalyse adaptation responses that are sensitive to the reality of NSW's regional communities, environments and economies. Building on regional climate vulnerabilities identified through the IRVA, in a series of workshops, key regional decision-makers devise long term strategies to address systemic vulnerabilities and plan potential adaptation pathways at a regional and sub-regional scale. This process has been successfully piloted in the South East region of NSW in the Enabling Adaptation in the South East (EASE) and Enabling Adaptation in the Australian Capital Territory (EnAACT) projects (Jacobs et al. 2014a). EASE consisted of two main components: sub-regional adaptation process benchmarking and development of strategic adaptation pathways.

- **Process benchmarking**: Monitoring and evaluation of climate change adaptation is acknowledged as challenging and complex (Bours et al. 2013). At regional scale many difficulties, such as attribution of outcomes, are amplified because of the multiplicity of actors and actions that operate across scales, sectors and responses. Moreover, the absence of a central governing authority ensures a general lack of strong policy instruments to 'enforce' climate adaptation responses through planning and implementation. For example, the South East region of NSW is comprised of 12 separate local government administrations with additional governance at regional scale through NSW government agencies, all contributing to some extent to climate change adaptation action.

  EASE trialled the use of process benchmarking as an alternative, soft policy instrument to determine what and where improvements in adaptation action may be made to drive enhanced regional performance (Huggins 2010). Best practice benchmarking is a technique commonly used in local government in Australia for a range of purposes from community engagement to financial performance (e.g. Productivity Commission 2012). Adaptation processes were defined according to Hansen et al.'s (2013) adaptation process cycle. Adaptation activity was measured through a survey of workshop participants as the proportion of participants that answered in the affirmative that their organisation had completed activity under a process (e.g. impact assessment) (Fig. 2.3). Scores tended to decline with progression around the adaptation process cycle. Earlier processes (such as impact assessment, vulnerability assessment and planning) in most sub-regions tended to show higher levels of activity than later processes (in particular monitoring and evaluation).
- **Strategic Adaptation Pathways**: Recent approaches to dealing with decision-making under deep uncertainty have emphasised the use of adaptive pathways (Hasnoot et al. 2013). We take a strategic approach to regional adaptation that conceptualises a region as a complex evolving system. The regional system in turn encompasses multiple sub-systems (or the region as a system-of-systems, Katina et al. 2014). With climate change as a system driver, adaptation can be envisioned as a series of transitions that emerge autonomously from the need to move away from business-as-usual in key regional sub-systems (Thorn et al. 2015). Incremental change occurs through exploration of the space of possibilities with a focus on the adjacent possible, which is one step away from what already exists (Mitleton-Kelly 2003) and can be realised by using existing resources in a novel way (a concept which underpins adaptive capacity, e.g. Ellis 2000). A 'successful' pathway is promoted through increasing returns and positive path dependency leading to 'sticky' change (Levin et al. 2012). A transformed system will ultimately emerge through

Sub-region	Impact Assessment	Vulnerability Assessment	Planning	Capacity Building	Implementation	Monitoring & evaluation
Alpine						
Tablelands						
Coast						
Scale	No activity	Very Low	Low	Medium	High	Very high

Fig. 2.3 Prototype reporting format of levels of activity benchmarked across adaptation processes for each sub-region of South East NSW. *Source*: authors

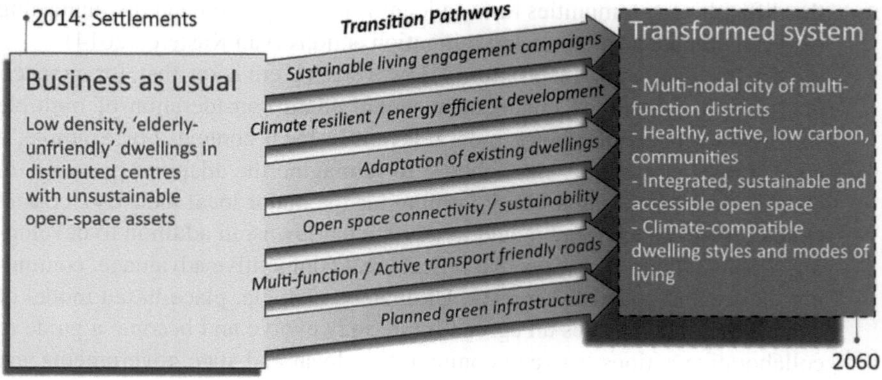

Fig. 2.4 An example of a model of change for settlements in the Australian Capital Territory. *Source*: Jacobs et al. (2014a)

knowledge sharing of success and self-organisation of the social system (Mitleton-Kelly 2003). In our view the role for government is to establish conditions that facilitate such changes by developing an understanding of community behaviour alongside careful long-term planning in consultation with the community, to reduce the risk of disruption to society from abrupt transformation. The strategic adaptation pathways process seeks to develop a structured process for government, at the local and regional scale, to consider and implement adaptation actions in regional decision-making to achieve desirable transformation.

Through Enabling Adaptation in the South East a suite of 'change models' (Fig. 2.4) was developed that described transition pathways to a transformed future for 12 regional subsystems. These models included large regional towns, agricultural service centres, potable water supply, emergency management, extensive grazing, dairy farming, coastal development, coastal ecosystem management, off-reserve biodiversity conservation, public land management, alpine tourism and beach tourism. Together they represent strategic pathways that provide state agencies, local government, industry and the community with a mosaic of potential future transitions required to adapt to climate change.

(c) **Dialogue to Enable Local Transformative Adaptation (DELTA)** Fazey et al. (2015) suggest that any process that seeks to envision the future must be legitimate with respect to assumptions, values and principles that future embodies; avoid lock in traps (Xu et al. 2015) that could reinforce existing social inequalities; and, understand how past change influences system transformation. DELTA, which is to be piloted in the NSW South East region in 2016 will engage local communities in the co-production of place based adaptation pathways to improve the relevance, saliency and legitimacy (Cash et al. 2003) of an agreed future. By allowing community members to engage with and reimagine local versions of region scale models we hope to overcome the difficulties communities have in identifying adaptation and distinguishing between climate adaptation and mitigation actions (van Kasteren 2014).

DELTA is a work in progress. However, we envision that for a specific location community engagement would entail consideration of multiple subsystem models of change most relevant to local context. For example, a coastal rural community may choose to reimagine the adaptive pathways in change models relating to dairy farming (as the major local industry), coastal development, beach tourism and large regional towns, in addition to developing supplementary pathways that reflect local competitive advantage, community desire and adaptive capacity. Ultimately, a single, place-based model of change that incorporates an agreed future may evolve and become a guide to collaborative actions between communities, local and state governments and business.

The Supporting Adaptation Architecture

Extensive adaptation architecture has been constructed by the NSW Government to support the ERA process (Fig. 2.5).

The key components of the adaptation architecture include:

(a) **Capacity building**: consists of three main sub-components.

- **Community of Practice**: a climate change community of practice (CoP) provides networking and information exchange opportunities across the NSW agency with responsibility for climate change adaptation (i.e. Office of Environment and Heritage). The CoP operates via an online social media platform on the staff intranet and provides information updates and notices of events (e.g. research seminars) and other climate change communications.
- **Newsletters**: Regionally-focused newsletters are issued to regions undergoing the vulnerability assessment process (i.e. IRVA). These highlight regionally relevant events, opportunities or contacts related to regional adaptation. An overarching, state-wide, interactive Adaptation Newsletter

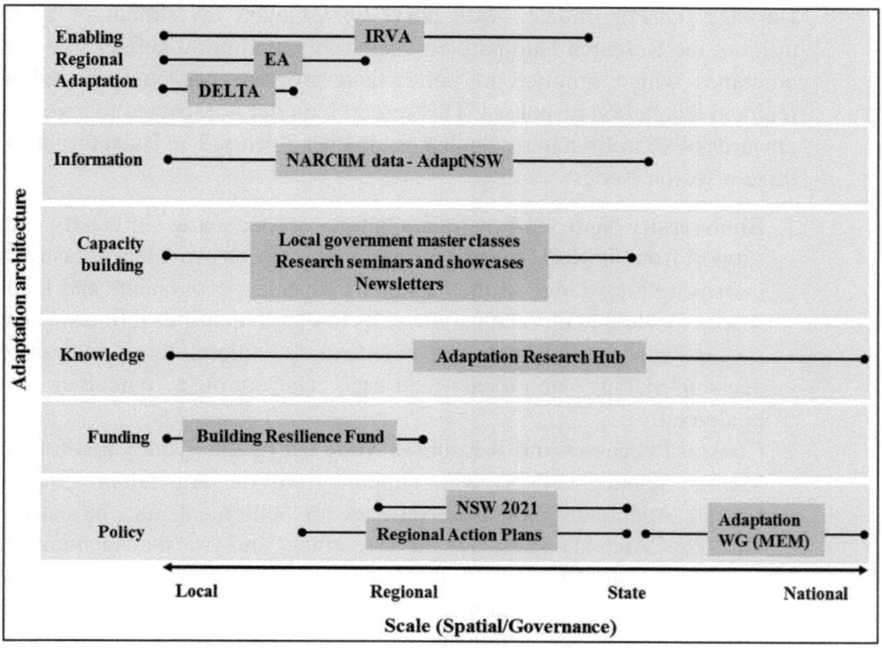

Fig. 2.5 Key components of the NSW Government's climate adaptation architecture matched to the scale over which individual programs operate. *Source*: authors

is also issued to both current and past participants in ERA to stay abreast of emerging climate change research and practice.

- **Master classes**: a series of hands-on discussions of issues emerging from current research and practice on climate adaptation specifically targeting local government. For example, three 'Urban Greening Masterclasses' have been delivered where local government practitioners present to their peers on success stories and challenges associated with their urban greening projects. Private sector support (from Horticulture Industry Australia's Nursery and Garden Industry Association) helps to cover the costs of these forums and also extends their influence beyond traditional government considerations.

(b) **Information dissemination**: AdaptNSW (OEH 2015) is a web portal containing information and links to all components of the NSW Government's climate impact and adaptation program. This micro site was launched in December 2015 logging 25,000 post-launch hits and 54 media articles. The site allows access to the latest regional climate projections for NSW.

(c) **Knowledge generation**

- **Research partnerships** through the NSW Adaptation Research Hub. The Hub is a collaboration between leading NSW universities and experts in climate-change and adaptation science, and the Office of Environment and

Heritage (OEH). Initiated with NSW Government investment of $2.75 million, the Research Hub partners match this funding to deliver research outcomes which are tied to policy and program priorities, including regional adaptation responses. This research model is expected to leverage upwards of $6 million in research activity over 3 years. The Hub consists of three research Nodes:

1. **Biodiversity Node** led by Climate Futures at Macquarie University with support from CSIRO (Australia's national research provider) focuses on increasing knowledge of the capacity of species, ecosystems and landscapes to adapt to climate variance. Its research identifies refuges where species can survive extreme events and explores how integrated decision-making on local land-use can optimise outcomes for biodiversity.
2. **Coastal Processes and Responses Node** led by the Sydney Institute of Marine Science (SIMS) with support from the Australian Climate Change Adaptation Research Network for Settlements and Infrastructure (ACCARNSI). It studies the assessment and risk management of, and adaptation responses to, the impacts of climate change on coastal and estuary zones. The aim is to improve the knowledge base that will inform decisions and actions taken by local communities.
3. **Adaptive Communities Node** is led by the Institute for Sustainable Futures at the University of Technology Sydney and supported by CSIRO. The focus is research into how urban and rural communities can best adapt to climate change and the ways government can provide support to local communities to build resilience. The ERA program emerged primarily through the applied research of this Node.

- **Down-scaled climate modelling**. The NSW Government has developed and released high resolution climate projections at a scale (10 × 10 km) to support local decision makers through the NSW and ACT Regional Climate Modelling (NARCliM) Project. This multiagency research partnership between the NSW and ACT Governments and the Climate Change Research Centre at the University of NSW provides an extensive dataset of more than 100 meteorological variables, which are also presented in regional summaries, including current climate and likely changes in climate (temperature and rainfall, severe fire weather, hot days and cold nights) by 2030 and 2070. Further impact information and access platforms are currently being prepared for release.

(d) **Dedicated funding for implementation**: a contestable grants scheme, 'Building Resilience to Climate Change', provides dedicated funding to local government for resilience projects to address identified climate change risks and vulnerabilities facing NSW councils. This $1 million program is funded through OEH and the NSW Environmental Trust and is administered by the peak local government association in NSW. The program aims to encourage:

1. Enhanced consideration of climate change impacts in local and regional decision making.
2. Delivery of projects that minimise climate change impacts for local and regional decision makers.
3. Implementation of climate change adaptation beyond current projects and programs.
4. Adaptive capacity in Local Government through a community of practitioners across professional disciplines with direct experience in implementing adaptation responses across NSW.

(e) **Policy environment**

- **State-wide policy**: 'NSW 2021' is a 10 year plan that sets the State Government's agenda for building the economy, focussing on government service delivery, and improving infrastructure that is designed to strengthen local environments and communities. This includes a specific commitment to assist local government, business and the community build resilience to future extreme events and hazards by helping them understand and minimise the impacts of climate change (NSW Government 2015). For the agency charged with delivering this goal (OEH) the corporate objective is to build resilience to climate change and environmental hazards and risks.
- **Regional policy**: Regional Action Plans (RAPs) detail the Government actions in each of the NSW regions, in line with the strategic policy directions established in NSW 2021 (NSW Government 2015). RAPs integrate planning for land use, transport and infrastructure investment aligned with higher level objectives of NSW 2021. The RAPS include 19 distinct actions across eight NSW regions to "develop information and tools to assist local communities address climate risk, identifying cross-sectoral vulnerabilities and opportunities to respond".
- **Federal policy linkages**: Since 2008 there has been an Adaptation Working Group facilitating engagement between the Australian and the State and Territory governments through different Federal reporting structures, which has enabled dialogue on related research programs and specific initiatives. In December 2013 the Council of Australian Governments (COAG) streamlined its Ministerial Council system and abolished the Standing Council on Environment and Water. In a positive development the National Environment Ministers Meeting (MEM) decided to revive this Working Group in February 2015 as one of a series of matters "requiring resolution on a national collaborative basis". However the "temporary" status of this working group, and the lack of a formal Ministerial Council for reporting may hamper its capacity for coordinated and efficient governance of cross-scale and cross-disciplinary adaptation policy and actions.

Conclusions

Successful climate change adaptation requires governments to establish a comprehensive architecture that operates over the long-term to support local- and regional-scale knowledge generation, capacity building and innovation (Jordan and Huitema 2014). We believe the core projects and architecture established by the NSW Government serve as an example of what is required to promote adaptation. To date the core projects within the ERA process have engaged 720 regional decision-makers through 33 participatory workshops and assessed adaptation in five NSW planning regions covering 75 % of the State's population and 64 % of Local Government Areas. However, a far greater number of regional and local decision makers have been engaged and informed about climate change through the adaptation architecture that supports ERA. Furthermore, we see evidence in the actions taken by the NSW Government through the Office of Environment and Heritage of the emergence of climate adaptation as a discrete policy field albeit with links to all areas of government (Massey et al. 2014).

Adaptation planning in NSW is ongoing. Planning is underway for three further state planning regions to commence ERA in the next 12 months, with plans to expand the Government's adaptation architecture through the placement of regional staff in key locations across NSW to act as the knowledge brokers, supporters and active participants to facilitate collective regional adaptation responses.

References

Australian Bureau of Statistics (ABS) (2015). http://www.abs.gov.au/ausstats/abs@nrp.nsf/lookup/1Main+Features12007-2011. Accessed 1 May 2015

Bours D, McGinn C, Pringle P (2013) Guidance note 1: twelve reasons why climate change adaptation M&E is challenging. SEA Change CoP, UKCIP. http://www.seachangecop.org/node/2728. Accessed 1 May 2015

Brooks N, Adger W (2005) Assessing and enhancing adaptive capacity, Technical Paper 7 of the adaptation policy framework. http://www.undp.org/climatechange/adapt/apf.htm. Accessed 1 May 2015

Cash D, Clark W, Alcock F, Dickson N, Eckley N, Guston D, Jager J, Mitchell R (2003) Knowledge systems for sustainable development. Proc Natl Acad Sci U S A 100(14): 8086–8091

Department of Water Resources (1994) New South Wales: water, people, places. Department of Water Resources, Parramatta

Ellis F (2000) Rural livelihoods and diversity in developing countries. Oxford University Press, Oxford, UK

Environment Protection Authority (EPA) (1997) NSW state of the environment report. Environment Protection Authority, Chatswood

Fankhauser S, Smith J, Tol R (1999) Weathering climate change: some simple rules to guide adaptation decisions. Ecol Econ 30(1):67–78

Fazey I, Wise RM, Lyon C, Câmpeanu C, Moug P, Davies TE (2015) Past and future adaptation pathways. Clim Dev 1–19 (ahead-of-print) doi:10.1080/17565529.2014.989192

Hansen L, Gregg RM, Arroyo V, Ellsworth S, Jackson L, Snover A (2013) The state of adaptation in the United States: an overview. A report for the John D. and Catherine T. MacArthur Foundation. http://www.ecoadapt.org/data/library-documents/TheStateofAdaptationintheUnitedStates2013.pdf. Accessed 1 May 2015

Hasnoot M, Kwakkel J, ter Matt J (2013) Dynamic adaptive policy pathways: a method for crafting robust decisions for a deeply uncertain world. Glob Environ Chang 23(2):485–498

Huggins R (2010) Regional competitive intelligence: benchmarking and policy-making. Reg Stud 44(5):639–658

Jacobs B, Boronyak-Vasco L (2014) Natural resources planning for climate change: extreme climate events and communities. Institute for Sustainable Futures. Available at: https://www.uts.edu.au/sites/default/files/JacobsBoronyakVasco2015extremeclimateevents.pdf. Accessed 1 May 2015

Jacobs B, Boronyak-Vasco L, Mikhailovich N (2014a) Enabling adaptation in the Australian capital territory. Institute for Sustainable Futures. Available at: http://www.environment.act.gov.au/cc/what-government-is-doing/climate-change-adaptation-and-resilience. Accessed 1 May 2015

Jacobs B, Lee C, O'Toole D, Vines K (2014b) Integrated regional vulnerability assessment of government services to climate change. Int J Clim Chang Strateg Manage 6(3):272–295

Jordan A, Huitema D (2014) Policy innovation in a changing climate: sources, patterns and effects. Glob Environ Chang 29(2014):387–394

Katina PF, Despotou G, Calida BY, Kholodkov T, Keating CB (2014) Sustainability of systems of systems. Int J Syst Syst Eng 5(2):93–113

Leach M, Scoones I, Stirling A (2010) Dynamic sustainabilities: technology, environment, social justice. Earthscan, New York

Leach M, Rokstrom J, Raskin P, Scoones IC, Stirling AC, Smith A, Thompson J, Millstone E, Ely A, Around E, Folke C, Olsson P (2012) Transforming innovation for sustainability. Ecol Soc 17(2):11–16

Levin K, Cashore B, Bernstein S, Auld G (2012) Overcoming the tragedy of super wicked problems: constraining our future selves to ameliorate global climate change. Policy Sci 45(2):123–152

Massey E, Biesbroek R, Huitema D, Jordan A (2014) Climate policy innovation: the adoption and diffusion of adaptation policies across Europe. Glob Environ Chang 29(2014):434–443

Mitleton-Kelly E (ed) (2003) Complex systems and evolutionary perspectives on organisations: the application of complexity theory to organisations, Advanced series in management. Elsevier Science, Oxford

New South Wales Government (2015). http://www.nsw.gov.au/2021/. Accessed 1 May 2015

Office of Environment and Heritage (OEH) (2015). http://www.climate.change/environment.nsw.gov.au. Accessed 1 May 2015

Pelling M (2011) Adaptation to climate change: from resilience to transformation. Routledge, London

Productivity Commission (2012). http://www.pc.gov.au/projects/study/regulation-benchmarking/local-government/report. Accessed 3 May 2015

Smith JB, Vogel JM, Cromwell JE III (2009) An architecture for government action on adaptation to climate change. An editorial comment. Clim Chang 95(1–2):53–61

Stirling A (2010) Keeping it complex. Nature 468(7327):1029–1031

Thorn J, Thornton TF, Helfgott A (2015) Autonomous adaptation to global environmental change in peri-urban settlements: evidence of a growing culture of innovation and revitalisation in Mathare Valley Slums, Nairobi. Glob Environ Chang 31:121–131

van Kasteren Y (2014) How are householders talking about climate change adaptation? J Environ Psychol 40(2014):339–350

Xu L, Marinova D, Guo X (2015) Resilience thinking: a renewed system approach for sustainability science. Sustain Sci 10(1):123–138

Chapter 3
Vulnerability Is Dynamic! Conceptualising a Dynamic Approach to Coastal Tourism Destinations' Vulnerability

Jillian Student, Bas Amelung, and Machiel Lamers

Abstract Coastal regions and islands are among the most popular tourist destinations. They are also highly vulnerable to climate change. Much of the literature on vulnerability, including IPCC reports, states that vulnerability is dynamic. However, vulnerability conceptualisations in the tourism realm have so far taken a static perspective. Static conceptualisation underestimates inherent uncertainties stemming from actor interactions (with one another and their environment) and processes. The interactions and processes are important for developing adaptive strategies in a dynamic world. Hence, frameworks for analysing tourism vulnerability as a dynamic phenomenon are urgently needed. This paper outlines the first steps taken towards a dynamic approach for analysing vulnerability of Caribbean coastal tourism. The approach consists of (1) a conceptual framework focusing on human-human and human-environment interactions at the actor level and (2) an evolutionary methodology. The methodology engages both Caribbean climate change experts and regional actors. Regional actors both respond to and help develop the framework through interactive, or companion, modelling. By focusing on interactions and processes, the approach is expected to yield key insights into the development of vulnerability through time, crucial information for adaptive management.

Keywords Dynamic vulnerability approach • Companion modelling • Agent-based modelling • Adaptive capacity • Vulnerability • Coastal tourism

---

J. Student (✉)
Environmental Policy Group, Wageningen University, P.O. Box 47, 6700 AA Wageningen, Netherlands

Environmental Systems Analysis Group, Wageningen University, P.O. Box 47, 6700 AA Wageningen, Netherlands
e-mail: jillian.student@wur.nl

B. Amelung
Environmental Systems Analysis Group, Wageningen University, P.O. Box 47, 6700 AA Wageningen, Netherlands

M. Lamers
Environmental Policy Group, Wageningen University, P.O. Box 47, 6700 AA Wageningen, Netherlands

W. Leal Filho (ed.), *Innovation in Climate Change Adaptation*, Climate Change Management, DOI 10.1007/978-3-319-25814-0_3

Introduction

Have we adapted to climate change? The simple answer is "No". The current array of research on climate change, scenarios, vulnerability and vulnerability assessments is a clear signal that humans have not adapted to climate change. If the answer were or could be "Yes", then it would presume that one set of (already executed) actions is sufficient to deal with future problems. However, climate change is a complex and dynamic process influenced by human mitigation and adaptation strategies where uncertainty persists.

Tourism is a useful setting to study the global impacts of vulnerability and adaptive capacity in a local context. Tourism is a complex human-environment system (e.g. Moreno and Becken 2009) with many interdependencies (e.g. on resources, among tourism operators) and many crossovers with other industries (Csete and Szécsi 2015). At the same time, different individuals are exposed to different types of harm and cope with these types of harm in different ways (Turner et al. 2003). For example, tourists and local tourism operators are exposed to different types of harm and coping mechanisms. On the one hand, tourists have a relatively high coping abilities (UNWTO-UNEP-WMO 2008) as they can stay away from destinations or engage in other activities (e.g. boat cruises instead of fishing and diving excursions). On the other hand, local tourism operators and the surrounding community have low coping abilities as they cannot easily change the tourism activities they offer or change their location (e.g. Kaján and Saarinen 2013; Moreno and Becken 2009). Some types of tourism and geographic regions are more vulnerable than others. Coastal tourism is highly vulnerable to extreme weather events, sea-level rise, beach erosion and ocean acidification (Moreno and Becken 2009; Shakeela and Becken 2014). For example, Caribbean islands are a 'hotspot' for climate change impacts due to high exposure levels and economic dependency on tourism (UNWTO-UNEP-WMO 2008). Tourism contributes 14.0 % to regional GDP and 12.3 % to regional employment in the Caribbean (World Travel and Tourism Council 2013).

Adaptive capacity and vulnerability are intricately interconnected. In order to analyse and develop adaptive capacity, vulnerability needs to be understood. Vulnerability to climate change is a global issue illustrated by international research collaborations of the IPCC and UN political alliances. Nevertheless, it requires local action. The IPCC definition of vulnerability is the "degree to which a system is susceptible to and is unable to cope with adverse effects (of climate change)" and uses exposure, sensitivity and adaptive capacity of a system as key components (Adger 2006, p. 269). The IPCC definition does not specifically describe the three components nor the relationships among the components (Eakin and Luers 2006). Many definitions of vulnerability exist with different disciplines focusing on different factors e.g. climate scientists focus on likelihood of occurrence and social scientists focus on socio-economic indicators (e.g. Adger 2006; Brooks 2003; Füssel 2007). Along with type of vulnerability, there are different temporal focuses: historical, present and future (Füssel 2007). These different definitions and emphases result in a range of conceptual frameworks trying to clarify the meaning of vulnerability. A complete review of the concept and development of the term vulnerability is beyond the scope of this paper (for reviews see Adger 2006; Eakin and Luers 2006; Schroter et al. 2005). The scope of this paper is to provide evidence for the need of dynamic approaches to vulnerability and adaptive capacity and outlines initial steps on how to operationalise a dynamic approach in the context of Caribbean coastal tourism.

Adaptation strategies are influenced by the static approaches to vulnerability. Adaptation strategies are typically ad hoc, short-term focused, reactive (e.g. repairing damaged items) and event specific (Kaján and Saarinen 2013). Adaptive capacities are dynamic processes made up of actions at the household, regional and national level (Smit and Wandel 2006). Adaptation often requires collaboration at social, political and spatial levels and adjustments to the local context (Csete and Szécsi 2015). Thus, future-looking vulnerability approaches and adaptive strategies are required in order to move beyond reactive short-term measures (Kaján and Saarinen 2013).

This paper explores the application of a dynamic approach to a dynamic problem. A growing body of literature recognises that vulnerability is dynamic (e.g. Adger 2006; Turner et al. 2003) and context dependent (e.g. Brooks 2003; Füssel 2007). Scientific approaches, however, do not match the definition of vulnerability- static approaches are applied to a dynamic problem. Climate change and tourism destination vulnerability continually shift; thus adaptation measures must be continuous and flexible (e.g. Brown et al. 2012; Csete and Szécsi 2015). Challenges for adaptation strategies are understanding interconnections, translating understanding into action, focusing on the long-term future and considering local levels and context (Turner et al. 2003). Moreover, current methodologies have not helped many local actors identify the importance of emerging vulnerability challenges. Local actors often do not understand scientific conceptual frameworks and are uncertain on how and whether they may personally be affected (Klein and Juhola 2014). This uncertainty lowers the importance of understanding vulnerability and delays decision-making and implementation. Therefore, engaging local actors in designing a dynamic conceptualisation of vulnerability is fundamental for developing long-term adaptive strategies.

Current Frameworks and Their Limitations

Vulnerability, in the context of tourism, is traditionally assessed using a top-down approach of a tourist destination's exposure, sensitivity and adaptive capacity to climate change (e.g. Moreno and Becken 2009; Perch-Nielsen 2010; Polsky et al. 2007; Schroter et al. 2005). Research has focused on specific events (e.g. severe storms and sea level rise) with predictable consequences (Csete and Szécsi 2015). These approaches analyse individual pieces of the system. In tourism, the most common adaptive strategies involve diversifying destinations activities and product portfolio (Kaján and Saarinen 2013). All of these approaches have useful ideas for inventorying the risks and hazards, but they do not provide a framework to understand how people and the environment interact with each other and with emerging risks and hazards. Moreover, local actors are not often included and represented in the process of making and analysing vulnerability assessments and considering adaptive capacity.

Eakin and Luers (2006) identify three main streams that have emerged in the debate on vulnerability definitions and assessments: (1) biophysical risk/hazard, (2) political ecology and/or political economy and (3) ecological resilience. Classic approaches resulting from the risk/hazard stream include determinism (nature causes hazards) and mechanistic engineering (technology reduces vulnerability) (Füssel 2007). This stream takes an instrumental natural science based perspective. For example, Disaster Risk Reduction (DRR) (Thomalla et al. 2006) takes an engineering approach focused on singular events, exposure and technological solutions, but does not focus on interactions among people. Typically, this approach takes a historical perspective (Mercer 2010) and aggregates known hazards and impacts (Füssel 2007). Relying on a risk-based understanding of vulnerability provides a limited perspective on adaptation because interconnectedness is not taken into account (Kaján and Saarinen 2013). A further limit of this approach is that it does not provide increased understanding of the different impacts on the system and its sub-sets nor what adaptive measures may be applied (Turner et al. 2003). Moreover, adaption involves a mixture of tools, the specific mixture is location and context specific (Csete and Szécsi 2015), and requires the buy-in of local actors.

The political ecology definition focuses on people (individuals, households, communities, etc.). The definition and approach asks how and which people are affected and what are the causes and outcomes of the heterogeneous adaptive capacities (resulting from different entitlements and capabilities) (Eakin and Luers 2006). This definition does consider agency, the capacities of individuals to act and effect change. However, it does not look at the broader scope of vulnerability in settings such as a coastal beach destination nor does the definition focus on what actions can be taken and what capacities are needed to reduce future vulnerabilities (ibid).

Ecological resilience, in contrast, focuses on a coupled human-environment system (Turner et al. 2003) and is informed by complexity theory. This definition and framework asks the questions why and how systems change (Eakin and Luers 2006). Ecological resilience focuses on thresholds and tipping points and is future-looking. Although the ecological resilience perspective does consider the interactions between humans and the environment, the perspective is less decisive on human-human interactions. The human dimension of adaptation involves actions, processes and outcomes and adjusting to changing conditions (Smit and Wandel 2006). These changes also come about because of interactions among actors (Csete and Szécsi 2015). Preferences, adaptation mechanisms and strategies of individuals and groups influence each other (Kaján and Saarinen 2013). Combined, these streams take into account agency, broader risks, human-environment interactions, thresholds and future scenarios. However, none of these three streams consider what defines a desirable state (Eakin and Luers 2006), which requires local actors to co-design the objectives and conceptual framework.

Static approaches identify different parts of the systems, but do not encourage systems thinking. Academic discussions have circled around dynamic approaches, however operationalising this thinking has been difficult (Becken 2013b). For example, current frameworks help identify actors, possible adaptation activities (Csete and Szécsi 2015), possible hazards and indicators. According to UNWTO-UNEP-WMO (2008), adaption strategies types are technological, management, behaviour, education and/or political. The current frameworks focus more on technological adaptation strategies and focus less on the other four adaptation strategies. In order to get beyond this scientific challenge of defining vulnerability in the local context, local peoples' tacit knowledge needs to be combined with scientific knowledge. Participatory approaches with local actors help relate science to the societal issue in a process of joint knowledge production. Analysing vulnerability and how it changes, gives opportunity to build adaptive capacity and limit harm to local people.

Operationalising a Dynamic Approach

A dynamic definition of vulnerability suggests the need for a dynamic approach. The approach and the conceptualisation of vulnerability and adaptive capacity are dynamic. Dynamic suggests a focus on the interactions among different variables. Thus, a dynamic approach is process-oriented, transdisciplinary and iterative. A dynamic approach for dynamic problems requires the use of different range of tools than are currently being applied to vulnerability issues in tourism. The tools currently applied provide insights on key variables (actors, biophysical challenges, possible scenarios, potential risks and extreme events predictions) (e.g. Moreno and Becken 2009; Perch-Nielsen 2010). Understanding system interactions in a local context requires local knowledge and participation. Transdisciplinary research endeavours to provide a holistic approach involving multiple disciplines and local participation to improve system understanding. Many levels and forms of participation exist (e.g. Barreteau et al. 2010; Hegger et al. 2012). The role of actors in this process is different than what has been done to date for vulnerability in tourism. This participatory process sees local actors not just as the end users or informers of the system, but also as actively involves individuals in the process of learning, co-creating, modifying and analysing the process. The following sections describe a means to operationalise a dynamic approach for this dynamic problem.

This paper responds to the need for new approaches to study the complex relations between tourism and climate change (Becken 2013a) by asking how dynamic vulnerability can be conceptualised in a coastal tourism context, what are the implications of this framework and how it can inform adaptive governance strategies. In this study, interactive modelling refers to two-way communication and learning between stakeholders (experts and local actors) and researchers through modelling and simulations. Simulations developed through role playing games and agent-based modelling (ABM) will be used because ABM provides an actor-oriented modelling environment for analysing the emergent properties of actor interactions over time.

Implications for Process

Dynamic approaches require learning and iteration. Adaptation studies have thus far limited focus in community perceptions (Kaján and Saarinen 2013). Many methodologies exist for studying vulnerability (e.g. economic modelling, surveys, Delphi surveys, workshops) (Becken 2013a). However, interactive modelling approaches are new to the tourism domain and provide different tools to include community perceptions. The process designed for studying dynamic vulnerability and adaptation of Caribbean coastal tourism destinations is inspired by companion modelling (e.g. Etienne 2011), which engages local actors in problem definition, determining the objective and forming the conceptual framework. The process appears linear, but is in fact made up of feedback loops in which the original objective, conceptual model and modelling tools can be altered as a result of interactions with local actors, altered objectives and better system understanding (Etienne 2011). The continual feedback loops operationalise researchers' suggestion that vulnerability approaches include built-in reflexivity (Hegger et al. 2012). The two main objectives of companion modelling processes are to (1) create knowledge of the system (interactions, interdependencies, patterns, etc.), in this case understand emerging vulnerabilities and the implications for adaptation, and (2) enhance decision-making by analysing what processes are available or could be considered to address these challenges. All companion modelling approaches explore objective one, but some do not include objective two. The objectives of companion modelling are in line with what researchers have identified as the information gap of vulnerability- lack of understanding of the system and limited decision-making capabilities.

The first phase focuses on inventorying existing knowledge: understanding the context, local actors' objectives and identifying relevant actors. The second phase involves co-constructing the conceptual framework using a combination of scientific, technical and local knowledge. The third phase involves operationalising the framework in the form of a role-playing game and/or computer simulations such as ABM. The fourth phase involves exploring different scenarios with local actors and the fifth phase involves monitoring and evaluation (adapted from Etienne 2011).

As local context is crucial for analysing vulnerability and adaptive capacity, case studies on two separate Caribbean islands are used. The first destination case study shows the learning process of joint knowledge production and what the implications are of a dynamic process on improving decision-making. The second case study demonstrates what has been learned by the process of the first case study and offers a comparison study to analyse what the similarities and differences are in the process of understanding local vulnerabilities and adaptive capacities.

Implications for Tools

The interactive process is supported by a range of tools. In the earliest stages, a fuzzy cognitive model provides a rough understanding of the scientific understanding of the system and the possible interactions and interdependencies. The conceptual framework combines scientific knowledge of the system and uses earlier frameworks to identify key actors and biophysical variables. A panel made up of experts on climate change (in the tourism context) provides information on the Caribbean coastal tourism context. Role-playing games executed in focus groups help make the problem, how the system interacts and how other actors behave more tangible. Role-playing games have the added benefit of being more approachable than computer simulations and help remove the perception of dealing with a black box, which is a common complaint of computer simulations (e.g. Barreteau et al. 2000). This enables actors to more easily contribute to monitoring and evaluating the system and its emergent properties.

Operationalising the same conceptual framework as role-playing games, computer simulations can help show how individual decisions result in different macro patterns. ABM is a useful simulation type as it is designed to describe heterogeneous and autonomous actors' interactions with each other and their local environment while offering a flexible platform to explore global tourism and climate change scenarios within a local tourism destination context (Bonabeau 2002). ABM has seen limited applications in tourism. A few examples of ABM applications in tourism include Balbi et al. (2013), Johnson and Sieber (2011) and Soboll and Schmude (2011). The computer simulation can be done one-on-one or can have multiple users. One-on-one interviews using simulations can look at multiple scenarios. Computer simulations enable applying multiple scenarios (climate change, tourist projects) and collecting data from different scenarios in a short period of time. Moreover, they help identify how individual decisions, actions and different practices can affect the system.

Participation

A gap exists between research on vulnerability, adaption, decision-making and actions taken at a local scale (Klein and Juhola 2014). Climate change is not one of the main vulnerabilities that locals respond to (Shakeela and Becken 2014). One explanation is that climate change at a local scale is difficult to conceptualise (Klein and Juhola 2014). By involving actors throughout the process, their real concerns about their local environment can become more explicit and they can actively engage in learning about their problems and the process. Moreover, involving local actors throughout the process provides opportunities to share their tactic knowledge (Hegger et al. 2012). The companion modelling approach provides guidelines for involving actors. The companion modelling charter states: equal accounts of identified actors' knowledge and perspectives; transparency of ideas used; iterative and adaptive processes; and evaluation of learning outcomes (evolution of actors' perspectives and interactions) as well as technical outcomes (Etienne 2011).

Local actors are heterogeneous- they have different roles in the community, different perceptions of climate change, different vulnerabilities and varying abilities to adapt (Scott et al. 2012). In the context of tourism, little is known on which actors must be involved in participatory processes (Hegger et al. 2012). As a baseline for participation, two types of actors should be involved in the participatory process: individuals affecting and affected by destination vulnerability and individuals who can make decisions to address vulnerabilities or develop adaptive capacity. Hegger et al. (2012) suggest success conditions for joint knowledge production to facilitate a productive participatory process. In terms of actor selection, they recommend the broadest actor involvement feasible within strategic and practical limits. Actor identification matrixes, scientific literature, snowballing techniques aid in identifying relevant participants for coastal tourism vulnerability. Involving experts helps to get access to existing scientific and policy information on the study.

Discussions

This study has indicated that vulnerability is dynamic and that current scientific approaches for tourism are static. If decision-makers and researchers want to understand who and what is vulnerable and how these vulnerabilities are attenuated or amplified through human-human and human-environment interactions and what can be done to limit vulnerabilities, a dynamic approach that considers diverse and complex interactions is essential (Turner et al. 2003). A few similar studies within tourism indicate the potential of exploring interactions and involving actors. For example, Balbi et al. (2013) used ABM to explore various actor strategies and climate scenarios to study the effects on tourism in the Italian alps and Soboll and Schmude (2011) explored the supply side of tourism ski areas and adapted their agent-based model to analyse human-environment interactions. Outside of the tourism domain are examples of the companion modelling approach. For example, in a study of Senegalese farmers, Barreteau et al. (2000) indicated the usefulness of combining ABM with role-playing games to improve coordination among local actors in a companion modelling approach.

Static approaches offer a sense of control and clarity by developing indicators and measurements for evaluating risk. By looking at specific events or assuming that new vulnerabilities do not emerge (that current vulnerabilities are an indication of future vulnerabilities), the variety of adaptive measure taken are limited and promotes choosing and supporting/maintaining one best solution. With individual events assessments, it is very likely that eventually the event will occur and that the approach predicts that event and how individuals can prepare for that single event. However, critical events are only a part of the vulnerability that local actors experience. The approach also assumes that the people, resources and abilities available at the current moment of time will be available in the future and during the critical event.

Dynamic processes are not focused on prediction. Rather, the focus is on improving understanding of the system. A dynamic approach is not reinventing the wheel. Instead, it takes aspects that static approaches have taught us and puts them in motion. Transdisciplinarity improves the adaptive process as it enables a different range of solutions and approaches. Vulnerability does not affect the biophysical environment nor people in isolation. Rather, vulnerability affects the interactions among people and their environment. One knowledge domain is insufficient to understand these interactions. Transdisciplinary approaches incorporate a wider body of knowledge, which helps assess the transdisciplinary challenge of climate change (e.g. Hegger et al. 2012). Interactive modelling through tools such as companion modelling aids transdisciplinary collaboration. The joint conceptual framework and exploration of role-playing games and computer simulations leads to an understanding of the underlying processes.

Despite including major human-human and human-environment interactions, it is not feasible to comprehensively consider the whole system and all its interactions (Turner et al. 2003). It remains important to be aware that each system is complex, involves stochasticity and is nonlinear. Dynamic future-focused research necessarily deals with uncertainty. However, uncertainty should not paralyse decision making processes (Kaján and Saarinen 2013; Scott 2011). Understanding of where uncertainties lie and how they can develop is more helpful than non-understanding of the unknown.

Taking a dynamic approach necessitates researchers giving up control of the end of product and sharing ideas with non-experts. Nonetheless, focusing on vulnerability's dynamic nature enables more flexibility in thinking. Both actors and researchers learn more about the system during interactive processes. Aggregate information is less useful for decision-making in a local context. Vulnerability approaches are more effective in understanding vulnerabilities and improving adaptive capacities when the local context is considered, when some of the interactions and complexity are identified and when the approaches provide a means for improving decision-making and implementation (Turner et al. 2003).

Conclusions

Vulnerability and adaption are continual processes as they affect and are affected by human-human and human-environment interactions. Research has shown that vulnerability and adaptive capacity are dynamic and demonstrate a growing need for new approaches to study this challenge. This research responds to the need for a dynamic conceptualisation of vulnerability as aggregate vulnerabilities are not enough for us to understand who and what is vulnerable and how these vulnerabilities emerge. This paper has argued that a dynamic problem (vulnerability) requires a dynamic approach. Understanding interactions is crucial, but how to approach this problem is less clear. The study has identified possible ways to operationalise a dynamic approach using an interactive modelling and engaging local actors. Interactive modelling (using a companion modelling approach) is a promising tool to conceptualise vulnerability and adaptive capacity as dynamic phenomena. Engaging local actors and experts throughout the process of conceptualising improves understanding of the system for both researchers and local actors.

Moreover, this study also aligns with previous research that suggest that tourism destinations' adaptive capacity deserves more focus (Kaján and Saarinen 2013). Few studies have focused explicitly on adaptation for coastal tourism and most climate change tourism literature to date has focused on North America, Western Europe, New Zealand and Australia (Becken 2013a). Moreover, transdisciplinary studies are limited (ibid). By engaging local actors in the process, both researchers and those affected by climate change gain a better understanding of the macro problem and the interconnectedness. Diverse transdisciplinary approaches help manage complex questions such as vulnerability (Eakin and Luers 2006). Improved understanding of vulnerability may lead to new insights for adaptive management in tourism destinations.

This process will be further adjusted to coastal tourism and will focus on a local tourism destination. Nonetheless, wider applications of this approach exist. First, by adjusting and applying the approach to other local Caribbean destinations, similar and/or distinct patterns and interactions can be identified. This dynamic process even has applications outside of tourism science. Companion modelling has largely been used for agricultural human-environment systems, but can be adjusted to analyse vulnerability in other human-environment contexts.

References

Adger W (2006) Vulnerability. Glob Environ Chang 16(3):268–281. doi:10.1016/j.gloenvcha.2006.02.006

Balbi S, Giupponi C, Perez P, Alberti M (2013) A spatial agent-based model for assessing strategies of adaptation to climate and tourism demand changes in an alpine tourism destination. Environ Model Softw 45:29–51

Barreteau O, Bousquet F, Attonaty J (2000) Role-playing games for opening the black box of multi-agent systems: method and lessons of its application to Senegal River. J Artif Soc Soc Simul 4(2):5

Barreteau O, Bots P, Daniell K (2010) A framework for clarifying "participation" in participatory research to prevent its rejection for the wrong reasons. Ecol Soc 15(2):1. http://www.ecologyandsociety.org/vol15/iss2/art1/

Becken S (2013a) A review of tourism and climate change as an evolving knowledge domain. Tour Manage Perspect 6:53–62. doi:10.1016/j.tmp.2012.11.006

Becken S (2013b) Developing a framework for assessing resilience of tourism sub-systems to climatic factors. Ann Tour Res 43:506–528. doi:10.1016/j.annals.2013.06.002

Bonabeau E (2002) Agent-based modeling: methods and techniques for simulating human systems. Proc Natl Acad Sci U S A 99(Suppl 3):7280–7287

Brooks N (2003) Vulnerability, risk and adaptation: a conceptual framework. Tyndall Centre Clim Chang Res Working Paper 38:1–16. http://www.tyndall.ac.uk/sites/default/files/wp38.pdf

Brown C, Ghile Y, Laverty M, Li K (2012) Decision scaling: linking bottom-up vulnerability analysis with climate projections in the water sector. Water Resour Res 48(9), W09537 1–12. doi:10.1029/2011WR011212

Csete M, Szécsi N (2015) The role of tourism management in adaptation to climate change – a study of a European inland area with a diversified tourism supply. J Sustain Tour 23(3):$ 32#477–496. doi:10.1080/09669582.2014.969735

Eakin H, Luers A (2006) Assessing the vulnerability of social-environmental systems. Annu Rev Environ Resour 31(1):365–394. doi:10.1146/annurev.energy.30.050504.144352

Etienne M (2011) Companion modelling – a participatory approach to support sustainable development. Quæ, Versailles. doi:10.1007/978-94-017-8557-0

Füssel H (2007) Vulnerability: a generally applicable conceptual framework for climate change research. Glob Environ Chang 17(2):155–167. doi:10.1016/j.gloenvcha.2006.05.002

Hegger D, Lamers M, Van Zeijl-Rozema A, Dieperink C (2012) Conceptualising joint knowledge production in regional climate change adaptation projects: success conditions and levers for action. Environ Sci Policy 18:52–65. doi:10.1016/j.envsci.2012.01.002

Johnson P, Sieber R (2011) An agent-based approach to providing tourism planning support. Environ Plann B 38(3):486–504. doi:10.1068/b35148

Kaján E, Saarinen J (2013) Tourism, climate change and adaptation: a review. Curr Issues Tour 16(2):167–195. doi:10.1080/13683500.2013.774323

Klein R, Juhola S (2014) A framework for Nordic actor-oriented climate adaptation research. Environ Sci Policy 40:101–115. doi:10.1016/j.envsci.2014.01.011

Mercer J (2010) Policy arena disaster risk reduction or climate adaptation: are we reinventing the wheel? J Int Dev 22:247–264. doi:10.1002/jid

Moreno A, Becken S (2009) A climate change vulnerability assessment methodology for coastal tourism. J Sustain Tour 17(4):473–488. doi:10.1080/09669580802651681

Perch-Nielsen S (2010) The vulnerability of beach tourism to climate change-an index approach. Clim Chang 100:579–606. doi:10.1007/s10584-009-9692-1

Polsky C, Neff R, Yarnal B (2007) Building comparable global change vulnerability assessments: the vulnerability scoping diagram. Glob Environ Chang 17(3–4):472–485. doi:10.1016/j.gloenvcha.2007.01.005

Schroter D, Polsky C, Patt A (2005) Assessing vulnerabilities to the effects of global change: an eight step approach. Mitig Adapt Strateg Glob Chang 10:573–596

Scott D (2011) Why sustainable tourism must address climate change. J Sustain Tour 19(1):17–34. doi:10.1080/09669582.2010.539694

Scott D, Simpson M, Sim R (2012) The vulnerability of Caribbean coastal tourism to scenarios of climate change related sea level rise. J Sustain Tour 20(6):883–898. doi:10.1080/09669582.2012.699063

Shakeela A, Becken S (2014) Understanding tourism leaders' perceptions of risks from climate change: an assessment of policy-making processes in the Maldives using the

social amplification of risk framework (SARF). J Sustain Tour 23(1):1–20. doi:10.1080/09669582.2014.918135
Smit B, Wandel J (2006) Adaptation, adaptive capacity and vulnerability. Glob Environ Chang 16(3):282–292. doi:10.1016/j.gloenvcha.2006.03.008
Soboll A, Schmude J (2011) Simulating tourism water consumption under climate change conditions using agent-based modeling: the example of ski areas. Ann Assoc Am Geogr 101(5):1049–1066. doi:10.1080/00045608.2011.561126
Thomalla F, Downing T, Spanger-Siegfried E, Han G, Rockström J (2006) Reducing hazard vulnerability: towards a common approach between disaster risk reduction and climate adaptation. Disasters 30(1):39–48. doi:10.1111/j.1467-9523.2006.00305.x
Turner B, Kasperson R, Matson P, McCarthy J, Corell R, Christensen L, Schiller A (2003) A framework for vulnerability analysis in sustainability science. Proc Natl Acad Sci U S A 100(14):8074–8079. doi:10.1073/pnas.1231335100
UNWTO-UNEP-WMO (2008) Climate change and tourism: responding to global challenges climate change and tourism responding to global challenges. Madrid. Retrieved from http://sdt.unwto.org/sites/all/files/docpdf/climate2008.pdf
World Travel and Tourism Council (2013) Travel and tourism economic impact 2013 Caribbean. World Travel and Tourism Council, London

Chapter 4
Climate Injustice in a Post-industrial City: The Case of Greater Manchester, UK

Aleksandra Kazmierczak

Abstract Whilst weather extremes are currently rarely experienced in Greater Manchester, UK, under the changing climate the temperatures are projected to rise and heatwaves are likely to become more frequent. This may be particularly dangerous to people considered to be vulnerable to excessive heat: those in poor health, young or old age, and those isolated from others because of cultural differences or sparse social networks. The risk of harm to people caused by high temperatures may be exacerbated by the urban morphology of the post-industrial conurbation, including the distribution of green spaces and the housing conditions.

This paper explores the risk of high temperatures to vulnerable communities in Greater Manchester, UK. It investigates the spatial distribution of the factors contributing to social vulnerability and the neighbourhood qualities affecting exposure to high temperatures in relation to urban heat island. The results suggest that more diverse communities and people living in rented accommodation and in poor quality housing are likely to be at the greatest risk of high temperatures. The paper concludes by proposing neighbourhood-level adaptation measures targeting the physical environment that could address this climate injustice.

Keywords Climate justice • Social vulnerability • Heatwaves • United Kingdom • Urban areas

Introduction

As the climate change becomes reality rather than distant threat, and the frequency and magnitude of extreme weather events intensifies, the researchers and policy makers turn their attention to the impacts the changing climate is likely to

A. Kazmierczak, M.Sc., Ph.D. (✉)
School of Planning and Geography, Cardiff University, Glamorgan Building, King Edward VII Avenue, Cardiff CF10 3WA, Wales, UK
e-mail: KazmierczakA@cardiff.ac.uk

W. Leal Filho (ed.), *Innovation in Climate Change Adaptation*, Climate Change Management, DOI 10.1007/978-3-319-25814-0_4

have on the people, in particular where the effects are going to be felt and who is going to be affected by the changing climate. The recent Intergovernmental Panel on Climate Change report notes that climate change risks are unevenly distributed and are generally greater for disadvantaged people and communities in countries at all levels of development (IPCC 2014). Thus, the issues of social justice in relation to climate change start to emerge: are some individuals and communities likely to be more affected than others? Who are the most vulnerable and the most affected people? Can we create climate justice—can we ensure that collectively and individually we have the ability to prepare for, respond to and recover from climate change impacts by considering existing vulnerabilities, resources and capabilities (Preston et al. 2014)?

This paper aims to contribute to the climate justice debate by investigating how the effects of increasingly high summer temperatures are likely to be spatially distributed among the urban communities in Greater Manchester, UK, taking into account their vulnerability, and how the urban environment can be changed at the neighbourhood level to address the risks of high temperatures. It starts from an overview of the literature relating to the impact of heatwaves on human health under the changing climate, and the factors influencing the vulnerability of communities to high temperatures. It then goes on to introduce the case study area and describe the methods used in the paper. The findings presents the analysis of the spatial distribution of vulnerable communities against urban heat island, green spaces and types of housing. Then the possible adaptation responses at the neighbourhood level are suggested and the paper concludes with recommendations for action and further research.

Literature Review

Climate change is seen as the biggest threat to public health this century (Costello et al. 2009). Amongst other effects, exposure to extreme and prolonged heat can be deadly; in post-industrial societies nearly 95 % of recorded human deaths that result from natural hazards are attributed to extreme temperatures (McGeehin and Mirabelli 2001). However, the association between high temperatures and human mortality varies greatly by climatic zone. In tropical regions, where temperatures are higher for longer periods of time, the temperature extremes do not significantly impact on weather-related mortality. On the contrary, severe but infrequent temperature fluctuations in temperate region during periods of generally milder weather conditions are associated with increases in weather-related mortality (McGeehin and Mirabelli 2001). This is why heatwaves, understood in the UK context as periods of at least two consecutive days with maximum temperatures exceeding 30 °C, and the temperature on intervening night above 15 °C (Met Office no date) are so dangerous. For example, the heatwave experienced in Europe in August 2003, with peak temperatures ranging from 38.5 °C in England to 47 °C in Portugal (Poumadere et al. 2005), caused an estimated 70,000 additional deaths across

Europe (Robine et al. 2008). In England alone, between 4 and 13 August 2003 there were 2091 (17 %) more deaths (ONS 2005). The projected changes in climate suggest that high summer temperatures and heatwaves will occur more frequently, both globally and in the UK (Meehl et al. 2007; Jenkins et al. 2009). For example, average summer night-time temperatures in the UK in the 2050s are projected to be around 2–3 °C higher than today (based on the central estimate of the medium emissions scenario relative to 1961–1990; Capon and Oakley 2012). This is likely to have implications for human comfort and health.

High temperatures do not affect all people in the same manner; some people are more susceptible to harm than others due to their different capacities to deal with hazards. Firstly, poor health can affect the body's ability to sweat and regulate its temperature (Semenza et al. 1999; NHS 2009). For example, high temperatures have been found to increase heat-related emergency admissions to London hospitals for respiratory disease and renal disease (Kovats et al. 2004). Other conditions that affect an individual's ability to adapt their behaviour to keep cool include nervous system disorders, Parkinson's or Alzheimer's diseases, epilepsy, having a disability or being bed-bound and unable to care for themselves (Semenza et al. 1999; NHS 2009), as well as mental-health illnesses (Kaiser et al. 2001).

Secondly, children and the elderly are more at risk during hot weather spells; those aged under 4 or over 65 are the most frequently admitted to hospitals during heatwaves (Knowlton et al. 2009). This is because the old and the very young tend to have reduced heat-regulating mechanisms (Center for Disease and Control Prevention 1993). They are more likely to have restricted mobility and/or cognitive capacity that may result in diminished control over their environments, including access to fluids and ability to open the windows (McGeehin and Mirabelli 2001). During the 2003 heatwave, deaths in England for people over the age of 75 increased by 23 % compared to the expected number (ONS 2005). Older people are more prone to heat-related illnesses for physiological reasons, such as impaired thermoregulation, reduced cardiovascular fitness or kidney functions (Semenza et al. 1999; Kovats and Ebi 2006). As a consequence, the elderly have been the most numerous victims of heatwaves (Hajat et al. 2006; Semenza et al. 1996; NHS 2009). Infants and small children are also more prone to heat-related illnesses (Kovats et al. 2004), due to their limited ability to take adaptive actions (NHS 2009). Children with certain predisposing illnesses such as diarrhoea, respiratory tract infections, and neurologic defects will be more at risk of hyperthermia during extreme heat (McGeehin and Mirabelli 2001).

Thirdly, the material situation of people seems to have an effect on temperature-related mortality (Kovats and Ebi 2006). One explanation is that populations of lower socio-economic status may not have access to air-conditioning because of the cost of an unit or utility bills (Semenza et al. 1999). Also, the fear of crime in poorer neighbourhoods may prevent people from opening the windows in their homes to provide appropriate ventilation, particularly during the night (Lindley et al. 2011).

In addition, individuals and communities unable to speak or read the official language, and thus not able to understand the heatwave warnings or guidance, may be particularly vulnerable to extreme weather events (McGeehin and Mirabelli

2001). In addition, cultural differences may hamper support from emergency management services due to misunderstandings over their nature and intent (Thrush et al. 2005). Finally, living arrangements and the availability and proximity of social networks affect vulnerability to high temperatures. People living on their own tend to be more vulnerable during heatwaves (McGeehin and Mirabelli 2001); during the 2003 heatwave 92 % of victims in France lived alone (Poumadere et al. 2005).

How badly the vulnerable people are affected by heatwaves depends also on the quality of their environment. In cities, the temperatures are usually intensified by the urban heat island phenomenon, whereby urban areas exhibit higher temperatures relative to their surroundings. This is due to the modification of energy balances through the complex topography and mass of buildings, the emission of heat from anthropogenic activities, and the replacement of vegetation with hard surfaces (Smith et al. 2011). Thus, the intensity of the urban heat island and therefore the effects of high temperatures of vulnerable communities can be mediated by the characteristics of the urban environment, such as the presence of green space or the characteristics of housing.

Urban parks have been proven to be on average 1 °C cooler than built-up areas, with larger parks having a greater cooling effect (Bowler et al. 2010); measurements taken around Kensington Gardens (London) showed temperature reductions of up to 4 °C and a cooling effect being felt up to distances of around 400 m (Doick et al. 2014). The cooling effect is pronounced during the night-time when heatwaves can be particularly problematic. Also, small-scale interventions such as green roofs and street trees can lower the temperatures in densely built-up urban areas with limited opportunities to establish new green spaces (Speak et al. 2013; Skelhorn et al. 2014); tree shading has been found to reduce surface temperatures by 15–20 °C, and the air temperature by 5–7 °C, significantly improving human comfort (Ennos 2011).

The type of houses that people live in also affect their exposure to high temperatures. Heat rises, and is easily transferred through thin roofs, so people residing on the top floor of apartments blocks experience higher rates of heat-related morbidity and mortality than those living on lower floors (Semenza et al. 1999). In Paris, just over half the victims of the 2003 heatwave lived on the two highest floors in traditional 'service rooms', commonly occupied by the elderly (Poumadere et al. 2005). Those living in rented accommodation may be more vulnerable as they usually are not permitted to make changes to the properties they live in and rely on the landlords to provide temperature regulation measures. Further, houses in the private sector in the UK tend to tend to have lower energy ratings than socially rented or owner-occupied ones (CLG 2013). This may mean that they are more prone not only to excess cold but also to overheating.

The measures that can be used to mitigate overheating in houses focus on reducing heat gain through windows and through the warming of external surfaces. These measures, the majority of which can be used as retrofits for existing buildings, include: planting deciduous trees to the south-west and south-east of buildings for shade; using low-energy double-glazing, awnings, shutters or blinds for the windows; roof and loft insulation; increased reflectivity of roofs and facades through light colours; wall insulation; green roofs (Porritt et al. 2010; EST 2005; EPA no date; Arup 2008). Porritt et al. (2010) modelled the impact on internal temperatures of some of these options and found that external wall insulation was the most effective intervention, followed by external window shutters, internal wall insulation, and painting the walls a lighter colour to reduce heat absorption. Therefore, relatively small changes to the buildings can increase human comfort.

This paper explores firstly the extent to which the spatial distribution of vulnerable communities coincides with the physical characteristics of the environment that can mitigate or exacerbate the impact of high temperatures on vulnerable communities using the Greater Manchester as the case study area. It then goes on to discuss the physical adaptation measures that can be used in a vulnerable neighbourhood to better adapt it to the changing climate.

Methods

Greater Manchester as the Case Study Area

The conurbation of Greater Manchester (GM) in the North West of England covers 1276 km^2 and is a home to over 2.5 million people. It is located on a river basin flanked by the Pennine hills in the north and east and stretching to lowland areas to the south and west. GM evolved during the industrial revolution in the nineteenth century from several towns. The conurbation continued to grow until 1960s until, following the global economic turn, it went through the process of post-industrial change, including significant job losses, rises in unemployment and poverty and an increase in the magnitude of social problems. In recent decades, urban regeneration in GM has made it the largest cluster for finance, law, media, research and higher education in England outside of London. However, due to the out-movement of wealthier residents to the suburbs and peri-urban areas, the conurbation is characterised by great income inequalities. Moreover, throughout its history, GM has been a multi-ethnic centre for many groups, more recently including South Asian, Caribbean and Chinese. Therefore, the socio-economic characteristics of this conurbation make it a case study relevant to other British and European post-industrial cities (Kazmierczak and Cavan 2011).

GM currently has a temperate climate, with mean annual temperature of 7.5–9.4 °C, depending on the altitude. Summer mean daily maximum temperature ranges between 16 and 20 °C. The warmest day in summer reaches between 25 and 27 °C; and the warmest night varies between 15 and 18 °C (Cavan 2011). However, under the high emissions scenario by the 2050s, the increase in the temperature of the warmest day of the summer is very unlikely to be less than 1.5 °C and very unlikely to exceed 6 °C; whilst it is unlikely that there will not be any days exceeding 30 °C, there is a small likelihood that even up to 25 days in the summer may reach over 30 °C. Similarly, there is a small likelihood that increase in the

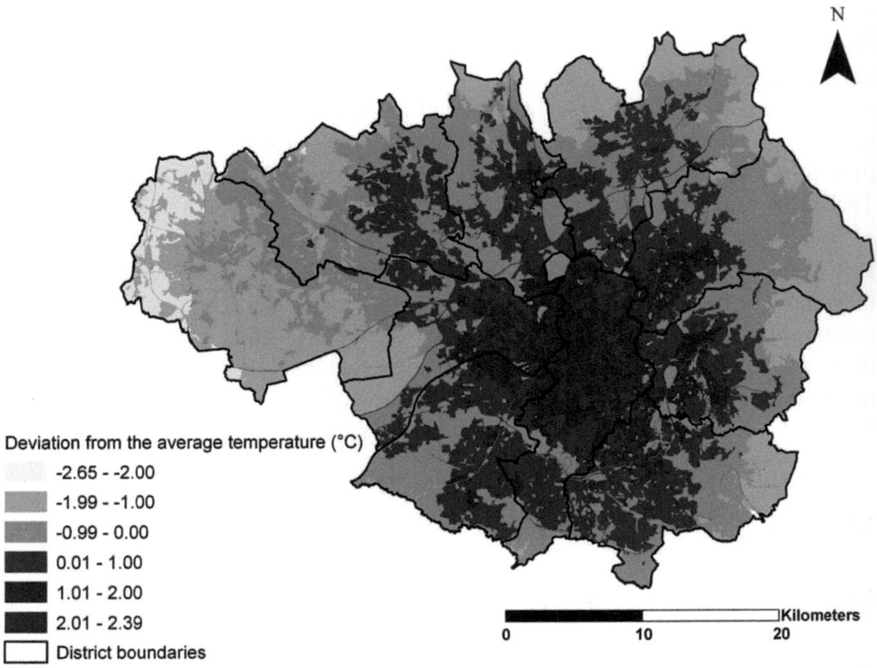

Fig. 4.1 The extent and intensity of UHI in GM: the deviation of surface temperatures from the average surface temperature in GM (Smith et al. 2011). Base map is © Crown Copyright/database right (2009). An Ordnance Survey/EDINA supplied service

temperature of the warmest night of the summer will increase up to 20–22 °C (Cavan 2011), thus offering no respite from the heat during the day. Thus, by the 2050s, a heatwave may be expected nearly every summer; there is a small likelihood that there will be several heatwave events a year (Cavan 2011). Therefore, in the future, the risk of thermal discomfort, morbidity and mortality, associated with high temperatures, is highly likely to increase.

The urban heat island intensity is used in this paper as a basis of understanding where heatwaves are likely to be exacerbated by the urban environment and thus have the greatest effect on people. The extent and intensity of the urban heat island (UHI) was modelled in GM based on the measurements of air and surface temperatures, and taking into consideration land use types; distances from urban centres; and building geometry (Smith et al. 2011) (Fig. 4.1).

Vulnerability of Communities in Greater Manchester

Taking into consideration the multiple factors affecting people's ability to deal with high temperatures, Kazmierczak and Cavan (2011) investigated the vulnerability of communities in GM to extreme weather events by carrying out Principal Component Analysis of a number of indicators related to people's age, health, living arrangements, ease of access to information and extent of social networks. The analysis of vulnerability was carried out for the spatial unit of Lower Super Output Areas (LSOA), which are compact areas of homogenous socio-economic characteristics and an average population of around 1500 people, constrained by the boundaries of the electoral wards used by the Office of National Statistics to report small area statistics across England and Wales (ONS 2008). More details about the analysis and methodology can be found in Kazmierczak and Cavan (2011).

Four different underlying aspects of vulnerability were identified by Kazmierczak and Cavan (2011). These were: (1) poverty and poor health; (2) diverse, dense and transient communities; high proportions of (3) children or (4) elderly in the population. Poorer and more diverse communities tend to concentrate in the highly urbanised areas close to the centre of the conurbation (although there are pockets of material deprivation in more remote locations). High proportions of the elderly and children are associated with more suburban locations (Fig. 4.2).

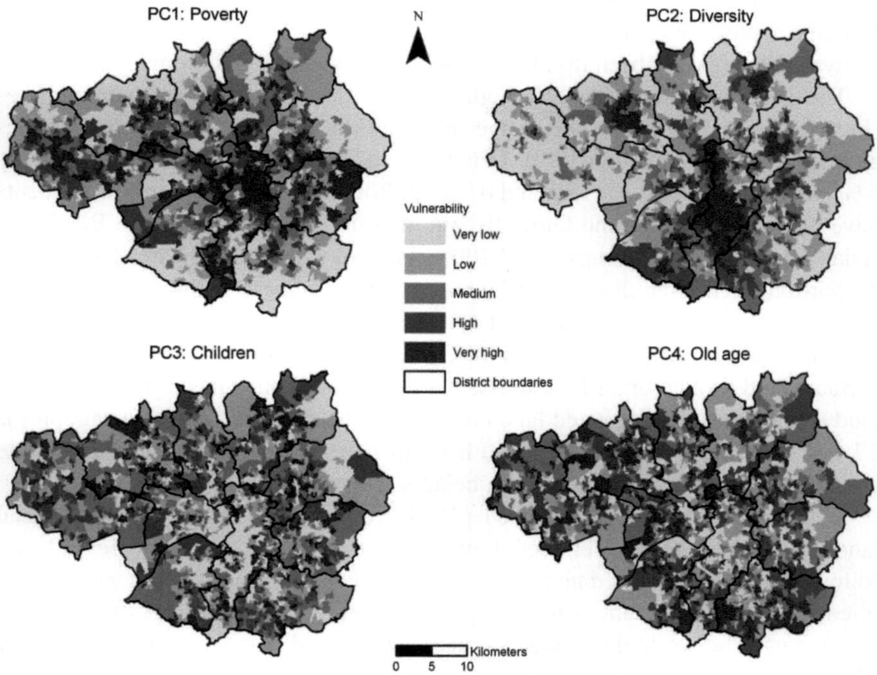

Fig. 4.2 Spatial distribution of different aspects of vulnerability of people and communities in GM (Kazmierczak and Cavan 2011)

Spatial Analysis of Associations Between Vulnerability and the Characteristics of the Urban Environment

The spatial data on vulnerable communities was overlaid with the intensity of urban heat island in order to analyse where the heatwaves associated with climate change are likely to have the greatest impact. Further, the vulnerability level for LSOAs was juxtaposed with the characteristics of the urban environment that may offset or exacerbate the impact of high temperatures on vulnerable communities: the proportion of green space (estimated based on the Urban Morphology Types, Gill et al. (2007) and the type of housing: percentage of flats (ONS 2002) and percentage of houses in poor quality (CLG 2008).

An Urban Neighbourhood as a Focus for Adaptive Actions

Following the analysis at the conurbation scale, an urban neighbourhood classed as vulnerable and located within the UHI extent was chosen for further investigation. The selected neighbourhood area covers around 1.96 km^2 (1.4 × 1.4 km). It includes various types of inner-city residential areas; major roads and quieter residential roads; parks and informal green spaces (Fig. 4.3). An analysis of the maximum surface temperatures (Smith et al) suggests that the deviations in the case study area from the average temperature in GM can reach up to 2.3 °C. For the purposes of the analysis, the area has been divided into 49 grid squares of 200 m by 200 m.

The area is characterised by high vulnerability of people and communities (Fig. 4.4). This is mainly due to the high cultural diversity of the communities residing in the area and high levels of poverty and poor health. The area is in the top 5 % most deprived areas in England (CLG 2008). On average, 28 % of the residents have been born outside the Great Britain, and nearly 55 % belong to Black and Asian minority ethnic groups (ONS 2002). Parts of the case study area have a high proportion of children that further increases vulnerability to high temperatures.

The area on average has a good provision of green space, although it varies between different locations (Fig. 4.8). One-fifth of the area is covered by natural surfaces, and another one-fifth is classed as gardens (although this signifies a type of land use, rather than vegetated land cover). The mean coverage of tree canopies is 17.4 % (Red Rose Forest 2011), with up to half of the trees located in private gardens, and a significant proportion being street trees.

Housing is predominantly rented (21 % from private and 50 % from social landlords; ONS 2002). Terraced houses are the main dwelling type (73 %), followed by semi-detached houses 14 %) and flats (11 %; Fig. 4.9). Since building orientation is an important factor in exposure to overheating, the proportion of walls facing particular world directions was also investigated (based on data provided by EPSRC SCORCHIO EP/E017428/1, Newcastle University). The walls exposed to south, south east and south west form 30.9 % of the total length of the residential building walls. The buildings potentially most exposed to sun, and thus potentially

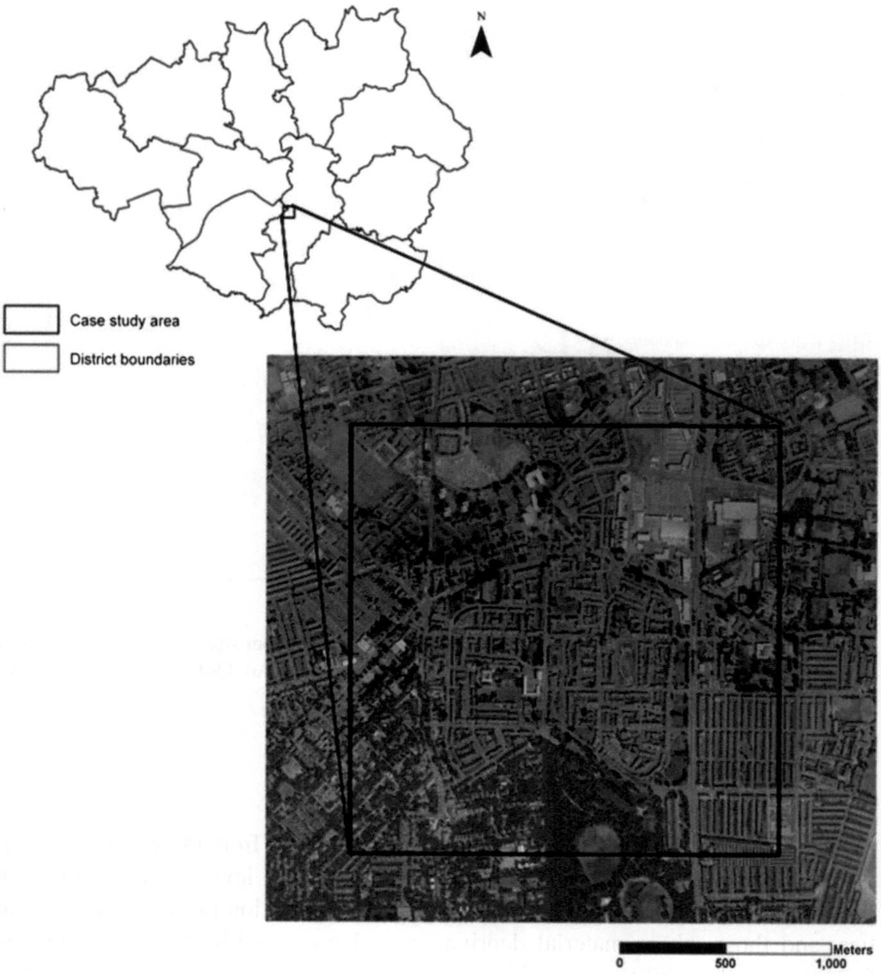

Fig. 4.3 The urban neighbourhood selected for further investigation. Aerial photography from Cities Revealed © The GeoInformation Group, 2008. Base map is © Crown Copyright/database right (2009). An Ordnance Survey/EDINA supplied service

prone to overheating, are present in the west part of the case study area (Fig. 4.9). The physical characteristics of the neighbourhood have been taken into consideration when suggesting potential adaptive actions reducing the risks from high temperatures to vulnerable communities.

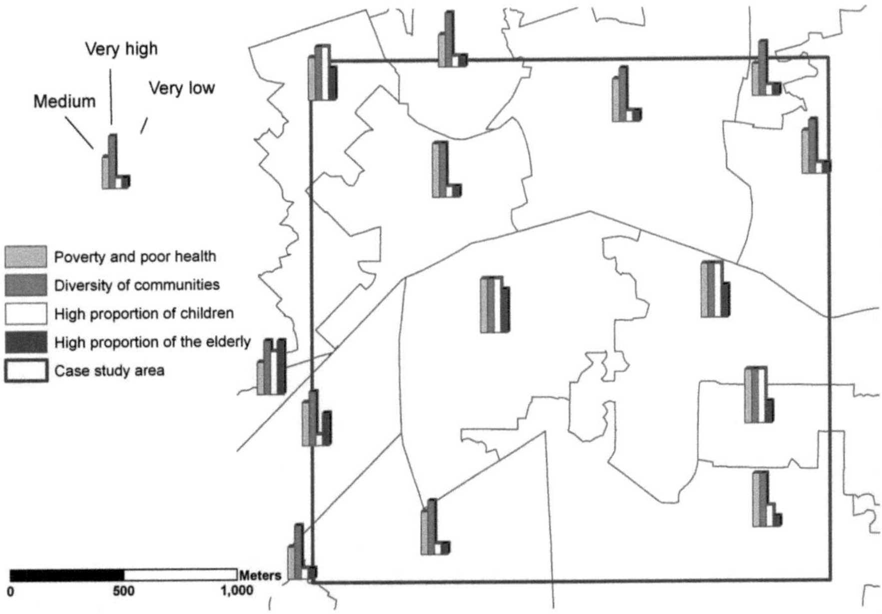

Fig. 4.4 Vulnerability of the communities in the Lower Super Output Areas within the case study area. Base map is © Crown Copyright/database right (2009). An Ordnance Survey/EDINA supplied service

Findings

Figure 4.5 presents the associations between the deviation from the average temperature in GM (the intensity of urban heat island) and the level of vulnerability of people and communities per LSOA. The spatial distribution of diverse communities, and those where material deprivation and poor health are the dominating factors of vulnerability, is positively associated with increasing temperatures. In particular, those areas that are vulnerable due to community diversity tend to be located in highly urbanised areas close to city and town centres and are more exposed to high temperatures. On the contrary, the exposure to the UHI of communities with high proportions of the elderly in their populations decreases slightly with the increasing vulnerability. No trends are present in relation to the vulnerability associated with a high proportion of children in the population.

There is a generally inverse relationship between the proportion of greenspace and the intensity of the urban heat island. Consistently with the higher concentrations of poorer and more diverse communities within the urban heat island, increasing vulnerability due to these factors is associated with the decreasing proportion of green space in the area where they live (Fig. 4.6). No such associations can be found in the case of areas with a particularly high proportion of the very young or older people. This suggests that areas inhabited by communities that are vulnerable due to socio-economic factors such as income and cultural diversity

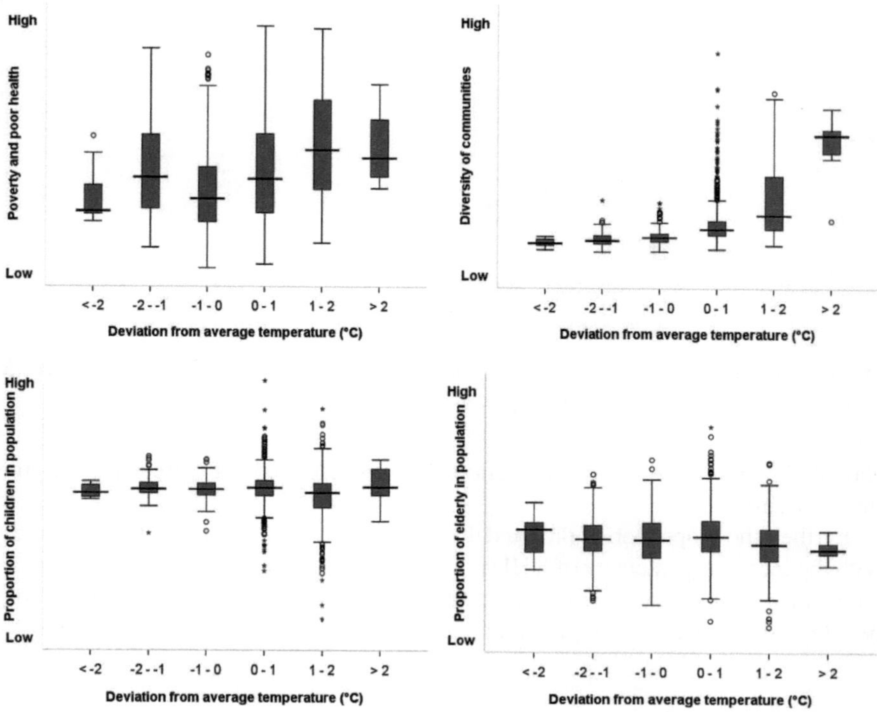

Fig. 4.5 The associations between the different aspects of vulnerability of communities and the intensity of urban heat island. The boundaries of the *box* are Tukey's hinges. The median is identified by a *line* inside the *box*. The length of the *box* is the interquartile range (IQR) computed from Tukey's hinges. Values more than 1.5 IQRs but less than 3 IQRs from the end of the *box* are labelled as outliers (*circle*). Values more than three IQRs from the end of a *box* are labelled as extreme outliers (asterisk)

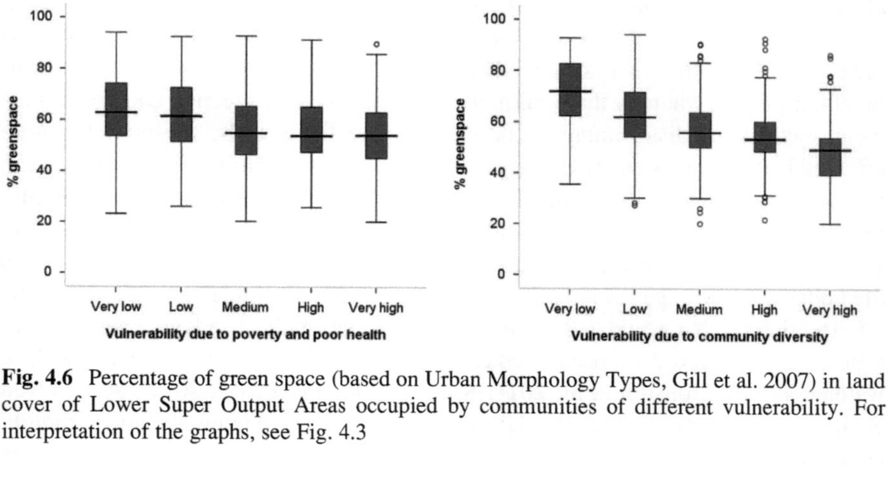

Fig. 4.6 Percentage of green space (based on Urban Morphology Types, Gill et al. 2007) in land cover of Lower Super Output Areas occupied by communities of different vulnerability. For interpretation of the graphs, see Fig. 4.3

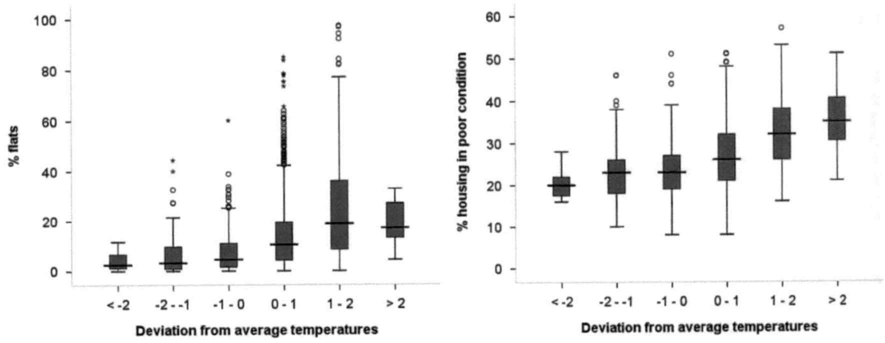

Fig. 4.7 Associations between the proportion of flats and houses of poor quality and the intensity of the UHI. For interpretation of the graphs, see Fig. 4.3

should be prioritised for adaptation actions focused on greening of the urban environment.

Further, the proportion of flats and dwellings in poor condition rises significantly with the increasing intensity of UHI (Fig. 4.7). This indicates that housing improvements may be necessary in order to ensure that the houses inhabited by vulnerable people remain cool in the high temperatures and offer refuge during heatwaves.

Discussion

Climate Injustice in Greater Manchester

The previous sections identified that the future higher temperatures are likely to affect people residing in GM, in particular the more deprived and diverse sectors of the society living in rented houses and flats of poorer quality, in the more urbanised locations with little green space. This clearly represents an example of environmental injustice, whereby the wealthier communities enjoy greener, cooler environments, whilst the disadvantaged communities are likely to be substantially more affected by heatwaves due to the physical qualities of their immediate environment. It reflects the general UK trend, where poorer communities and those with a higher proportion of people from Black and Ethnic Minority backgrounds tend to live in areas with less green space (CABE 2010). More specifically, the paper also adds to the body of knowledge investigating the exposure of vulnerable communities in the UK to climate change-related events, whereby materially disadvantaged communities tend to be more exposed to coastal flooding (Walker et al. 2006) and surface water flooding (Houston et al. 2011; Kazmierczak and Cavan 2011). Also, Oven et al. (2012) found that in the UK many areas experiencing the most rapidly changing hazards coincided with the places where the proportion of older people was projected to increase, thus potentially further increasing the risks to older populations.

Therefore, this paper emphasises the extent of social inequalities in relation to climate change impacts, in the context of a post-industrial metropolitan area, which may reflect the situation in other British and European cities of similar history. It also calls for actions reducing the potential impacts of high temperatures on vulnerable communities. This could be done, firstly, by addressing the underlying causes of vulnerability. However, the tangled social, economic and demographic issues contributing to vulnerability in this deprived and culturally mixed conurbation may take decades to resolve. On the contrary, the measures targeting the physical aspects of the urban environment, which could provide a protective buffer between the vulnerable communities and the high temperatures, are relatively easy to implement. Therefore, next section proposes a number of such 'easy fixes' at the scale of the urban neighbourhood that can make the area better prepared for the future with higher temperatures.

Adaptation Responses at the Neighbourhood Level

The suggested adaptation responses take into account the physical characteristics of the urban neighbourhoods and are targeted at the land cover (Fig. 4.8) and buildings (Fig. 4.9). The first set of actions focuses mainly on the improved management and accessibility of current green spaces, such as parks, and providing additional greening in the densely built-up environment in the form of trees in private gardens, street trees and green roofs. Another adaptation measure suggested is a gradual introduction of reflective road and pavement surfaces (for example, when scheduled rod works take place), in particular in the residential areas. It has been estimated that every 10 % increase in solar reflectance could decrease surface temperatures by 4 °C (EPA no date).

The second set of actions is about adapting the built environment. The measures that can be used to mitigate summer overheating in houses focus on reducing heat gain through windows and through the warming of external surfaces by increasing reflectivity of walls and roofs through painting them a lighter colour, loft insulation, external and internal wall insulation, providing double glazing and shading through shutters, blinds and awnings (Porritt et al. 2010; EST 2005; Arup 2008). In areas where the houses are particularly exposed to the sun during the hottest hours due to their orientation, planting of vegetation for shading is recommended.

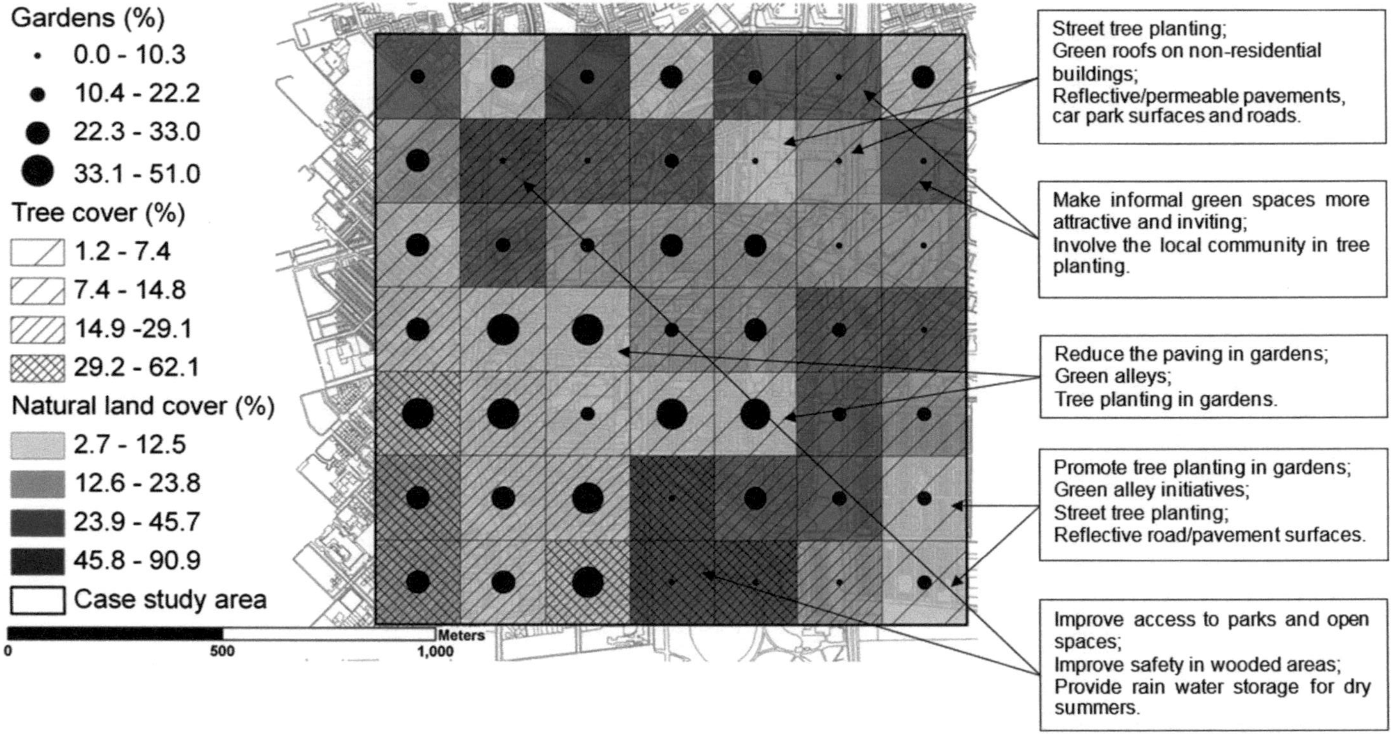

Fig. 4.8 Proposed adaptation responses in relation to land cover management. Tree cover data after Red Rose Forest (2011). Base map is © Crown Copyright/ database right (2009). An Ordnance Survey/EDINA supplied service

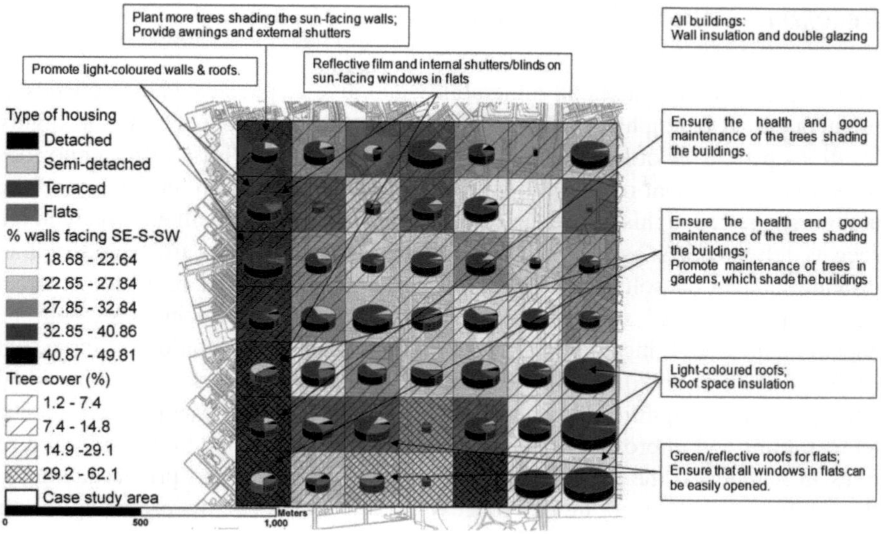

Fig. 4.9 Suggestions of adaptation responses at the building level. Type of housing and wall orientation based on data provided by EPSRC SCORCHIO EP/E017428/1, Newcastle University. Tree cover after Red Rose Forest (2011). Base map is © Crown Copyright/database right (2009). An Ordnance Survey/EDINA supplied service

Nested Scales of Climate Injustice

The investigation into the characteristics of an urban neighbourhood guiding the choice of adaptation measures has revealed a high diversity of housing types and varying green space provision (Figs. 4.8 and 4.9). This draws attention to the fact that social inequalities affecting the climate change impacts on urban communities are present at different spatial scale. This is certainly true at the international level, where the poorest countries are bearing the brunt of climate change despite having the lowest greenhouse gases emissions (IPCC 2014). Within individual countries, regional inequalities are present: in the UK, there is a North-South divide associated with patterns of flood disadvantage as the northern locations in the UK tend to have higher levels of deprivation as well as having wetter climates (Houston et al. 2011; Lindley et al. 2011), whilst the South is disproportionately affected by water poverty linked to potential water scarcity coinciding with pockets of deprivation (Benzie et al. 2011). Finally, this paper indicates that inequalities in vulnerability of communities to high temperatures are also firmly present at the scale of a conurbation; and that even within one neighbourhood the physical characteristics of the environment can mean that some people are affected by the climate change hazards more than others. This stresses the urgency of actions to reduce these inequalities.

Recommendations for Action and Research

Multiple underlying causes of vulnerability and diverse physical characteristics of urban areas also emphasise that planning adaptive actions needs to be highly location-specific in order to provide suitable solutions. Whilst local authorities have a very important role to play in climate change adaptation, in the case of the actions suggested in this paper, the changes in land cover and built environment are seen as delivered best as a collaborative effort. Local authorities are in a position to provide some of the solutions (e.g. maintenance of green spaces, provision of street trees, replacing road surfaces with reflective materials). Businesses could be encouraged through incentives or regulations to provide vegetation on large, flat roofs (Kazmierczak 2014). Maintenance of trees on private land by individual homeowners can significantly contribute to the shading in the area to help reduce temperatures and improve human comfort, as suggested by the high proportion of trees in gardens in the investigated neighbourhood. Social and private landlords should be also involved in the process of improvements by providing better insulation of the houses or providing appropriate greening.

One of the limitations of this research is its reliance on the socio-economic data representing the current vulnerability. Whilst future climate estimates are available, there are limited projections of how the socio-economic situation affecting vulnerability is likely to change in the future. It is thus recommended that future research focuses on prognoses of the future vulnerability in conjunction with climate projections in order to estimate the extent of the future climate change risks. Further, this research is based on the data that has been averaged for spatial units, which may not reflect best the individual vulnerabilities: not all people living in areas classed as vulnerable are vulnerable themselves; and, on the contrary, there would be vulnerable individuals living in areas not classed as vulnerable. Thus, extending the research to assess the vulnerability of individual people and households would be a valuable addition to the exploration of the nested scales of vulnerability.

Conclusions

This paper explored the risk of high temperatures to people and communities in GM. It has identified the types of communities at risk and investigated the characteristics of their environment that may mitigate or exacerbate the impacts of high temperatures. The results suggest that those at the highest risk to high temperatures are disadvantaged due to their material situation, cultural origin, advanced age, poor health, living in rented accommodation and in highly urbanised areas with little green space. This proves that climate injustice is present at the scale of a conurbation. However, the social justice issues can be at least partially addressed by implementing relatively simple changes in the physical environment occupied by the vulnerable communities to reduce their exposure to high temperatures. This requires a collaborative, multifaceted action by local authorities, residents, landlords and the private sector. Further research is required to provide a better understanding of the inequalities associated with climate injustice present at different spatial scales.

Acknowledgements This paper draws on research carried out by the author at The University of Manchester between 2009 and 2012 within the project Ecocities: The Bruntwood Initiative for Sustainable Cities. The author acknowledges Ordnance Survey for the use of MasterMap data. The paper has used data generated by the Sustainable Cities: Options for Responding to Climate Change Impacts and Outcomes (EPSRC SCORCHIO EP/E017428/1) project including the empirical model of the urban heat island in Greater Manchester (Smith et al. 2011); building wall orientation; and building type. The author thanks Dr Sarah Lindley for sharing this data. The author also acknowledges the Red Rose Forest for the use of the Greater Manchester Tree Audit dataset.

References

Arup (2008) Your home in a changing climate. Retrofitting existing homes for climate change impacts. Report for policy makers. Greater London Authority, London

Benzie M, Harvey A, Burningham K, Hodgson N, Siddiqui A (2011) Vulnerability to heatwaves and drought: case studies of adaptation to climate change in south-west England. Joseph Rowntree Foundation, York

Bowler DE, Buyung-Ali L, Knight TM, Pullin AS (2010) Urban greening to cool towns and cities: a systematic review of the empirical evidence. Landsc Urban Plan 97:147–155

CABE (2010) Urban green nation: building the evidence base. Commission for Architecture and the Built Environment, London

Capon R, Oakley G (2012) Climate change risk assessment for the built environment sector. Department for Environment, Food and Rural Affairs, London

Cavan G (2011) Climate change projections for Greater Manchester, EcoCities. The University of Manchester, Manchester, UK

Center for Disease and Control Prevention (1993) Heat-related deaths—United States. Morb Mortal Wkly Rep 42:558–560

CLG (2008) The English indices of deprivation 2007. Communities and Local Government, London

CLG (2013) English housing survey HOUSEHOLDS 2011–12. Office of National Statistics, London

Costello A, Abbas M, Allen A, Ball S, Bell S, Bellamy R, Friel S, Groce N, Johnson A, Kett M, Lee M, Levy C, Maslin M, McCoy D, McGuire B, Montgomery H, Napier D, Pagel C, Patel J, Puppim de Oliveira JA, Redclift N, Rees H, Rogger D, Scott J, Stephenson J, Twigg J, Wolff J, Patterson C (2009) Managing the health effects of climate change. Lancet 373:1693–1733

Doick KJ, Peace A, Hutchings TR (2014) The role of one large greenspace in mitigating London's nocturnal urban heat island. Sci Total Environ 493:662–671

Ennos R (2011) Quantifying the cooling and anti-flooding benefits of green infrastructure. Presentation at the Green Infrastructure workshop, University of Manchester, 1st November 2011

EPA (no date) Reducing urban heat islands: compendium of strategies. Environmental Protection Agency, Washington, DC

EST (2005) Energy efficiency best practice in housing. Reducing overheating – a designer's guide. Energy Saving Trust, London

Forest RR (2011) Greater Manchester tree audit dataset. Red Rose Forest, Manchester

Gill SE, Handley JF, Ennos AR, Pauleit S (2007) Adapting cities for climate change: the role of green infrastructure. Built Environ 33:115–133
Hajat S, Kovats RS, Lachowycz K (2006) Heat-related and cold-related deaths in England and Wales: who is at risk? Occup Environ Med 64:93–100
Houston D, Werritt A, Bassett D, Geddes A, Hoolachan A, McMillan M (2011) Pluvial (rain-related) flooding in urban areas: the invisible hazard. Joseph Rowntree Foundation, York
IPCC (2014) Climate change 2014: synthesis report. Contribution of Working Groups I, II and III to the 5th assessment report of the intergovernmental panel on climate change [Core Writing Team, Pachauri RK, Meyer LA (eds)]. IPCC, Geneva, 151 pp
Jenkins GJ, Perry M, Prior J (2009) The climate of the United Kingdom and recent trends. UKCIP, Oxford, UK
Kaiser R, Rubin C, Henderson A, Wolfe M, Kieszak S, Parrott C, Adcock M (2001) Heat-related death and mental illness during the 1999 Cincinnati heatwave. Am J Forensic Med Pathol 22(3):303–307
Kazmierczak A (2014) Innovative ways of supporting the establishment of green infrastructure in cities: collaboration of local authorities with investors and property owners. In: Bergier T, Kronenberg J, Lisicki P (eds) Sustainable development applications 4. Nature in the city – solutions. Sendzimir Foundation, Warsaw, pp 98–109
Kazmierczak A, Cavan G (2011) Surface water flooding risk to urban communities: analysis of vulnerability, hazard and exposure. Landsc Urban Plan 103(2):185–197
Knowlton K, Rotkin-Ellman M, King G, Margolis HG, Smith D, Solomon G, Trent R, English P (2009) The 2006 California heatwave: impacts on hospitalizations and emergency department visits. Environ Health Perspect 117(1):61–67
Kovats RS, Ebi KL (2006) Heatwaves and public health in Europe. Eur J Pub Health 16(6): 592–599
Kovats RS, Hajat S, Wilkinson P (2004) Contrasting patterns of mortality and hospital admissions during hot weather and heatwaves in Greater London, UK. Occup Environ Med 61(11): 893–898
Lindley S, O'Neill J, Kandeh J, Lawson N, Christian R, O'Neill M (2011) Climate change, justice and vulnerability. Joseph Rowntree Foundation, York
McGeehin MA, Mirabelli M (2001) The potential impacts of climate variability and change on temperature-related morbidity and mortality in the United States. Environ Health Perspect 109: 185–189
Meehl GA, Stocker TF, Collins WD, Friedlingstein P, Gaye AT, Gregory JM, Kitoh A, Knutti R, Murphy JM, Noda A, Raper SCB, Watterson IG, Weaver AJ, Zhao Z-C (2007) Global climate projections. In: Solomon S, Qin D, Manning M, Chen Z, Marquis M, Avery K, Tignor M, Miller HL (eds) Climate change 2007: the physical science basis. Contribution of working group I to the fourth assessment report of the intergovernmental panel on climate change. Cambridge University Press, Cambridge
Met Office (no date) Heat-health watch. http://www.metoffice.gov.uk/public/weather/heat-health/#?tab=heatHealth
NHS (2009) Heatwave plan for England. NHS, London
ONS (2002) Census data 2001 dataset. Office of National Statistics, London
ONS (2005) Horizons. The facts behind the figures. Issue 33. Office for National Statistics, London
ONS (2008) Names and codes for super output area geography. Office for National Statistics, London
Oven KJ et al (2012) Climate change and health and social care: defining future hazard, vulnerability and risk for infrastructure systems supporting older people's health care in England. Appl Geogr 33(1):16–24
Porritt, S.M., Shao, L., Cropper, P.C., Goodier, C.I. (2010) Building orientation and occupancy patterns and their effect on interventions to reduce overheating in dwellings during heatwaves. In: Proceedings of conference: IESD PhD conference: energy and sustainable development. Institute of Energy and Sustainable Development, De Montfort University, Leicester, UK, 21 May 2010

Poumadere M, Mays C, Le Mer S, Blong R (2005) The 2003 heatwave in France: dangerous climate change here and now. Risk Anal 25(6):1483–1494

Preston I, Banks N, Hargreaves K, Kazmierczak A, Lucas K, Mayne R, Downing C, Street R (2014) Climate change and social justice: an evidence review. Joseph Rowntree, York

Robine J-M, Cheung SLK, Le Roy S, Van Oyen H, Griffiths C, Michel J-P, Herrmann FR (2008) Death toll exceeded 70,000 in Europe during the summer of 2003. Epidemiology 331(2): 171–178

Semenza JC, Rubin CH, Falter KH, Selanikio JD, Flanders WD, Howe HL, Wilhelm JL (1996) Heat-related deaths during the July 1995 heatwave in Chicago. N Engl J Med 335:84–90

Semenza JC, McCullough JE, Flanders WD, McGeehin MA, Lumpkin JR (1999) Excess hospital admissions during the July 1995 heatwave in Chicago. Am J Prev Med 16:269–277

Skelhorn C, Lindley S, Levermore G (2014) The impact of vegetation types on air and surface temperatures in a temperate city: A fine scale assessment in Manchester, UK. Landsc Urban Plan 121:129–140

Smith CL, Webb A, Levermore GF, Lindley SJ, Beswick K (2011) Fine-scale spatial temperature patterns across a UK conurbation. Clim Chang 109:269–286. doi:10.1007/s10584-011-0021-0

Speak AF, Rothwell JJ, Lindley SJ, Smith CL (2013) Reduction of the urban cooling effects of an intensive green roof due to vegetation damage. Urban Clim 3:40–55

Thrush D, Burningham K, Fielding J (2005) Flood warning for vulnerable groups: a review of literature. DEFRA/Environment Agency, London

Walker G, Burningham K, Fielding J, Smith G, Thrush D, Fay H (2006) Addressing environmental inequalities: flood risk. Environment Agency, Bristol

Chapter 5
Reforms that Integrate Climate Change Adaptation with Disaster Risk Management Based on the Australian Experience of Bushfires and Floods

Michael Howes

Abstract Responding to disasters such as floods and bushfires demands immediate action, while adapting to the impacts of climate change requires a consistent long-term policy commitment. Systems of government need to be able to do both at the same time, despite all the other demands on scarce public resources. This chapter summarises the findings of a project that searched for opportunities to improve the situation by integrating disaster risk management and climate change adaptation. The research was based on comparative case studies of recent extreme bushfires and floods in Australia. The paper offers some practical recommendations for reform that consist of changes to agencies and funding to empower local communities as well as improve collaboration within and between sectors of society. This entails: focussing on a common policy goal of building resilience; empowering communities with local resilience building grants; promoting institutional learning by embedding climate change experts within disaster risk management organisations; and, facilitating interagency collaboration through improved networking at all levels. Such reforms will help to build resilience to both disasters and climate change.

Keywords Climate change adaptation • Disaster risk management • Policy integration • Australia

Introduction

Australia presents a fascinating case study for research into both climate change adaptation and disaster risk management because of its high level of vulnerability and multi-level system of government. In terms of geography, it is a large island

M. Howes (✉)
Griffith School of Environment, Urban Research Program, Griffith University, Southport, QLD 4222, Australia
e-mail: m.howes@griffith.edu.au

W. Leal Filho (ed.), *Innovation in Climate Change Adaptation*, Climate Change Management, DOI 10.1007/978-3-319-25814-0_5

continent covering 7.7 billion square kilometres with a coastline that stretches for 25,760 km (CIA 2015; Howes 2005). The climate ranges from a wet tropical north to a temperate south and east, but the continent is dominated by a large arid centre and west. The climate is highly variable as it is influenced by the cycles of El Nino and La Nina events and can swing from long periods of drought to intense rainfall in the course of a year (AAS 2015; IPCC 2014a). Although it is a wealthy nation (with an annual GDP per capita of US$46,600 adjusted for purchasing power parity) the economy still relies heavily on primary industries, such as agriculture and mining, as well as eco-tourism based on natural assets such as the Great Barrier Reef (CIA 2015; Howes 2005). Most of the population of 23 million is located in cities and towns that are spread along low-lying areas around the coast (DCC 2009). These factors make the country, its economy, and its population highly vulnerable to disasters such as bushfires or floods and this situation will worsen as the climate changes (IPCC 2012; COAG 2011, 2007).

On the other side of the equation, the multi-level Australian system of government is based on a constitution that was drafted in the 1890s that combined the conventions of the UK parliament with the federal structure of the USA (Howes 2005). This has led to a three tiered system that consists of: (1) The Commonwealth (national) government; (2) Six state and two territory governments; and, (3) Some 560 local councils. Within each level are a plethora of agencies and departments with different portfolios of responsibility that have an interest in both climate change adaptation and disaster risk management (e.g. environment, emergency services, transport, housing, health, etc.). The underlying dynamic of this system is an on-going power struggle within and between levels of government over jurisdiction and funding (Howes 2005; Toyne 1994). This makes developing a whole of government response to the challenges posed by climate change adaptation and disaster risk management difficult (Howes et al. 2013).

The Need for This Work

The latest Intergovernmental Panel on Climate Change (IPCC) Assessment Reports confirm that the climate is changing, that this change is being driven by human activity, and that there are significant environmental, economic and social consequences (IPCC 2013). Such findings are supported by independent research from scientific organisations around the world (e.g. AAS 2015; Royal Society 2014; NOAA 2013). While there has been a great deal of public debate on mitigation and reducing greenhouse gas emissions, there is also the somewhat less high profile issue of how to adapt to the impacts that cannot be avoided (IPCC 2014a, b). These impacts will impose significant costs on all sectors of society (Stern 2005). The problem of climate change is therefore multi-faceted, complex, global, long-term, serious and urgent. In short it is a classic example of a 'wicked problem' that is difficult to address (Howes et al. 2013; Head 2008; Rittel and Webber 1973).

The IPCC has also revealed that climate change will increase the intensity, duration and/or frequency of extreme events such as floods or bushfires, and this will be a significant problem for highly vulnerable countries like Australia (IPCC 2012, 2014a; AAS 2015). Disaster risk management puts additional pressure on the existing Australian system of government, particularly during major events that require an immediate large scale responses from different branches and levels of government (Howes et al. 2013, 2015; Handmer and Dovers 2007; Prosser and Peters 2010; Cronstedt 2002). As the risk of disasters such as bushfires and floods increases due to climate change, more pressure will be placed on these systems and their resources at a time when governments at all levels are attempting to reduce expenditure and lower the level of public debt. Integrating disaster risk management and climate change adaptation allows for the sharing of knowledge, resources and facilities, hence providing the opportunity for a more efficient, effective and appropriate response (Howes 2015). It could also lead to a more proactive approach to disaster risk management by shifting the focus to prevention and preparation for the increased impacts to come (Handmer et al. 2011).

Research Performed

Australia's climate variability and vulnerability can best be illustrated by two recent extreme events. The Millennium Drought that lasted 10 years (2001–2010) was induced by successive El Nino events and led to water restrictions in every major city and town as well as an increase in bushfires. The drought ended in the summer of 2010–2011 when the east coast of the country experienced extensive flooding due to a deluge from an unusually strong La Nina event (Howes et al. 2013). This paper is based on the findings of a research project that investigated three case studies related to these events. First, was the extreme 2009 'Black Saturday' bushfires that burnt out a large section of the state of Victoria, killed 173 people and damaged or destroyed 2133 homes (Victorian Bushfires Royal Commission 2010a, b, c). Second, was the extensive 2011 Brisbane Floods in Queensland that led to the deaths of 35 people and the inundation of 20,000 homes (Queensland Floods Commission of Inquiry 2012). Third, was the intense 2011 Perth Hills bushfires in Western Australia that destroyed 71 homes and damaged 39 (Government of Western Australia 2011). These case studies are geographically spread from the south to north and east to west. They also involve three different state governments and several different local councils. Further, they are the types of extreme events that are being influenced by climate change. This suggests that the results of this research should have more general implications (Howes et al. 2013).

An investigation into these three case studies was conducted by a collaborative team of researchers from Griffith University in Brisbane (in the state of Queensland) and RMIT University in Melbourne (in the state of Victoria) in 2012–2013. The project was funded by a grant from the National Climate Change Adaptation Research Facility (NCCARF) and supported by the Queensland Department of Community Safety. The research was conducted in a series of stages. First, a literature review was undertaken to identify the key policy issues relating to climate change adaptation and disaster risk management in Australia. Second, the final reports of the official inquiries into these three events were analysed (Victorian Bushfires Royal Commission 2010a, b, c; Queensland Floods Commission of Inquiry 2012; Government of Western Australia 2011). Third, a series of semi-structured interviews were conducted with key stakeholders within the relevant state organisations and community groups. Finally, forums were held in Brisbane (in Queensland), Melbourne (in Victoria) and Perth (in Western Australia) with a broader range of stakeholders to provide a peer-review process by practitioners that could test and refine the findings (Howes et al. 2013).

The first two stages of this project led to the identification of four themes that could help to integrate climate change adaptation and disaster risk management: (1) Focus on building resilience; (2) Empower communities; (3) Promote institutional learning; and, (4) Facilitate interagency collaboration and communication. These themes were then explored in semi-structured interviews with key stakeholders that led to the synthesis of three proposed reforms: a new collaborative funding model, the creation of local community resilience grants, and the embedding of climate researchers within disaster risk management agencies. These proposals were then taken to the stakeholder forums where they were put through a peer-review process by practitioners. During these forums a set of institutional changes emerged that were grouped together to make a fourth proposal for reform to improve interagency collaboration via networking (Howes 2015). The findings on these four themes and their corresponding proposals for reform are outlined in turn below along with additional material from related research conducted by the author. A more detailed analysis of these case studies, the methods and findings can be found in Howes et al. (2013, 2015) and Heazle et al. (2013).

Results

Focus on Building Resilience

The research literature reviewed in stage one of this project revealed that identifying a common goal is one of the most effective ways of getting different agencies, levels of government, and even sectors working together to address a wicked problem (Howes et al. 2013; Head 2008). This is not an easy task given the fragmented multi-level organisational structures and ingrained competition of the Australian system of government (Howes et al. 2013; Howes 2005; Toyne 1994). In the 1990s there was some success in having sustainable development adopted as a common goal, both in Australia and around the world, although the pursuit of this goal has subsequently proved difficult (Howes 2005). Over the last decade the idea of building resilience has gained prominence globally as a worthy objective for all sectors of society (Aldunce et al. 2014; Alexander 2013; Davoudi 2012).

The participants in this study agreed that all levels of government and their agencies should be focussed on building community resilience to both climate change and disasters. They also agreed that the community and businesses should share this goal. As one respondent put it: "*A resilient community is one which pulls together. A not so resilient community would be one that just evaporates and people go their own separate ways.*" There was recognition, however, that the other more disparate goals of organisations may prove a stumbling block, so there needs to be some incentive for change (Howes et al. 2013).

One proposed reform that could assist was the idea of a collaborative funding model for the public sector that would generate an ongoing financial incentive for cooperation to build resilience (Howes et al. 2015). Under the current system of funding, each agency is allocated a set of resources (money, staffing, powers, equipment, etc.) that is jealously guarded and 'turf wars' may break out if one agency suspects another of encroaching on its jurisdiction. It was proposed that part of each budget be placed in a common pool. A competitive grant application process would then encourage agencies to form collaborations across the public sector and bring in stakeholders from the community and business to work on specific projects. Proposals would only be funded if they could deliver effective, efficient and appropriate outcomes. There is some precedent for this kind of arrangement, the NCCARF grant that funded this research, for example, led to the collaboration of two universities and a government department. This proposal would not only help to promote the common goal of resilience building, it would also help to improve interagency collaboration and community empowerment.

Empowering Communities

One of the key findings of this research was the confirmation that governments are not able to meet all the needs and demands of its citizens. The research literature suggested that there will increasingly be a need to be a shift from government to governance, where new partnerships between agencies, the community and business are used to increase the capacity for community-based climate change adaptation and disaster risk management (O'Brien et al. 2006; Dovers 1998). Obviously the state should not vacate the field as it has vital resources and response capacities that are well beyond those of other sectors. It can, however, expand its role to act as a facilitator that enables the other sectors to build their own capacities (Howes et al. 2013).

The participants in this study agreed with proposals to empower communities. As one participant said in relation to climate change adaptation: "*there's certainly room for a greater commitment and greater effort and resources to be put into community engagement and other stakeholder engagement.*" One of the key issues here is to there are different kinds of communities and each needs to be approached in different ways. A geographical community such as a suburb, for example, can be invited to face-to-face meetings, whereas an on-line community would need a different approach. Communities of young people are easier to reach by social media, while elderly residents are more likely to listen to announcements on the local radio station. The participants were clear that engagement has to go beyond simple public education campaigns about the level of risk or what to do in an emergency, it had to enable the community to participate in building its own resilience. A good place to start would be to forge cooperative relationships with existing community networks (Howes et al. 2013).

On this point, a second proposed reform put to the stakeholder forums was the idea of creating local community resilience building grants (Howes 2015). Most local councils already offer small grants to community organisations (e.g. funding for surf life-saving groups to help run their beach patrols). This funding pool might be increased and some money set aside to fund community resilience building initiatives. Local residents could propose simple measures, such as establishing a network of volunteers to check on elderly residents in an emergency situation. A town hall meeting, both face-to-face and on-line, would give each proposal a hearing and let residents vote on which projects to fund. This would raise awareness about the risks as well as empower the community to take ownership of building resilience.

Promoting Institutional Learning

The research literature reviewed in phase one of this project suggested that agencies need to be able to learn from their experience and be willing to actively seek to expand their body of knowledge and expertise (Waugh and Streib 2006). This might entail a continuous review of the current understanding of risks as new research findings comes to hand (Birkmann and von Teichman 2011). There is also a great deal of knowledge and expertise spread across a range of different agencies that could be of mutual benefit if was shared across the public sector (Howes et al. 2014). In short, there needs to be a built-in capacity for institutional learning and improvement.

Participants in this project agreed with this idea. Those involved in disaster risk management agencies were particularly keen to find out more about climate change and what the implications are for their operations. Likewise, those working on climate change adaptation wanted to know more about disaster risk management. An interesting point that was raised is that the aftermath of a major disaster creates an environment where governments and their agencies are open to change. As one participant stated: "*there's a window of opportunity after any major event in a place to say, this is what we have to embed in the corporate knowledge and understanding, ... We've only got this little window of opportunity to do that in, otherwise people forget. ... the priorities get overtaken by the next most important thing.*" So timing is important when setting up a strategy to improve institutional learning, but agencies also need to be prepared to make the most of such opportunities (Howes et al. 2013).

The third proposal for reform put to the forum participants was the idea of embedding climate change researchers within disaster risk management agencies (Howes et al. 2015). This would allow them to have some input into the development of risk management strategies and also to learn about the work of the emergency services. For larger agencies this would simply mean adding experts to existing research teams. For smaller agencies, this might involve establishing a collaborative link with a relevant research organisation. Such arrangements are already supported by the Australian Research Council *Linkage Grant* scheme that funds projects where researchers team up with external organisations to work on improving the understanding of a problem and seek solutions. This proposed reform would expand this approach as well as embed researchers permanently within larger agencies.

Facilitating Interagency Collaboration

A common theme that emerged from the research literature is that problems such as climate change and disaster risk management demand a whole of government approach to policy formulation and implementation (Mitchell et al. 2010; Waugh and Streib 2006). This challenges the organisational architecture of governments, its bureaucratic operating routines, and underlying models of policymaking, particularly in Australia (Heazle et al. 2013; APSC 2007; Head 2008; Garnaut 2008; Toyne 1994).

This is something that the participants in this project were particularly concerned about. Many of them were working within agencies and all had experience relating to the difficulty of communicating and collaborating with other agencies. As one participant pointed out collaboration needs to be: "*working in partnership, recognising the skills of the various agencies and how they can actually complement each other but having a common goal.*" It is difficult to achieve this when agencies have been set up to work in 'administrative silos' that are narrowly defined by the limits of their delegated powers and allotted portfolio domains (Howes et al. 2013).

During the forums, a set of practical changes emerged that were put together to form a fourth proposed reform based on the idea of networking (Howes 2015). First, was the need to start at the top, using organisations like the Council of Australian Governments, ministerial councils, and inter-agency executive committees to get all levels of government to commit to collaboration from the top down. Next, was a proposal for groups of senior officers drawn from different agencies to negotiate common day-to-day management changes that would facilitate increased collaboration. Finally was the idea of a network of collaboration champions with members spread across all agencies. These would be front-line officers who were keen to establish good working relations with their counterparts in other agencies.

Discussion

These four reforms would enable the various agencies and departments at all levels of government to make more effective and efficient use of the scarce public resources that have been allocated to them and for which they compete. They would build capacity within the system of government as well as help to empower communities to take ownership of the process of building their own resilience to both disasters and climate change. In this sense they would help to shift the focus from government to governance by establishing networks of partnerships within and across the different sectors of society.

Comparing these findings to the research literature reviewed in phase one of the project suggests that the implications go well beyond the immediate situation of Australia and these specific issues. As a wicked problem, climate change and the increased risk of climate-related disasters that it brings expose some of the inadequacies of the existing systems of government around the world as well as the theoretical models on which they were built (Heazle et al. 2013; Mata-Lima et al. 2013; Howes et al. 2013; Head 2008; Rittel and Webber 1973). First, the basic organisational architecture of modern democracies was developed in the eighteenth and nineteenth centuries and was not designed to deal with such complex twenty-first century problems (Howes 2005; Toyne 1994; Beck 1992). Second, public sector procedures developed in the early twentieth century were deliberately designed to disaggregate policy responses into smaller specialised tasks and therefore have difficulty in developing the whole of government approach that climate change demands (Howes et al. 2015; APSC 2007; Gerth and Mills 1998). Third, the different approaches to policymaking offered by competing rational comprehensive and incremental schools in the mid-twentieth century fail to generate a practical strategy for responding to such wicked problems that were thrust onto the political agenda in the 1980s (Heazle et al. 2013; Lindblom 1979; Dror 1964).

Obviously the four reforms proposed in the results section of this paper will not solve these broader issues. They can, however, form part of a strategy to provide some immediate relief for already stressed systems of government by enabling them to make better use of resources. Such a move could be a useful first step, but significant structural transformation may be necessary to consolidate these gains. Such changes are, however, beyond the scope of this chapter.

Conclusions

This paper has summarised the findings of a research project that investigated opportunities for integrating climate change adaptation and disaster risk management in Australia. It is based on three case studies of severe climate-related disasters: the 2009 Victorian Black Saturday Bushfires; the 2011 Perth Hills Bushfires; and, the 2011 Brisbane Floods. Four themes emerged and four associated reforms were proposed. First, creating a pool of public money that can be used to fund collaborations within and between sectors will help to focus on building community resilience. Second, reallocating some local government resources to a resilience fund will empower communities to take more of the initiative on climate change adaptation and disaster risk management. Third, linking climate researchers to disaster risk management agencies, either by embedding them or via some collaborative arrangement, will increase the much needed institutional learning and improvement. Finally, a set of procedural changes that promote networking at all levels can improve inter-agency communication and collaboration. These reforms should enable governing systems to make more effective, efficient and appropriate use of public resources. While these proposals have been derived in the Australian context, they may have broader implications for other countries as well as the underlying design of governing institutions and their policymaking procedures. Such implications could be the focus of future research.

Acknowledgements The author would like to thank his colleagues whose support helped to provide the foundation for this paper: Deanna Grant-Smith (Queensland University of Technology), Kim Reis, Peter Tangney, Michael Heazle, and Paul Burton (Griffith University), and Darryn McEvoy and Karyn Bosomworth (RMIT University). The author would also like to thank the financial support provided for his research through various grants over many years by the Australian Government, the National Climate Change Adaptation Research Facility, the Queensland Government, the Griffith Climate Change Response Program, and the Urban Research Program at Griffith University.

References

Aldunce P, Beilin R, Handmer J, Howden M (2014) Framing disaster resilience. Disaster Prev Manage 23(3):252–270

Alexander D (2013) Resilience and disaster risk reduction: an etymological journey. Nat Hazards Earth Syst Sci 13:2707–2716

Australian Academy of Science (AAS) (2015) The science of climate change: questions and answers. AAS, Canberra

Australian Public Service Commission (APSC) (2007) Tackling wicked problems: a public policy perspective. Australian Government, Canberra

Beck U (1992) Risk society: towards a new modernity. Sage, London

Birkmann J, von Teichman K (2011) Integrating disaster risk reduction and climate change adaptation: key challenges – scales, knowledge, and norms. Sustain Sci 5(2):171–184

Central Intelligence Agency (CIA) (2015) World Factbook. CIA, Washington, DC

Council of Australian Governments (COAG) (2007) National climate change adaptation framework. COAG, Canberra

Council of Australian Governments (COAG) (2011) National strategy for disaster resilience: building our nation's resilience to disasters. COAG, Canberra

Cronstedt M (2002) Prevention, preparedness, response, recovery: an out dated concept? Aust J Emerg Manage 17(2):10–13

Davoudi S (2012) Resilience: a bridging concept or a dead end? Plann Theory Pract 13 (2):299–307

Department of Climate Change (DCC) (2009) Climate change risks to Australia's coasts: a first pass national assessment. DCC, Canberra

Dovers S (1998) Community involvement in environmental management: thoughts for emergency management. Aust J Emerg Manage 13(Winter):6–11

Dror Y (1964) Muddling through: 'science' or inertia? Public Adm Rev 24(3):153–157

Garnaut R (2008) Garnaut climate change review: final report. Cambridge University Press, Cambridge

Gerth H, Mills C (1998) From Max Weber: essays in sociology. Oxford University Press, New York

Government of Western Australia (GWA) (2011) A shared responsibility: the report of the Perth Hills bushfire review. GWA, Perth

Handmer J, Dovers S (2007) Handbook of disaster and emergency policies and institutions. Earthscan, London

Handmer J, McLennan B, Towers B, Whittaker J, Yardley F, McKellar R, Fellegara I (2011) Emergency management and climate change: national climate change adaptation research plan – an updated review of the literature. National Climate Change Adaptation Research Facility, Gold Coast

Head B (2008) Wicked problems in public policy. Public Policy 3(2):101–118

Heazle M, Tangney P, Burton P, Howes M, Grant-Smith D, Reis K, Bosomworth K (2013) Mainstreaming climate change adaptation: an incremental approach to disaster risk management in Australia. Environ Sci Policy 33:162–170

Howes M (2005) Politics and the environment: risk and the role of government and industry. Allen & Unwin/Earthscan, Sydney/London

Howes M (2015) Disaster risk management and climate change adaptation: a new approach. In: Palutikof JP, Boulter SL, Barnett J, Rissik D (eds) Applied studies in climate adaptation. Wiley, Oxford

Howes M, Grant-Smith D, Reis K, Bosomworth K, Tangney P, Heazle M, McEvoy D, Burton P (2013) Rethinking disaster risk management and climate change adaptation. National Climate Change Adaptation Research Facility, Gold Coast

Howes M, Tomerini D, Dodson J, Eslami-Endargoli L, Mustelin J (2014) Integrating pollution reporting, climate change adaptation, disaster risk management and land-use planning: a case study of the 2011 Brisbane floods. In: Burton P (ed) Responding to climate change: lessons from an Australian hotspot. CSIRO Publishing, Melbourne, pp 85–96

Howes M, Tangney P, Reis K, Grant-Smith D, Heazle M, Bosomworth K, Burton P (2015) Towards networked governance: improving interagency communication and collaboration for disaster risk management and climate change adaptation. J Environ Plan Manag 58 (5):757–776

Intergovernmental Panel on Climate Change (IPCC) (2012) Managing the risks of extreme events and disasters to advance climate change adaptation: a special report of working groups I and II of the intergovernmental panel on climate change. Cambridge University Press, Cambridge

Intergovernmental Panel on Climate Change (IPCC) (2013) Climate change 2013: the physical science basis, working group I contribution to the 5th assessment report of the intergovernmental panel on climate change. Cambridge University Press, Cambridge

Intergovernmental Panel on Climate Change (IPCC) (2014a) Climate change 2014: impacts, adaptation, and vulnerability, working group II contribution to the 5th assessment report of the intergovernmental panel on climate change. Cambridge University Press, Cambridge

Intergovernmental Panel on Climate Change (IPCC) (2014b) Climate change 2014: mitigation of climate change, working group III contribution to the 5th assessment report of the intergovernmental panel on climate change. Cambridge University Press, Cambridge

Lindblom C (1979) Still muddling, not yet through. Public Adm Rev 39(6):517–526

Mata-Lima H, Alvino-Borba A, Pinheiro A, Mata-Lima A, Almeida J (2013) Impacts of natural disasters on environmental and socio-economic systems: what makes the difference? Ambiente Soc 16(3):45–64

Mitchell T, van Aalst M, Villanueva P (2010) Assessing progress on integrating disaster risk reduction and climate change adaptation in development processes. Institute of Development Studies, University of Sussex, Brighton
National Oceanic and Atmospheric Administration (NOAA) (2013) State of the climate report. NOAA, Washington, DC
O'Brien G, O'Keefe P, Rose J, Wisner B (2006) Climate change and disaster management. Disasters 30(1):64–80
Prosser B, Peters C (2010) Directions in disaster resilience policy. Aust J Disaster Manage 25 (3):8–11
Queensland Floods Commission of Inquiry (QFCI) (2012) Queensland floods commission of inquiry: final report. Queensland Floods Commission of Inquiry, Brisbane
Rittel H, Webber M (1973) Dilemmas in a general theory of planning. Policy Sci 4:155–167
Royal Society (2014) A short guide to climate science. Royal Society, London
Stern N (2005) Stern review on the economics of climate change. HM Treasury, London
Toyne P (1994) The reluctant nation: environment, law and politics in Australia. ABC Books, Sydney
Victorian Bushfires Royal Commission (VBRC) (2010a) Final report: summary. Parliament of Victoria, Melbourne
Victorian Bushfires Royal Commission (VBRC) (2010b) Final report: volume 1: the fires and the fire-related deaths. Parliament of Victoria, Melbourne
Victorian Bushfires Royal Commission (VBRC) (2010c) Final report: volume 11: fire preparation, response and recovery. Parliament of Victoria, Melbourne
Waugh W, Streib G (2006) Collaboration and leadership for effective emergency management. Public Adm Rev 66:131–140

Chapter 6
The CityTree: A Vertical Plant Filter for Enhanced Temperature Management

Peter Sänger and Victor Splittgerber

Abstract Today, already over 50 % of the world's population is living in cities. Climate change is an enormous challenge both for residents and for the vegetation. The situation is worsened by the decreasing air quality. This causes that dwellers are suffering from syndromes such as stress, cancer and allergies caused by heat, noise and air pollution. For this reason a vertical plant filter, the CityTree has been developed. It is a vertical plant structure that cleans the air, cools the surroundings, holds back water and reduces noise. Together with its potential to display advertisements, it is a marketing tool for companies. As a result, clean and cool air can be provided economically profitable.

In a lot of metropolitan areas worldwide façade greening is used widely. In contrast in Europe it is still an exception. The paper shows the advantages of such type of greening and underlays it with data. This allows to achieve specific aims such as cooling, stress reduction and air cleaning. Moreover the design is an important feature so that the structure can be integrated into European style cities. The free standing solution offers more independence to deliver the advantages of vertical greening in heat and air pollution hot-spots. Thus a more sustainable and environmental friendly city with a higher standard of living can be created.

Keywords Climate protection • Climate change adaptation • Air pollution • Fine dust • Nature-based-solutions • CityTree • Sustainability • Vertical plant wall • Moss • Façade greening

Introduction

Climate change is an enormous challenge both for residents and for the vegetation. Between 2002 and 2012 numerous extreme weather events caused 80,000 deaths and economic losses of more than 95 billion euros, whereas the deaths are mainly caused by extreme temperatures (European Commission 2014). Currently 73 % of the European population lives in cities (UN 2014). This proportion is expected to

P. Sänger • V. Splittgerber (✉)
Green City Solutions GmbH & Co. KG, Andreas-Schubert-Str. 23, 01069 Dresden, Germany
e-mail: v.splittgerber@mygcs.de

W. Leal Filho (ed.), *Innovation in Climate Change Adaptation*, Climate Change Management, DOI 10.1007/978-3-319-25814-0_6

rise to 82 % by 2050, resulting in more than 36 million new urban residents (UN Population Division 2010). In this respect, climate mitigation must be accompanied by concepts of climate change adaptation (IPCC AR5 2014). Facade greening, as a Nature-Based-Solution (NBS), is widespread due to the known beneficial effects on the urban climate, air and noise pollution. NBS are innovative applications of knowledge of nature or inspired by nature and support, maintain and improve natural capital and human welfare. Putting the right plants into the right place and support those with technical engineering can create enormous benefits for cities and its dwellers. But so far there are only few examples of a consistent mixture of plants, engineering and business strategy. To reveal the full potential of NBS this paper provides an overview of technical solutions, scientific evidence and broader application opportunities for one type of NBS.

Heat and Cooling Vegetation

The heat load in cities can be reduced with green walls and green roofs—such as in the Mediterranean region by up to 10 K [ambient daytime temperature by an average of 0.94 °C (Bowler et al. 2010)]. These approaches can also help to reduce the energy demand in buildings (by 10–15 %) (Bigham 2011) and to improve the quality of life [energy savings from green roofs around 10–15 % (Zinzi and Agnoli 2011; Alexandri and Jones 2008)]. In particular close greenery reduces diseases such as heart damage, obesity and depression (Forest Research 2010). Taking this into account, in England, the treatment costs were reduced by around £2.1 billion (Hartig et al. 2014a, b). This quality of life is a crucial factor for the future viability, vitality and competitiveness of a city, for example it improves the attractiveness of cities for residents and businesses, and thus it increases property values and the economic activity (Hartig et al. 2014a, b). At the same time nearby greenery is reducing the pressure on peripheral natural areas. Examples are the Promenade Plantée in Paris, where a railway line was converted into a park.

State of Knowledge

City Climate

Climate in cities usually differs significantly from the average local weather conditions. Typical features of the so-called urban climate are nocturnal heat islands, a heavily modified radiation budget and changing wind currents. Since the urban climate is directly related to the design of the environment, the local climate can be altered by changes in the city structure both for the better and for the worse. If the impact of a proposed change in city structure should be predicted, the use of numerical simulation models is essential. In the city, the most diverse climatic conditions can be found in confined spaces next to each other: a drafty, shady street just a few meters from a windless, sunny place. Nowhere are there so many different materials with different physical properties next to each other as in cities. Therefore the combination of different modeling methods, such as fluid mechanics, thermodynamics, or agricultural meteorology is required. In the late 1980s three-dimensional flow models were available (e.g. MISKAM), in which the city was modeled in three dimensions (Eichhorn 1989). However the climate was reduced to one component, the wind field. Even today such flow models offer an important aid in the simulation of the distribution of pollutants, however, for a comprehensive assessment of the urban climate as a whole they are not suitable. In light of this lack of "real" urban climate models, the model ENVI-met was developed (Bruse 1999). The model ENVI-met is a three-dimensional coupled flow energy balance model (Bruse and Fleer 1998). The physical fundamentals are based on the laws of fluid mechanics, thermodynamics and atmospheric physics. Complex structures can be matched by the combination of basic elements (e.g. cubes). Numerical flows are simulating the wind and illumination by the sun of the resulting structures.

Through the interaction of sun and shades as well as the different physical properties of the materials (specific heat, reflective properties ...) the surface temperatures evolve differently over a simulated day, which in turn releases (depending on the wind field) more or less heat to the air. In addition to the distribution of air temperature more meteorological variables such as humidity, turbulence and the various radiation fluxes are calculated in the model, which make up the entire microclimate of the studied structure. In order to simulate the interactions between vegetation and the atmosphere, the physiological behavior of the plants is modeled. This includes the opening and closing of stomata to control the water vapor exchange with the environment, the absorption of water through the roots or the change in leaf temperature during the day.

Vertical Greening: A Nature-Based Solution to Climate Change Adaptation (Nature-Based Solution/NBS)

As the population is concentrated in exposed areas, such as cities, the damage of climate change could reach unacceptable levels (UN 2014). This rapid urbanization has an impact on the availability of resources and poses challenges to economic growth (UN Population Division 2010). Due to the ongoing economic recession European cities fight to integrate both health and quality of life in their cities. In this respect, climate protection concepts need to be complemented by climate change adaptation measures, as stated in the Assessment Report of the IPCC. Nature-based Solutions (NBS) can offer such integration.

NBS are innovative applications of knowledge of nature or inspired by nature and support, maintain and improve natural capital. Nature-based solutions that are including vertical greening play an important role in order to integrate green, blue and gray infrastructure. In particular green infrastructures can reduce energy and resource requirements and costs. For example trees, green roofs and green walls are providing cooling and insulation, are reducing urban heat island effects, and have the ability to reduce the use of heating and air conditioning. In the Mediterranean region the use of green walls and green roofs in cities can reduce the heat load by up to 10 K (The increase of green area ratio can also reduce the ambient daytime temperature by an average of 0.94 °C (Bowler et al. 2010); with an average night reduction of 1.15 °C (Gill et al. 2007)). These approaches also reduce the risk of floods, air pollution and the energy demand in buildings (by 10–15 %) and improve the quality [For instance, energy savings of green roofs were estimated at approximately 10–15 % (Bigham 2011), with a reduction of 12 % of energy demand in the Mediterranean region (Zinzi and Agnoli 2011). In cities like Athens, the cooling demand in buildings can be reduced by 66 % (Alexandri and Jones 2008)]. Other benefits include better recreation. For example the availability of green areas are directly related to public health (Forest Research 2010). In England the benefits of urban green spaces are estimated to reduce health costs by £2.1 billion (Hartig et al. 2014a, b). These benefits support particularly vulnerable groups such as children, the elderly and people with low socioeconomic status. In addition NBS empower urban neighborhoods, social bonds and support networks and innovation centers with new technologies. Examples include the Promenade Plantée in Paris or the Parco Nord in Milan. In this context, those Living Labs can facilitate the development and testing of new approaches. Because of the numerous advantages the experience with Nature-Based Solutions should be shared and transferred.

Arrangements

Many plants cannot be found in the conventional urban landscape because the tolerance range for a possible plant growth is not guaranteed. Biological plant diversity is restricted increasingly to a small group of plants and trees that can survive in spite of the air pollution and global warming. Due to the special pot system, as shown in Fig. 6.1, with the "CityTree" it is also possible to cultivate endangered species of moss and plants such as Polytrichum longisetum, Philonotis Marchica and Ditrichum Pusillum. From appropriate regional nurseries endangered species such as Antennaria Dioica and Aremonia Agrimonoides can be acquired. The plant diversity creates staging for corresponding fauna like bees, beetles, spiders, butterflies and dragonflies. Because urban green spaces are often limited and separated, CityTrees provide a structure for networking habitats, as recommended in the statutes of the European NATURA 2000 program. On "green" distribution paths insects and other micro-organisms can overcome man-made barriers and new habitats are created. Especially because moss is representing the lower plants and contribute to an improved habitat structure for biodiversity.

Fig. 6.1 From *left* to *right*: Single pot system with moss, plant symbioses, laboratory test for growth, laboratory test of fine dust uptake

Air Pollution and Vegetation

A second aim is to minimize health risks as a consequence of air pollution, which is the cause of "...more than two million premature deaths each year..." (WHO 2006). Thus, dwellers are suffering from syndromes like stress, asthma allergies, bronchitis and cancer caused by heat, noise and air pollution. These environmental impacts are worsened by climate change. Therefore, the European Union (EU) has determined critical values (2008/50/EG). Air pollution in cities consists of 90 % fine dust and 10 % ozone. The directive of 2005 (which was amended in 2008 and 2010) determined the following critical values for fine dust: PM (Particulate Matter) 10: 40 $\mu g/m^3$, PM 2.5: 20 μm^3 (Umweltbundesamt 2014).

State of Knowledge

Fine dust implies health risks for humans (Kryzanowski et al. 2005). There are a multiple projects such as EUROCHAMP (www.eurochamp.org), BENEFITS OF URBAN GREEN SPACE (BUGS) or PROGREENCITY (www.progreencity.com) which demonstrate the effort of the European Union to reduce emissions. These arrangements expand the knowledge of air pollution. In addition some of them provide approaches for new solutions (Draheim 2005).

Arrangements

Leafs and other parts of the plants decelerate particles (Kappis et al. 2007). The potential for filtering fine dust depends on physiological and chemical characteristics of the plants and their composition (Roloff 2013). Moreover vegetation can build ventilation barriers and lead to a higher fine dust concentration (Pugh et al. 2012). The city vegetation suffers vital damage as a result of prevailing diseases, pollution load (not industrial-resistant groves) and climate change which have a negative influence on their potential to bind particulate matter. As a result of laboratory tests the experts found that moss is excellent in fixing pollutants (Frahm and Sabovljevic 2007). Due to its specific physiologic features moss is different from flowering plants. It does not have a distinctive circulation system to convey water from the soil into the plant. As a result moss needs to gather the required nutrients out of the air. Thus it is called "cation exchanger". The surface has the ability for electrostatic charging. Contrary to the moss fine dust carries a negative charge. As a result forces of attraction and bounding arise which allow the moss to accumulate the substances and convert them into phytomass. Moss has a bacterial film on its surface which absorbs inorganic compounds and accumulate them in their organic material (Frahm and Sabovljevic 2007). In laboratory tests this double-process was proven and in addition a retention rate for particulate matter could be measured (20 g PM/m^2/year) that is 10 times higher than conventional plants.

Description of Solution: The CityTree

The so-called CityTree (Fig. 6.2) has a permeable and flow-optimized structure. Because a ground anchoring is not required the construction can be placed at any location. To augment the natural capacities a sophisticated structure with automated irrigation, water and energy supply was created. This enables the positioning in areas with high air pollutant concentration. Depending on the prevailing wind direction, the exposure to pollutant emitters and the sun, the CityTree is aligned and planted. Therefore an algorithmic analysis is carried out to achieve an excellent effectiveness. By courtesy of intelligent selection and positioning of the used plant matter the adapted planting concept leads to a fine dust reduction of up to 25 % (Klippel and Jazbec 2009). Furthermore moss is used as an optimal substrate in which the vascular plants flourish. This way synergy effects and symbioses are created. The chosen structures of the vascular plants are able to reduce wind speed and PM 10-particles. The particles are attracted, bounded and converted into phytomass.

Fig. 6.2 The CityTree with a bench and an QR-code to use it as a marketing tool

Differentiation to the State of the Art

Façade greening is reducing the air pollution by fixing fine dust and nitrogen dioxide. The much greater problem are local hotspot areas. So far the vertical greening cannot be used effectively in the hotspot areas due to the city's ownership structure. The fully free-standing construction solves these problems. For the flow-optimized positioning flow analyses were conducted and already used by urban development. The basic component of the advancement in contrast to other flat irrigation systems is the permeable single-pot irrigation system. The CityTree combines all of the mentioned individual aspects by using plants to improve the air quality. Placed in urban areas the independent system is able to shape the airflow and reduce the air pollution. Geographic information systems provide the possibility to link various location based information. Therefore different layers were used for modelling a load situation for every type of place. The objective is to determine the best location and arrangement of the system.

Delivering a Cooling Plant Filter

The automated, patented and TÜV approved system is unique in the way it combines the functional planting, the necessary technical infrastructure and associated analysis technology. Due to more than 50 studies and research reports conditions were developed which embrace all emission sources. Especially studies from the Lancaster Environment Center, the University of Applied Science Dresden, the "Fachvereinigung Bauwerksbegrünung e.V." (FBB), the

"Forschungsgesellschaft Landschaftsentwicklung Landschaftsbau e.V." (FLL) and the Federal Ministry for the Environment, Nature Conversation, Building and Nuclear Safety were included. They all recommend vertical greening as an arrangement for air pollution control. For example the Lancaster Environment Center simulated that such models are able to reduce 30 % of fine dust in urban street canyons (Pugh et al. 2012). The University of Applied Science Dresden validated the fixation on leaf surfaces in outside test beds using ivy (Hedera helix) (Schröder et al. 2011). Building on previous studies and in cooperation with the professorships of innovative cultivation techniques, vegetable gardening and greenhouse management of the University of Applied Science Dresden, the Institute of Agricultural System Technology and the Institute of Building Theory of the University of Technology Dresden a fully free-standing planting system was developed—the so-called CityTree. As the surface can be used to display information it can inform about the relevance of nature in cities (as shown in Fig. 6.3).

Technically the CityTree is characterized by:

- Each single plant is growing in moss which serves as a substrate (no soil required). Thus a symbiosis is formed in each pot.
- The CityTree has a fully-automated water and nutrient supply system.

Fig. 6.3 The CityTree in the city of Jena, Germany with additional information on the metal surface

- The single pots exhibit a determined small distance to each other. Thus create a screen structure permeable to air which assist the sedimentation of fine dust and prevent aeration barrier.
- The irrigation of the single pot system with 1.682 ventilates supplies an excellent watering for every single plant.
- The supply system allows an easy data transfer and maintenance is reduced. The system integrates up to 72 sensors.
- The system is equipped with a photovoltaic system to guarantee the supply of the electronic components in the control and steering circuit.
- To prevent vandalism expanded metal was used to clad the system.
- On both sides a wooden bench was included. Thus the CityTree is an attractive street furniture.
- As demonstrated in recent studies natural green supports stress reduction. Evergreen plants determine an all-year recreation effect and provide to the formation of a sustainable society.
- The dimensions of the CityTree are 2.9 m × 3.75 m × 0.65 m.
- Due to the evaporation the surrounding is cooled by up to 17 K (as shown in Fig. 6.4).
- A sustainable rainwater management enables an expandable tank capacity.

Thus an effective contribution towards the adaptation to climate change is generated.

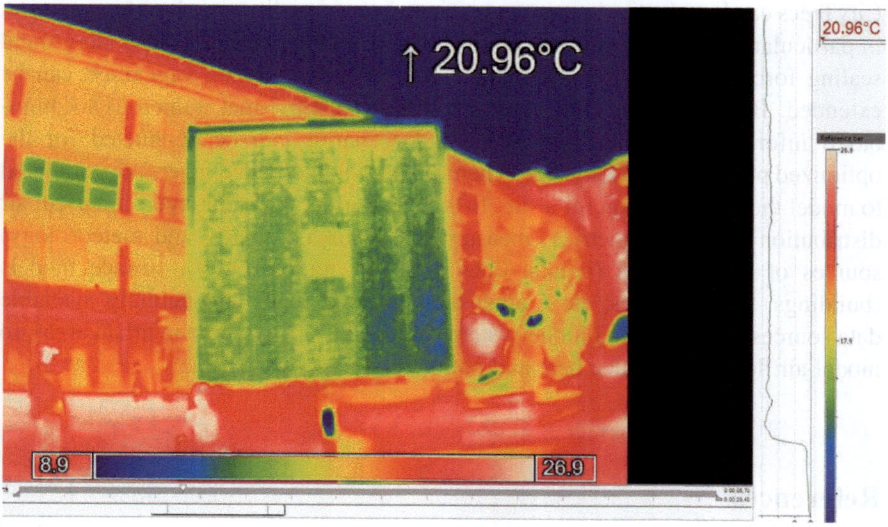

Fig. 6.4 Temperature reduction of up to 17 K by the CityTree (here in Jena)

Conclusions

The authors demonstrated that fine dust absorbing moss can be combined with covering plants to form a symbiosis to enhance the air cleaning capacity of vertical greening. Cultivated in a freestanding vertical structure the space efficient construction gives a higher degree of freedom for positioning than conventional facade greening. Thus this type of vertical plant structure can be positioned by the help of CFD analyses to optimize the transport of particles to the cleaning surface. This approach can contribute to a significantly better air quality in cities especially in "street canyons". Moreover the importance of climate change adaptation is increasing and the structures can be placed in areas that are heating up very quickly. Strategically placed these structures create "recreation islands" that are delivering shade and cool air. Hydroponic cultivation technology and the use of moss as a substrate results in a 50 % higher evaporation rate than conventional green spaces, whereby the environment is cooled by up to 17 °C (as shown in Fig. 6.4).

Especially in the hot and dry summers this allows a substantial cooling and creates a recreation area. As this affects the older population that suffers from climate change, a higher climate justice is achieved. This type of greenery not only uses a special symbiosis of moss and plants but by detaching vertical greening from preexisting constructions e.g. facades, the possibilities are also extended to a wide and new range of applications. In contrast to the previous approaches where existing vegetation is examined for its impact on the local climate situation, these vertical structures allow an immediate effect on the load and climate in a city street. CityTrees can be selectively used, e.g. to improve air quality or reduce temperature, in particular in critical locations. For instance if spatial conditions or underground sealing forbid conventional vegetation. This way the existing vegetation can be extended. By using combined climate models like Envi-met numerous location-based information can be networked and scenarios can be conducted for the optimized positioning of CityTree. Therefore various so-called layers are combined to model the load situation for a site. In the simulation, different layers are used: air distribution modeling, metrics (measuring stations air quality and meteorology) sources of air pollution (points, lines or areas, remote entry), surface models (buildings, trees, bridges, etc.), other data (general maps and publicly available data sources). So this stand-alone system can be placed anywhere in urban areas, to model air flows and reduce heat and pollution stress.

References

Alexandri E, Jones P (2008) Temperature decreases in an urban canyon due to green walls and green roofs in diverse climates. Build Environ 43(4):480–493

Bigham R (2011) The little details. Pollut Eng 43(4):7

Bowler DE et al (2010) Urban greening to cool towns and cities: a systematic review of the empirical evidence. Landsc Urban Plan 97(3):147–155

Bruse M (1999) Die Auswirkungen kleinskaliger Umweltgestaltung auf das Mikroklima. Dissertation, Ruhr-Universität Bochum, Bochum

Bruse M, Fleer H (1998) Simulating surface-plant-air interactions inside urban environments with a three dimensional numerical model. Environ Model Softw 13(1998):373–384

Draheim T (2005) Die räumliche und zeitliche Variabilität der PM 10-Schwebstaubkonzentration in Berlin unter Berücksichtigung der Großwettertypen. Geographischen Institut der Humboldt-Universität zu Berlin, Berlin

Eichhorn J (1989) Entwicklung und Anwendung eines dreidimensionalen mikroskaligen Stadtklimamodells. Dissertation, Universität Mainz, Mainz. http://www.staff.uni-mainz.de/eichhorn/docs/diss_eichhorn_1989.PDF

European Commission (2014) Communication from The Commission to The European Parliament, The Council, The European Economic and Social Committee and The Committee of the Regions. The post 2015 Hyogo framework for action: managing risks to achieve resilience. http://ec.europa.eu/echo/files/news/post_hyogo_managing_risks_en.pdf

Forest Research (2010) Benefits of green infrastructure. http://www.forestry.gov.u/pdf/urgp_benefits_of_green_infrastructure.pdf/$FIL/urgp_benefits_of_green_infrastructure.pdf

Frahm JP, Sabovljevic M (2007) Feinstaubreduzierung durch Moose. Nees Institut für Biodiversität der Pflanzen, Bonn

Gill SE et al (2007) Adapting cities for climate change: the role of the green infrastructure. Built Environ 33:115–133

Hartig T et al (2014a) Nature and health. Annu Rev Public Health 35:21.1–21.22

Hartig T, Mitchell R, de Vries S, Frumkin H (2014b) Nature and health. Annu Rev Public Health 35:207–228

IPCC (2014) Intergovernmental panel on climate change (IPCC) 5th assessment report: climate change 2014 (AR5). IPCC, Geneva

Kappis C et al (2007) Studie zum wissenschaftlichen Erkenntnisstand über das Feinstaubfilterungspotential (qualitativ und quantitativ) von Pflanzen. Berlin, pp15–18

Klippel N, Jazbec R (2009) Pflanzen filtern Feinstaub und Stickoxide. Umwelt Perspektiven, Illnau

Kryzanowski M, Kuna-Dibbert B, Schneider J (eds) (2005) Health effects of transport-related air pollution. Denmark. http://www.euro.who.int/__data/assets/pdf_file/0006/74715/E86650.pdf

Pugh TAM, MacKenzie RA, Wyatt DJ, Hewitt NC (2012) Effectiveness of green infrastructure for improvement of air quality in Urban Street Canyons, American Chemical Society: Lancaster Environment Centre, Lancaster University, Lancaster, U.K., LA1 4YQ. Environ Sci Technol 46(14):7692–7699

Roloff A (2013) Stadt- und Straßenbäume der Zukunft – welche Arten sind geeignet? Forstwiss Beitr Tharandt Beih 14:173–187

Schröder FG, Wolter S, Wolter A (2011) Abschätzung des Leistungspotentials vertikal begrünter Flächen, basierend auf einem nachhaltig ökologisch nutzbaren Pflanzensystem mit Regenwassermanagement. http://www.htw-dresden.de/fileadmin/userfiles/land/pdf/Gartenbau/Abschlussbericht-nachhaltige_Pflanzensysteme2011.pdf

UN (2014). http://esa.un.org/unpd/wup/Highlights/WUP2014-Highlights.pdf

UN Population Division (2010) cited in European Environment Agency, 2010. The European Environment, State and Outlook 2010, Living in an urban world; European Commission, 2011. Global Europe 2050. Executive summary. https://ec.europa.eu/research/social-sciences/pdf/global-europe-2050-summary-report_en.pdf

Umweltbundesamt (Federal Environment Agency) (2014) Feinstaub. http://www.umweltbundesamt.de/themen/luft/luftschadstoffe/feinstaub

World Health Organisation (WHO) (2006) WHO Air quality guidelines for particulate matter, ozone, nitrogen dioxide and sulfur dioxide, Global update 2005. Summary of risk assessment. Schweiz. http://whqlibdoc.who.int/hq/2006/WHO_SDE_PHE_OEH_06.02_eng.pdf

Zinzi M, Agnoli S (2011) Cool and green roofs. An energy and comfort comparison between passive cooling and mitigation urban heat island techniques for residential buildings in the Mediterranean region. Energy Build 55:66–76

Chapter 7
Mainstreaming Climate Change Adaptation into Development in the Gambia: A Window of Opportunity for Transformative Processes?

Hannes Lauer and Irit Eguavoen

Abstract Climate change adaptation (CCA) has emerged as a new paradigm of development politics. As adaptation has turned out to be less tangible than mitigation, controversies about the meaning and implementation have come up.

This paper is based on empirical research in The Gambia analyzing how CCA is mainstreamed into development strategies.

There is much political activism noticeable for translating the international idea of CCA to the local realities of The Gambia. These political efforts offer windows of opportunities for transformative processes. Many of these, however, are not seized due to country-specific and external factors. Despite this, some pragmatic and creative, approaches from the Gambian climate change network provide some adaptation and development co-benefits.

Keywords Climate mainstreaming • Governance • Institutions • West Africa • Gambia

Introduction

The Gambia, the smallest country of mainland Africa, finds itself confronted with the need to develop and adapt to climate change at the same time. A pragmatic way to do this is to "*address the two in an integrated way, through mainstreaming*" (Ayers et al. 2014). This poses an immense challenge because The Gambia has already been targeted by development cooperation for decades and still struggles to

H. Lauer (✉)
WASCAL, University of Bonn, Walter-Flex-Str. 3, 53113 Bonn, Germany
e-mail: HannesLauer@gmx.de

I. Eguavoen
Center for Development Research (ZEF), University of Bonn, Walter-Flex-Str. 3, 53113 Bonn, Germany
e-mail: eguavoen@uni-bonn.de

W. Leal Filho (ed.), *Innovation in Climate Change Adaptation*, Climate Change Management, DOI 10.1007/978-3-319-25814-0_7

meet the basic needs of its people. The challenge is also immense because adaptation is an intangible and still relatively vague concept with unresolved questions, opening up space for controversies (Pelling 2011). Given these challenges, it is necessary to investigate how adaptation is mainstreamed, as well as if this mainstreaming will have significant positive effects on the development agenda of The Gambia. Developing countries have witnessed the rise and fall of many approaches and paradigms proclaimed to be turning points around which funding has concentrated (Ireland 2012). All too often these paradigms have been absorbed into business as usual development frameworks, only presenting development in a new guise, without tackling underlying vulnerabilities (Pieterse 2010; Ireland 2012; Ireland and Keegan 2013). Great hopes are placed in adaptation now. Climate change adaptation (CCA) is expected to have a significant impact on the development discourse (Cannon and Müller-Mahn 2010) as it has stepped out of the shadow of mitigation and emerged as a new leading paradigm in the development sector. Adaptation is expected to be nothing less than "*an opportunity for social reform, for the questioning of values that drive inequalities in development and our unsustainable relationship with the environment*" (Pelling 2011).

It is against this backdrop that this paper analyses the ongoing process of mainstreaming CCA into the Gambian development strategies. Referring to the current academic debate on adaptation and transformation (Pelling 2011; O'Brien 2012; O'Brien and Sygna 2013; Eriksen 2013; IPCC 2014), the objective is to fathom if the mainstreaming process offers windows for transformative processes, which might go beyond a depoliticised implementation of adaptation as a climate proof add-on for the existing development strategies.

Much literature, many guidelines and toolkits neglect how governance works in Africa where adaptation must take place (Lockwood 2013). Hence, the second part of the paper presents empirical findings which provide some insights to climate governance network in the Gambia.

Adaptation at the Crossroad: Between Resilience and Transformation

The growing volume of funding mechanisms predicts a bright future for CCA. The Green Climate Fund alone is expected to provide 100 billion USD per year from 2020 onwards. This accounts for almost 80 % of official aid from member countries of the Organisation for Economic Co-operation and Development (OECD) (Ireland and Keegan 2013; Lockwood 2013). This global call for adaptation is finding its audience. Governments and organisations in West Africa are responding in order to bring themselves in position to engage with adaptation.

The crux of the matter is that adaptation is targeting uncertain future impacts. Experience about what actually makes an intervention an effective adaptation to climate change and how adaptation should look like in practice is rather rare, especially in Africa (Lockwood 2013). Adaptation is a process closely interwoven with multidimensional societal processes (Eriksen 2013). As a result, unlike mitigation, adaptation outcomes are difficult to measure.

Although "*there is now a large and increasing academic literature on adaptation and development*" (Lockwood 2013), decision-makers and implementers who are engaging with adaptation do so very much in a learning-by-doing attitude. The conceptualization of adaptation across multiple scales and its impact on the development discourse depends on power constellations between political actors (Pelling 2011; Eguavoen et al. 2015) and their willingness to facilitate change (O'Brien 2013). Accordingly, adaptation is at a critical point, where political decisions determine if adaptation can realize its potential to restart the quest for sustainable development. This development ideal bringing the environment, the economy and the social dimension under one umbrella is still to be achieved. The initial concept of sustainable development from the 1980s and 1990s "*has morphed into ecological modernization*" (Pelling 2011). Climate change, as a coevolution of development, offers room for reconfiguration, whereas "*first mitigation and now adaptation provide global challenges that call for a rethinking of development goals, visions and methods*" (Pelling 2011).

This explains why adaptation is increasingly debated in relation with transformative processes. CCA that seeks transformation as outcome builds on the conviction that, though there are many open questions and blind spots, *we know enough* about cause and effects of environmental- and climate change to recognize that only fundamental deliberate changes might create a livable future for subsequent generations (O'Brien 2013). However, it is "*not always clear what exactly needs to be transformed and why, whose interest these transformations serve and what will be the consequences*" (O'Brien 2012). Accordingly, transformation is opposed by notions of adaptation that prefer system-intern responses to occurring or expected adverse effects. These notions coalesced around resilience, a concept originally deriving from ecology and systems theory. Because the concept of resilience focuses on absorbing perturbations, such as shocks, to finally swing back and maintain the functioning of a system (Adger 2000; MacKinnon and Derickson 2013), it is criticized for being rather conservative as it is applied for social systems (Brown 2014). Although recently there has been engagement to strengthen the social dimension in resilience writing (ibid.), other than transformation, the mainstream notion of resilience takes social structures for granted. It offers a simplified understanding why certain countries, regions or social groups are vulnerable and presents rather technical solutions that allow to integrate adaptation into existing agendas, strategies and plans, even without changing or questioning them (Ireland 2012).

Transformation is rather a vision or a processual operation that requires practical (techniques and behaviors), political (system and structures) and personal changes (beliefs, values, worldviews and paradigms) (O'Brien and Sygna 2013). Thus the mainstreaming process in The Gambia is analyzed in search of processes that go beyond the utilization of adaptation as apolitical response to, or anticipation of a certain risk that threatens the system. As transformative processes imply change on a multitude of spheres and by a multitude of actors, the analysis is based on field research which adopted a methodological triangulation to examine the mainstreaming process from different angles. It consisted of an analysis of the relevant climate change policy papers and strategies, of expert interviews and of participant observation. The 17 in-depth expert interviews were conducted in a semi-structured manner with experts from all relevant government institutions, with national and international consultants and with three NGO representatives. The participant observation consisted of being an embedded intern at the leading national environment agency for 7 weeks. This cooperation provided insights in the routine work of a Gambian government institution and gave access to the Gambian climate network as it was possible to join to field trips and to take part in workshops and in countrywide conferences.

Climate Policy in The Gambia: A Historical Overview

The Gambia follows the pathway prescribed by the United Nations Framework Convention on Climate Change (UNFCCC) (see Fig. 7.1). The response strategy proposed by the UNFCCC consists of basic assessments which are followed by

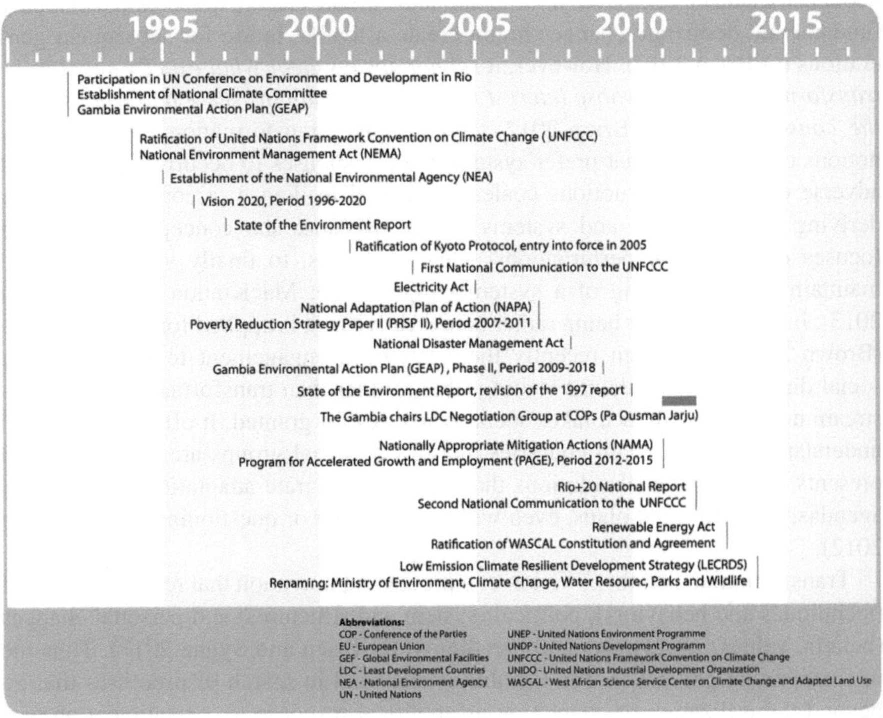

Fig. 7.1 Historical outline of the adaptation policy process in the Gambia. *Source*: authors, Design: J. Vajen

political strategy papers (Lamour 2013). The Gambia conducted National Capacity Self-Assessments with greenhouse gas inventories and vulnerability assessments. They build the base for the two National Communications to the UNFCCC, submitted by the Government of The Gambia (GoTG) in 2003 and 2012. The two main policy papers are the National Adaptation Plan of Action (NAPA) from 2007 and the plan for the National Appropriate Mitigation Actions (NAMA) from 2012. The NAPA sets the focus on improving the adaptive capacity of the country's key vulnerable sectors and regions against the main environmental stressors (GoTG 2007) and the NAMA presents a strategy to develop emission-intensive sectors more sustainable (GoTG 2011).

Climate change activities were accompanied by legal changes in various sectors, such as by the National Disaster Management Act (2008) or the Renewable Energy Act (2013). Additional sectoral policies were set in place by various institutions who engage with environmental management and issues such as biodiversity, water management or agricultural regulations.

The current political challenge is to manage and channel these various climate activities. The policy formulation process of the past 20 years has resulted in a "*myriad of existing climate change and development related strategies and reports*" (Lamour 2013). A first step of disentangling was the development of the medium term strategy Program for Accelerated Growth and Employment (PAGE) from 2011 which considers climate change as cross-cutting issue impacting on various sectors. The newly prepared Low Emission Climate Resilient Development Strategy (LECRDS) from 2014 takes another step as it hooks up on existing strategies to channel adaptation and mitigation into an integrated development strategy (Lamour 2013). Great expectations are also placed in the National Climate Change Policy. Its elaboration has recently been initiated by the government. The policy is expected to set the legal framework for future climate change policies.

Institutional Reorganization and Tight Network of Experts

Given the country's small overall population of less than two million people, the size of the political and academic elite is manageable. Experts in environmental policy and project implementation usually know each other personally. Workers of the ministries, the government agencies or researchers from the University of The Gambia, as well as donors and staff of international organizations form a tight social network that gets reinforced with every planning meeting or workshop. These events create a regular interface between different institutions working in the environmental sector.

The close-knit network character, however, does not necessarily imply a political comfort zone. The institutional framework for climate change governance in The Gambia has undergone some changes over the past decade. Political responsibilities were reallocated. Agencies and other organizations are constantly restructured or renamed according to the latest policy strategy. For these reasons, ministries and other organizations claim authority over climate change and partly compete with each other. This situation was described by an international consultant during the field research as: "*When climate change comes around every sector has a hat with climate change on*".

Institutional struggles for competence and the lucrative financial means for CCA projects became apparent in the National Climate Change Committee. This multi-stakeholder organization holds periodically meetings and functions as technical decision making body for climate change. It was reported that a central debate in the Committee concerns institutional mandates. In early 2014, there were four institutions present in the Committee that can be considered key players for the national climate change governance: (a) The Department of Water Resources (DWR) under the Ministry of Fisheries, Water Resources and National Assembly Matters, (b) The Ministry of Forestry and the Environment (MoFEN), (c) the National Environmental Agency (NEA) under the MoFEN, as well as (d) the National Disaster Management Agency (NDMA). The UNFCCC focal point and the chair of the National Climate Change Committee were automatically linked to the directorate of (a) but have to report to (b) as the ministry is the political body for environmental issues.

All interviewed experts working in government institutions are aware that institutional struggles are cumbersome and that existing structures are '*bubbled*'. They consistently expressed the need to sort out authority and responsibilities. Climate mainstreaming can be supportive in this regard when addressing "*issues of institutional architecture at national level [defining] which ministry of department is the nation's lead agency while simultaneously distributing responsibilities across sectors, encouraging dialogue, emphasising effective coordination and promoting systematic knowledge sharing*" (Jallow and Craft 2014).

This mainstreaming process is underway. The awaited National Climate Policy is intended to regulate responsibilities and a major institutional restructuring occurred in late 2014. The Ministry of Forestry and the Environment was renamed into the Ministry of Environment, Climate Change, Water Resources, Parks and Wildlife. The post of the minister was appointed to Pa Ousman Jarju, an internationally renowned expert on climate change. His achievements in climate diplomacy as chair of the Least Developed Countries (LDC) group at the UNFCCC Conferences of the Parties (COPs) and as special climate envoy (the first ever appointed from the LDC group) provide him strong legitimacy in leading the mainstreaming process in The Gambia. Mr. Jarju, as head of the DWR, had acted as the chair of the National Climate Change Committee and the UNFCCC focal point and was involved in the planning process of all climate change related documents about the Gambia.

It is an important observation that Mr. Jarju and a small number of other outstanding and internationally known environmental experts are very influential in the mainstreaming process. Together with some technically skilled people working in leadership positions of environmental institutions, they guide the policy formulation and implementation. The majority of staff below this level of experience, however, had rather limited technical knowledge on climate change. As a result, the climate change policy and mainstreaming process lays on the shoulders of a relatively small number of consultants and Gambian experts who are very active in driving the process.

Mainstreaming CCA in The Gambia

Political mainstreaming is a top-down process. It is often understood as the integration of a certain issue into political strategies and institutional agendas. However, more holistic approaches see mainstreaming as a long-lasting iterative process (Olhoff and Schaer 2010) that goes beyond the act of integration and is based on different pillars. Figure 7.2 presents a three-pillar scheme for such coherent mainstreaming (for comparable illustration see Ayers et al. 2014). It represents mainstreaming as a linear process that countries may follow step by step. But in practice mainstreaming is a process without clearly defined beginning and end. Awareness building on climate change, for example, is practiced by various actors and agencies and it is therefore difficult to determine whether it is part of the political mainstreaming process. Ayers et al. (2014) showed that there "*is no single best approach to doing mainstreaming*" and that frameworks and illustrations rather provide a starting point to understand the process.

Pillar 1: Sensitization and Capacity

The first pillar is considered the basis for political CCA mainstreaming. It includes understanding the impact of climate change on various sectors in a country and be

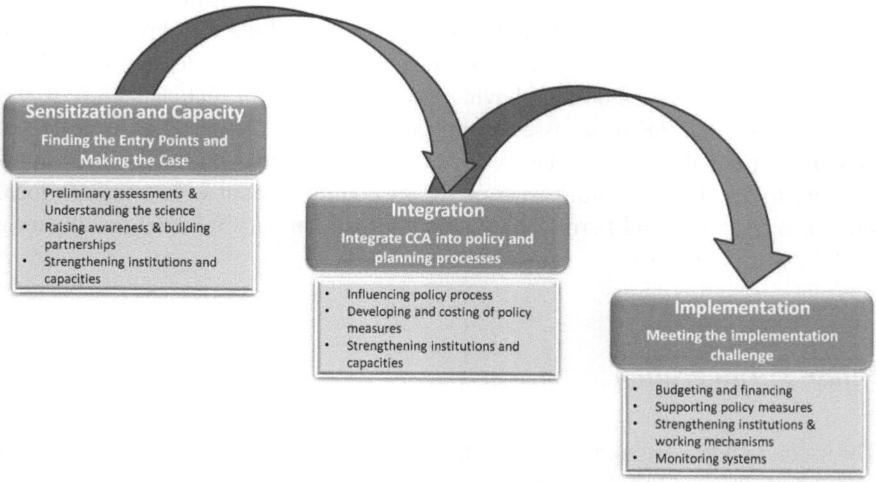

Fig. 7.2 The three pillars of mainstreaming. *Source*: authors, based on UNDP-UNEP (2009)

informed about possible adaptation options. This knowledge needs to be constantly shared within society and among policy-makers.

In the Gambia, the most discussed obstacle concerning the first pillar was the scarcity of experts who could support the mainstreaming process and back up the leading experts (see above). Though more technical capacity will certainly be needed, it would be an oversimplification to reduce governance problems to matters of missing knowledge (Lockwood 2013). Knowledge and skill would need to come with willingness for change or with, what an international expert had called, "*the culture of consequences*". It is therefore important to share knowledge and advocate for adaptation and mitigation beyond the border of the existing environmental and climate change network. Advocacy would need to convince political elites and the powerful Ministry of Finance and Economic Affairs. In the Gambia, there have been attempts to sensitize the whole cabinet during a series of working dinners. But still, it is most likely that CCA will be compromised when coming to monetary decisions.

But not only sensitization of politicians is needed. It is also important to translate climate science into a more comprehensive language in order to reach the public. In The Gambia many sensitization initiatives are ongoing. They include the introduction of climate change into the curriculum of basic and secondary schools, the establishing of environmental study programs by the University of the Gambia, the work of environmental reporters and the activities of various NGOs. The University of the Gambia hosts a West African Master of Science Program in Climate Change and Education (Eguavoen and Tambo 2015). But sensitization is a long-lasting process facing the difficulty that people might give preference to short-term strategies because gaining a livelihood is already difficult.

Missing sensitization of the wider public reveals a basic constraint for mainstreaming to be successful - the top-down nature of the whole process. This entails the risk that actions will not reach to the so-called *space of places* where people live. Particularly NGO representatives moaned that it is nice by the government to establish documents like the NAPA, but the whole top-down adaptation approach has the effect that measures will either never be implemented, or local people will misunderstand them, having the ultimate risk of maladaptation. It was also criticized that the needs assessments do not address the needs of local people properly. The main climate change documents rather focus on macro level interventions to develop basic infrastructure for the country's main sectors than on local structures. Agrawal and Perrin (2009) had come to similar findings by comparing NAPAs of 18 countries.

Pillar 2: Integration

The second pillar is often perceived as the actual mainstreaming. Integration mainly consists of policy formulation as well as developing and budgeting of adaptation measures.

Most of the climate related documents in the Gambia are detailed papers of high quality that follow UNFCCC guidelines. Policies and strategies have been written by teams of Gambian and international experts. There has been much political activity under the second pillar (see above). The climate change network has made sure that climate related documents are in line with the overall development goals of the country. Climate change documents overlap substantially with the country's flagship environmental and poverty reduction strategies. The "*economic structure and the development status, and the key role of weather and climate on physical, social and economic vulnerability*" (GoTG 2007) has the effect that most strategies aim to develop certain sectors, namely agriculture, forestry, energy and the coastal zone (erosion control of beaches and income diversification). This duplication shows that CCA (and also mitigation) is utilized as an opportunity to invest in the country's key sectors. The coastal zone is especially important for the Gambian tourist sector.

Following the top-down criticism mentioned under the first pillar, it is a crucial question for the policy formulation process how to downscale international and national policies and how to upscale local knowledge and communal adaptation (Vincent et al. 2013). We observed that constructive dialogue between the political sphere, NGOs and local people was missing on many occasions—be it in the daily routine, on conferences or on workshops. Though lip service was paid to integrate local knowledge in the adaptation planning process, in practice participation was often taking the form of consultation where parties presented their view without getting into discussions. For example on workshops and conferences the floor was open for question and answer sessions where local people very explicitly accused decision-makers for not addressing their needs, for sharing information to late and for having no voice in the decision-making process. NGO representatives complained that no practical consequences are drawn from such often lengthy question and answer sessions.

Pillar 3: Implementation

Making the final leap from the policy to the realization is ultimately the decisive step described as third pillar.

Whoever we had asked about what would be needed most to implement adaptation, named higher budgets and missing financial resources. During time of research, every CCA project investment and even the development of most strategy papers was financed by international donors. Nothing is implemented until foreign funding is available. Though the moral claim of LDCs for funding is legitimate, dependency on external funds is problematic, especially because the big money for CCA is not yet flowing. It had happened several times that The Gambia was missing out on funds because the country could not meet the funding requirements and provide the domestic contribution. As a result, only two NAPA projects have started implementation. The other eight projects are pending due to lack of funds.

An own domestic fund for adaptation seems, though expressed as a goal to be achieved, a future vision rather than a tangible option. Many respondents also suspected that if money for CCA implementation was available, it would be used inefficiently. Respondents complained that a lot of money was spent for international consultants, for conferences and workshops, but not much for concrete CCA measures.

Insights from the Gambian climate change network revealed how working routines and practices can be cumbersome for CCA implementation. Administrative hurdles are time-consuming in the daily work of the main agencies. Grinding paperwork and waiting for authorizations hinders effective workflows. Particularly the implementation of projects that involve a multiple actors is difficult, because different rules and regulations of the involved institutions have to be fulfilled. Making matters worse, the staff turnover inside and between the implementing institutions is very high. Employees up to the secretaries are shuffled from one position to the other within the institutions and beyond. It makes it difficult to establish long-term working relations and CCA expertise among the mid-level staff.

Conclusion

The new paradigm of CCA has introduced new political structures and financial mechanisms. Its implementation, however, struggles with similar problem constellations, structural prerequisites and obstacles like other development paradigms. Meaning that in The Gambia much of the progress being observable under the second pillar of the mainstreaming process, where experts are active in establishing policy frameworks, still misses out on translation into practice. What Lockwood (2013) emphasized is true for The Gambia, the adaptation policy process is hardly a rational and linear one, following guidelines and policy frameworks. Lack of staff capacity, and the missing 'so-called culture of consequences', aid dependency, the top-down approach, as well as issues concerning the governance structures for CCA (under the other two pillars) make the mainstreaming process difficult. CCA mainstreaming in The Gambia seems to offer an additional opportunity and funds to strengthen existing strategies and programs rather than to be a political act of changing underlying processes. This does neither imply that new projects and technologies will not bring improvements, nor that existing strategies for sustainable development were ineffective. Merging them with CCA, however, would imply a *business as usual* approach with additional financial sources.

To also present a positive outlook, it is vital to emphasize that The Gambia is actively working on adaptation and thereby creative in developing important economic sectors. Policy-makers have understood that integrated adaptation and mitigation strategies may support overall development targets. Climate change offers the opportunity to allocate funds and create infrastructures that allow tapping local potential, such as investments in sustainable farming or fishing schemes and livelihood strategies, or to overcome fossil dependency by investing in renewable energies. This pragmatic approach should not be interpreted as a deliberate transformation on multiple scales, but as a development opportunity for The Gambia. Finally, the active political role that The Gambia has been pursuing at global conferences helped to create pressure by the LDCs to make the heavily-polluting countries move faster towards transformations for sustainability and keep their financial promises. This generates hope that CCA can be more than a "*mobilization without political issue*" (Swyngedouw 2010).

Acknowledgement The study was conducted for the West African Service Center for Climate Change and Adapted Land Use (www.wascal.org) and funded by the German Federal Ministry of Education and Research. Special thanks are due to Prof. Müller-Mahn from the University of Bonn for supervision, as well to interview partners and the staff of the National Environmental Agency of The Gambia for their support and kind hospitality.

References

Adger N (2000) Social and ecological resilience: are they related? Prog Hum Geogr 24 (3):347–364

Agrawal A, Perrin N (2009) Climate Adaptation, local institutions and rural livelihoods. In: Adger N, Lorenzoni I, O'Brien K (eds) Adapting to climate change: thresholds, values, governance. Cambridge University Press, Cambridge

Ayers JM, Huq S, Faisal AM, Hussain ST (2014) Mainstreaming climate change adaptation into development: a case study of Bangladesh. WIREs Clim Chang 5:37–51

Brown K (2014) Global environmental change I: a social turn for resilience? Prog Hum Geogr 38 (1):107–117

Cannon T, Müller-Mahn D (2010) Vulnerability, resilience and development discourses in context of climate change. Nat Hazards 55:621–635

Eguavoen I, Tambo E (2015) Transformative learning for global change? Reflections on the WASCAL master programme in climate change and education in the Gambia. In: Dietz T, Tischler J, Haltermann I (eds) Cultural dimensions of climate change and the environment in Africa (working title) Brill, Leiden (Forthcoming)

Eguavoen I, Schulz K, de Wit S, Weisser F, Müller-Mahn D (2015) Political dimensions of climate change adaptation. Conceptual reflections and African examples. In: Filho WL (ed) Handbook of climate change adaptation. Springer, Berlin (Forthcoming)

Eriksen S (2013) Understanding how to respond to climate change in a context of transformational change: the contribution of sustainable adaptation. In: Sygna L, O'Brien K, Wolf J (eds) A changing environment for human security. Transformative approaches to research, policy and action. Routledge, London

Government of The Gambia (GoTG) (2007) National adaptation programme of action on climate change. Banjul. Available: http://unfccc.int/resource/docs/napa/gmb01.pdf

Government of The Gambia (GoTG) (2011) Nationally appropriate mitigation actions. Banjul. Available: http://unfccc.int/files/focus/application/pdf/nama_foc_prop_gambia.pdf

IPCC (2014) Summary for policymakers. In: Field CB, Barros VR, Dokken DJ, Mach KJ, Mastrandrea MD, Bilir TE, Chatterjee M, Ebi KL, Estrada YO, Genova RC, Girma B, Kissel ES, Levy AN, MacCracken S, Mastrandrea PR, White LL (eds) Climate change 2014: impacts, adaptation, and vulnerability. Part A: global and sectoral aspects. Contribution of working group II to the 5th assessment report of the intergovernmental panel on climate change. Cambridge University Press, Cambridge

Ireland P (2012) Climate change adaptation. Business-as-usual aid and development or an emerging discourse for change? Int J Dev Iss 11(2):92–110

Ireland P, Keegan P (2013) Climate change adaptation. Challenging the mainstream. In: Sygna L, O'Brien K, Wolf J (eds) A changing environment for human security. Transformative approaches to research, policy and action. Routledge, London

Jallow BP, Craft B (2014) Engaging effectively in climate diplomacy. Policy pointers from the Gambia. IIED Briefing. Iied, London. Available: http://pubs.iied.org/17246IIED.html

Lamour Y (2013) Low emission climate resilient development strategy (LECRDS) for the Gambia – Unpublished manuscript. Caori Consult, Banjul

Lockwood M (2013) What can climate-adaptation policy in Sub-Saharan Africa learn from research on governance and politics? Dev Policy Rev 31(6):647–676

MacKinnon D, Derickson K (2013) From resilience to resourcefulness: a critique of resilience policy and activism. Prog Hum Geogr 37(2):253–270

O'Brien K, Sygna L (2013) Responding to climate change: the three spheres of transformation. In: Proceedings of transformation in a changing climate, University of Oslo, Oslo, 19–21 June 2013

O'Brien K (2012) Global environmental change II: from adaptation to deliberate transformation. Prog Hum Geogr 36(5):667–676

O'Brien K (2013) Global environmental change III: closing the gap between knowledge and action. Prog Hum Geogr 37(4):587–596

Olhoff A, Schaer C (2010) Screening tools and guidelines to support the mainstreaming of climate change adaptation into development assistance – a stocktaking report. UNDP, New York

Pelling M (2011) Adaptation to climate change. From resilience to transformation. Routledge, London

Pieterse JN (2010) Development theory. Sage, London

Swyngedouw E (2010) Apocalypse forever? Post-political populism and the spectre of climate change. Theory Cult Soc 27(2–3):213–232

UNDP-UNEP (2009) Mainstreaming poverty-environment linkages into development planning: a handbook for practitioners. http://www.unpei.org/sites/default/files/dmdocuments/PEI%20Full%20handbook.pdf

Vincent K, Naess LO, Goulden M (2013) National level policies versus local level realities – can the two be reconciled to promote sustainable adaptation? In: Sygna L, O'Brien K, Wolf J (eds) A changing environment for human security. Transformative approaches to research, policy and action. Routledge, London

Chapter 8
Promoting Climate Smart Agriculture Through Space Technology in Nigeria

Idowu O. Ologeh, Joshua B. Akarakiri, and Francis A. Adesina

Abstract Agriculture is one of the sectors mostly affected by climate change. Nigerian farmers have been losing their harvests to the impacts of climate change leading to lower crop production and poorer livelihoods. Climate Smart Agriculture (CSA) is an adaptation strategy that helps rural farmers to be resilient to and cope with the effects of climate change. It can be improved through the use of space technology by empowering key actors, providing them with reliable weather forecasts at the right time.

This paper presents an assessment of already adopted space applications in Nigerian agricultural sector; the distribution of mobile phones to rural farmers by government for easy access to CSA information from extension workers. It is also a policy research on other unpractised space applications, especially the conversion of geo-data to relevant information on climate and hazards that can help local farmers, nourishing them with timely agricultural advice which enables them to have higher crop yields and a more efficient use of seeds, water and fertilizers. The farmers will also receive early warnings for drought, flooding and/or diseases on their mobile phones, thus maximizing its use. The results of this paper will be useful for crop production agencies and NGOs in Nigeria and Sub-Saharan Africa.

Keywords Climate smart agriculture • Space technology • Climate change • Food sustainability • Rural farmers

Introduction: Background to the Study

Food security and climate change are urgent and inter-related issues in the agriculture sector (Cattaneo and Arslan 2013). Food production varies from year to year, largely as a result of weather conditions especially in tropical areas indicating that climate has obvious and direct effects on agricultural production (Parry et al. 2001). Many developing countries especially in Africa are most vulnerable to climate change because their crop production is largely rain fed. They have inadequate

I.O. Ologeh (✉) • J.B. Akarakiri • F.A. Adesina
African Institute of Science Policy and Innovation, Obafemi Awolowo University, Ile-Ife, Nigeria
e-mail: iologeh@gmail.com; idowu.ologeh@dangoteprojects.com

W. Leal Filho (ed.), *Innovation in Climate Change Adaptation*, Climate Change Management, DOI 10.1007/978-3-319-25814-0_8

infrastructure to respond well to increasing variability in weather conditions associated with global climate change. Furthermore they sometimes have undeveloped capacities to invest in innovative adaptations (Lybbert and Sumner 2010). The most affected groups are those who depend wholly on agriculture especially the farmers and landless labourers and others mostly women who live on trading in agricultural products (Marrewijk 2011).

Women are most vulnerable because of their overwhelming gender role as home keepers and compelling socio-economic responsibilities in farming and trading (Schalatek 2009; CARE 2010). According to Annon (2006), women are responsible for carrying out 70 % of agricultural labour, 50 % of animal husbandry related activities and 60 % of food processing activities. Restricted access to land and livestock is really a constraint to women's ability to effectively cope with and contribute to societal adaptation to climate change in the agricultural sector (Mohammed and Abdulquadri 2012). Even though women are therefore disproportionately affected, they still play a crucial role in climate change adaptation and mitigation actions.

Nigeria is one of the countries in sub-Sahara Africa that is not self sufficient in food production, although it possesses substantial agricultural potentials which include, large array of plant and animal, naturally fertile and irrigable agricultural lands, water bodies and diverse agricultural climate (FAO 2002; Ojo and Adebayo 2012; Orefi 2012). Government agencies and officials connected with food production, food importation, and food distribution have been working hard to make the country self-sufficient in food production. They are concerned about the rising cost of food and raising awareness of the efforts being made to ensure that Nigeria is food-secured (FAO 1997; IFPRI 2002). Food shortage in Nigeria is as a result of some challenges that are hampering agricultural development and self-sufficient food production in Nigeria; the most prominent of these is inaccurate base line information on climate (Oni 2008; Ali 2013).

According to Intergovernmental Panel on Climate Change (IPCC) (2007), "Climate change is a change in the state of the climate that can be identified (e.g. using statistical tests) by changes in the mean and/or the variability of its properties, and that persists for an extended period, typically decades or longer. It refers to any change in climate over time, whether due to natural variability or as a result of human activity". The climate system warming is now evident; there are observations of increases in global average air and ocean temperatures, widespread melting of snow and ice and rising global average sea level. The temperature increase is global and is greater at higher northern latitudes. Average Arctic temperatures have almost doubled the global average rate in the past 100 years (IPCC 2007). Also, the degree of climate change impacts felt by individuals and nations depends on the ability of different societal and environmental systems to mitigate or adapt to change.

Climate change is impacting the society and ecosystems in a broad variety of ways; it can increase or decrease rainfall, influence agricultural crop yields, affect human health, cause changes to forests and other ecosystems, or even impact energy supply (USEPA 2014). Climate-related impacts are occurring across the six geo-political zones of the country and across many sectors of the economy. The federal government is already preparing for the impacts of climate change through "adaptation," which is planning for the changes that are expected to occur. One of the important adaptation strategies is climate-smart agriculture.

According to Marrewijk (2011) Climate-Smart Agriculture (CSA) "is a strategy to reduce rural communities' vulnerability for climate change." He further defined it as an important climate change adaptation strategy to help small (subsistence) farmers adapt to climate change by an intensification or diversification of their livelihood strategy, thereby reducing their vulnerability. CSA is important because it reduces the increasing pressure on the natural environment caused by growing world population by increasing food security in a sustainable manner (Cattaneo and Arslan 2013). CSA also is gaining much attention, and popularity in Nigeria, as it is considered a new climate change adaptation development strategy for poor subsistence farmers and therefore a perfect option for climate finance.

CSA can be promoted through the use of space technology. Space technology has been useful through the use of meteorological and remote sensing satellite applications in understanding the climate changing system (Yang et al. 2013). These satellites produced more than 30 years consistent solar and atmospheric temperature data and are an invaluable asset for climate monitoring and understanding the changing climate (Box et al. 2009). It is also an efficient approach for monitoring land cover and its changes through time over a variety of spatial scales. Space technology provides geo-data which is converted to relevant information on climate and hazards and is of innumerable use to local farmers, nourishing them with timely agricultural advice which enables higher crop yields and a more efficient use of seeds, water and fertilizers. Such data will be processed by relevant government agencies to produce early warnings for drought, flooding and/or diseases and conveyed to farmers through their mobile phones or through agricultural extension workers.

Statement of the Problem

In Nigeria, agricultural production remains the main source of livelihood for rural communities. With perceived long-term changes in climatic parameters particularly temperature and rainfall, farming activities are expected to be severely affected in many areas and this is already manifesting. Vulnerability to climate-related shocks, such as droughts and floods, varies by location, indicating the need for climate-smart agriculture and location-specific policy responses. In recent times, due to warmer climates, farmers have been experiencing crop failure. Maize farmers in South-West Nigeria experienced crop failure twice in 2013 (FAE 2013). A major reason for this problem is the missing link between farmers and the Ministry of Agriculture (MOA) on the one hand and the Nigerian Meteorological Agency (NIMET) and the National Space Research and Development Agency (NASRDA) on the other. There exists a large information gap among these critical stakeholders; the interactions among them are too weak to bring about innovations that can engender space application in climate smart agriculture. Climate information tools are needed to assist policy makers understand the impact of climate variability and change in different agro-communities and devise appropriate response strategies to mitigate climate change impact on agricultural production.

Farmers in Europe and America in particular are employing Space Technology as a major tool in coping with climate change because of its enormous potentials for generating real time data on the environment. If this is properly adopted in Nigeria, farmers will have access to critical resources that would help them make informed decisions to improve their agricultural businesses. The focus of this study is thus to create a platform for interactions among the critical stakeholders to stimulate rapid development of agricultural sector towards strengthening the adaptive capacities of farmers in the face of a rapidly changing climate through the adoption of space applications.

Objectives of the Study

The general objective of this study is to assess the extent and potential applications of space-based technology for climate-smart Agricultural development in Nigeria. The specific objectives of the study are to:

1. Identify the existing space-based technologies for Climate-Smart Agriculture in Nigeria.
2. Determine the potentials in the use of space-based technologies for Climate-Smart Agriculture in Nigeria.

Significance of the Study

This study will examine space technology as means of developing low cost coping and adaptation strategies through CSA. The information when made available to agricultural extension workers should be disseminated to the farmers through all available communication channels. This will help them to plan ahead and know when best to plant a crop in a year and what adaptation measures to adopt.

The study is also expected to provide valuable data on the extent of interaction between the MOA, NIMET, NASRDA and farmers, while identifying challenges and obstacles hindering such interactions. This will help both urban and rural farmers to practice climate smart agriculture thus minimizing loss and maximizing yield which on the long run will make the agricultural sector lucrative for investment, thus reducing unemployment problem in urban centres, coupled with related societal problems; incessant food shortage and crime. It is important to understand the spatial and temporal relationship of climate variability and agricultural productivity in planning for the economic development of the country, which relies mostly on agriculture. The study would assist the policy makers to formulate better policy towards the promotion of climate smart agriculture in Nigeria.

Methodology

Study Area

The study surveyed small scale holder farms clusters from the four geo-political zones within the guinea savannah belt of Nigeria. In each geo-political zone, the State with the highest food production was sampled for the study. After much literature review and academic consultations, the following states are chosen for this study.

North-Central Zone—Benue State
South-East Zone—Enugu State
North East Zone—Kaduna State
South-West Zone—Oyo State

Data Source and Analysis

Background information on indigenous climate change mitigation and adaptation strategies will be acquired through guided interview from farmers and extension workers alike. Six hundred questionnaires will be purposively distributed to 600 small holding farmers across the study area. Ten questionnaires will be administered to the technical experts in the Crop Production Department in the Ministry of Agriculture of each selected state to assess their level of relationship with NIMET and NASRDA. Ten technical experts each in NIMET and NASRDA will also be given questionnaires to assess their contribution to climate smart agriculture. The questionnaires were designed to collect data on farmers' biometric data, gender, farm size, types of crops cultivated, farm size, knowledge of climatic data, experiences on climate change and use of space data. The results are presented in percentages, tables and charts.

A pilot study was carried out in Oyo State to validate the research instrument.

Results and Discussion

Demographic and Socio-economic Characteristics of Respondents

In the four geo-political zones surveyed, 600 questionnaires were distributed to small scale farmers but 478 questionnaires were retrieved. The average age of respondents is 52 years with majority falling between the brackets of 45–50 years. This age distribution justifies while these farmers were most vulnerable; this is the stage in which their children are in tertiary/secondary education and needs financial support to remain in school. If their farms yield is low, it will affect their children education and subsequently their future.

Forty eight percent of respondents are female, while 52 % are male; in Oyo and Kaduna states, majority of the farmers (68 %) are men, but in Enugu and Benue, majority (64 %) are women. Enugu and Benue result are in line with the statement of Annon (2006) who said women are responsible for carrying out 70 % of agricultural labour. The result also shows that nearly 69 % of the women rented the lands; 10 % leased it while the remaining 21 % owned/inherited the lands. This result concurs with the work of Mohammed, 2012 who said women has limited access to land.

Majority of the farmers own one farm, but some farmers in Kaduna and Oyo states owned two to three farms in different locations. An average size of each farm is one hectare (six plots). The crops common to the farms surveyed are cassava, maize, yam, tomatoes, millet and sorghum.

Five technical experts each from the four states, NIMET and NASRDA are administered questionnaires. In total, 30 technical experts' opinions are obtained. Eighty-six percent of them are men; 60 % has masters degree while 7 % possess doctorate degrees.

Perceived Changes Due to Climate Change

When asked if they are affected by climate change, most farmers do not understand what climate change is. Climate change was explained to them in terms of increase or decrease in temperature, rainfall and humidity. Ninety-eight percent respondents signified that they are experiencing erratic rainfall, while 56 % indicated there is much increase in temperature. The farmers were given a table of ten parameters that are affecting their farms due to climate change, the three parameters that topped the list are crop failure, pests and diseases and drought. Figure 8.1 is a chart showing these parameters.

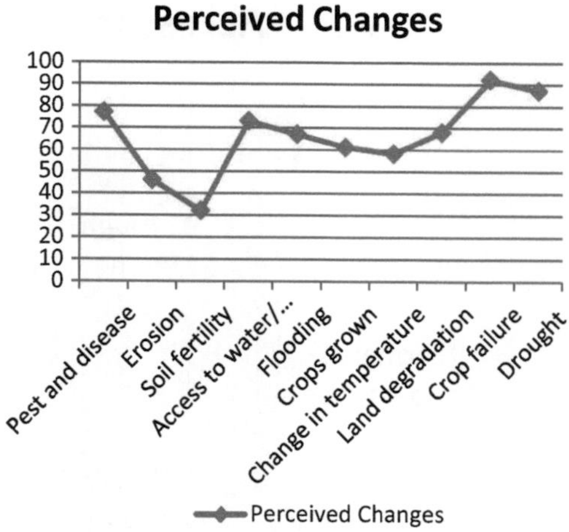

Fig. 8.1 Perceived changes due to climate change. *Source*: author

Access to Climatic Data

The exposure of the farmers to climatic data varies from one geo-political zone to another. Forty-seven percent of Oyo State farmers representing the South–west geo-political zone have heard or known climatic data, the figure decreases as we go up north. Their source of knowledge is displayed on Fig. 8.2. The three major sources of their information on climatic data are extension workers (89 %), radio programs (83 %) and through their mobile phones (66 %).

The result also shows that the farmers do not know much about the agencies in charge of these climatic data. Seventeen percent indicated they know about NIMET, while 6 % indicated they know NASRDA. This clearly make known that there is a gap between the activities of NIMET and NASRDA and that of extension workers. If they work together, the extension workers would have been promoting the activities of these agencies to the farmers.

Majority of the farmers (92 %) due to their ignorance of climatic data claimed they have not made used of any climatic data, neither did it affect their farming. The remaining 8 % indicated they made use of climatic data which is rainfall forecast they heard on radio but they cannot verify if it has any impact on their farming operations.

The technical experts from NIMET and NASRDA on the other hand claimed they are making climatic data available through publications, radio and television programs, workshops and seminars. They ascertain that agricultural technical experts and stakeholders are often invited to their workshops and seminars and this information are available to them. They claim, the Ministry of Agriculture is the one failing to relate the information to the farmers.

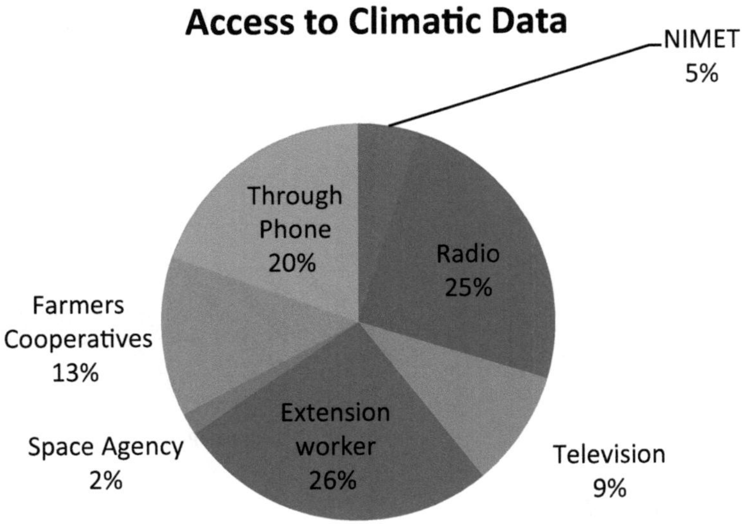

Fig. 8.2 Access to climatic data. *Source*: author

Indigenous Response to Climate Change

The farmers are asked to indicate how they respond to climate change effects indigenously. Fifty nine percent indicated they resulted to swamp farming when the rain is late coming, 23 % uses irrigation to augment rain-fed agriculture while 37 % rely on government intervention (the farmers that depend on government intervention though they come late are mostly poor and cannot practice any adaptation strategy if not assisted by the government). Some farmers diversify their cropping pattern to adapt to climate change effects; 76 % practice mixed cropping (if some crops fail, others may survive), 64 % practice crop substitution (replacing water loving crops with cassava which is climate resistant) and 19 % practice land fallowing. Land fallowing is not widely practiced because of restricted access to land. Majority of the farmers (73 %) uses pesticides to control pests on their farms, as they all experience increase in pests infection on their various farms (Table 8.1).

The Existing Space Based Applications for CSA Development in Nigeria

Based on the information retrieved from the 478 respondents (farmers) and 30 agricultural technical experts across the study area, the various existing space based technologies that are presently available and in use by Nigerian rural farmers are listed in the Table 8.2. The table list these practices and their efficacy; the efficacy

Table 8.1 Indigenous adaptation measures against climate change

Indigenous adaptation	Percentage (%)
Swamp farming	59
Irrigation	23
Mixed cropping	76
Crop substitution	64
Land fallowing	19
Use of pesticides	73
Government intervention	37

Source: author

Table 8.2 Existing space technology applications in use by Nigerian rural farmers

S/N	Space tool	Efficacy
1	Weather index-based insurance schemes for farmers	2
2	Mobile phones	4
3	Seasonal rainfall prediction (SRP) by NIMET	1
4	Meteorological open data policies and systems	1

Source: author

of these tools is rated on the scale of 1–5 with 1 as least efficient to 5 as most efficient. The efficacy is rated on how available, useful and satisfactory these tools are to the farmers. These technologies, their users, advantages, efficacy and limitations are also discussed below:

1. Weather Index-Based Insurance Schemes for Farmers: This space assisted tool is designed for farmers. It is a smart approach to control weather and climate risk because it uses a weather indicator, such as rainfall, to determine insurance disbursements. The weather data is provided by meteorological satellites; it is processed into climatic information through weather forecasting models to reduce some of the risks borne by farmers. This scheme improves farmers' livelihoods and enables them to invest in climate-smart technologies, which in turn secure the nation's food supply. The various effects of climate change reduce livelihood options for millions of small-scale farmers in Nigeria and commercial banks are unwilling to lend to farmers, if they do their interests are always high (21 %). As a result of their poverty, and in line with the sayings of Lybbert and Sumner (2010), majority of these vulnerable farmers are risk-prone. They cannot invest in improved seeds and other agricultural inputs which might help them to substantially increase their yields. Even if they do, and their efforts are affected by climate change, they may lose their crop, their livelihood and their food security and also incur a huge debt. To assist these farmers, the Ministry of Agriculture in conjunction with Bank of Industry provides weather-index based insurance. Only 14 % of surveyed farmers have access to this information and out of them, only 6 % have benefitted from the scheme. Though this scheme is useful and satisfactory for the beneficiaries, but because it is not easily accessible to all farmers, its efficacy is rated as 2.

2. Mobile Phones: This space operated tool is used by all stake holders. It can be used at any location based on network coverage. It is an efficient and effective means of reaching the farmers with information on distribution of subsidized seeds and fertilizers, climatic information, loans and government interventions. The information is sent to their mobile phones which are registered in the national farmer's database; 78 % of surveyed farmers use mobile phones. This tool is dependent on the quality of mobile network provider. Often, the network in the villages were most of these farms were located are poor, thus impairing communication. The efficacy of this tool is rated 4, because it is widely used, acceptable, useful and satisfactory for users, whose network coverage is good.
3. Seasonal Rainfall Prediction (SRP) by NIMET: This space promoted tool is for farmers and agricultural technical experts. Its aim is to provide seasonal (daily, weekly, monthly, quarterly) rainfall prediction using meteorological satellite data and appropriate weather forecasting models. The information provided will enlighten rural farmers on the usefulness of climatic data (rainfall forecasting) in agriculture, and how they can best plan their farming activities to avoid weather-related losses. A major setback of this tool is its poor publicity; SRP is mainly known to crop production experts and not to many rural farmers. None of the surveyed farmers indicated they are furnished officially with this information, only 17 % indicated they learnt about SRP through NIMET radio sponsored programs, thus its efficacy is rated 1.
4. Meteorological Open Data Policies and Systems: This space related tool is for use by all stakeholders. It is an open online database where enlightened farmers, researchers, policy makers and extension workers can assess meteorological data. This is a project which is yet to be published online for public use. None of the surveyed farmers know about it and its efficacy is rated 1.

The Potential Space Based Applications for CSA Development in Nigeria

Key potential space tools that are affordable and can be used to promote CSA in Nigeria include:

Remote Sensing

Although this technology is already employed by some large scale farms in Nigeria, it is far away from the reach of rural farmers. These rural farmers produce the bulk of the harvest in the country, thus this technology must be made available to them. Remote sensing promotes precision agriculture by using sensors on aerial or satellite platforms to record light from specific wavelengths that is reflected back from on-ground targets to determine variations in a farm (NASA 1999). The images when interpolated with soil maps and other global positioning system (GPS)-

referenced data can be used to develop geo-located maps of specific features on farms. The product can be used by farmers with the help of Geographic Information System (GIS) analysts and extension workers to determine the cause of the variations in the farm and solutions. It helps to determine exact pesticide and fertilizer application, the amount of water needed by plant (irrigation) when rainfall is erratic and can help formulate appropriate agricultural policies.

The Conversion of Geo-data to Relevant Climate Information

This technology is also used by large mechanized farms in Nigeria, but not available to the major food producers—the rural farmers. Geo-data are stored information about geographic locations that can be used with a geographic information system (GIS). Geo-data, when converted to relevant information on climate, weather and hazards, can help rural farmers generate information for customised and timely agricultural advice.

In 2013, NIMET predicted early rainfall which was so, but this rainfall stopped abruptly in May for about 2 weeks with intense temperature causing maize crop failure. If the information about the rainfall break has been passed across to rural farmers on time, they would have made provision for irrigation and saved their harvests. This information can be passed across to rural farmers through their phone for instant delivery or through extension workers for better understanding of the situation.

Also the data is useful for early warnings of drought, flooding and/or diseases which can help farmers plan their planting years appropriately. For example, in 2014, Ile-Ife town in South Western Nigeria experienced late rainfall; while their counterparts in other states were cultivating, they are waiting for rainfall; there was no one to inform them of the cause of delayed rainfall or how long they are to wait. If they have had access to relevant information, they would have practiced swamp farming and still have farm produce to sell by mid-year.

Satellite Radio

Satellite radio is a radio service broadcast from satellites with the signal broadcast that is nationwide, covering a much wider geographical area than terrestrial radio stations. This is another space based technology that can aid instant dissemination of remote sensing information and climatic data across to rural farmers in real-time. Other important information on government aid, loans, distribution of fertilizers etc. can also be passed through this medium. The advantage of satellite radio over mobile phones is that various programs and educational talks can be aired in the local dialects of the farmers. Radio drama series can also help paint clearer pictures of the information to farmers and repetition of this information over the radio will make it stick on the minds of the rural farmers. The Ministry of Agriculture need set up only one satellite radio station with programs in major Nigerian languages. The receivers are to be distributed to farmers bit by bit until it circulates and the exercise monitored so it will not be sabotaged.

Teleconferencing

A teleconference or videoconference is a live exchange of information among groups of persons that are remote from one another but linked by a telecommunications system. Rural farmers' cooperative or association meetings can be remotely linked through telecommunication. The farmers in the south-west may have solution to the problems being faced by the farmers in the middle belt. When these farmers are allowed to meet online, they can share their problems and get instant solution from fellow farmers who have solved that problem in their geo-political zone. Various adaptation and mitigation strategies against climate change that have been proofed can also be shared through this medium.

Conclusion

Rural farmers are already poor, and they are getting poorer due to waning crop yield. It is the role of the government through the Ministry of Agriculture at the national level and various state levels to assist these rural farmers through technologies that can improve their crop yield and subsequently livelihood.

The study has shown that rural farmers in Nigeria are aware that climate is changing with remarkable effects on their crop yield. It also explains that there is a weak relationship between the climatic data providers and the rural farmers, as the later are losing much benefit because of limited or no access to climatic data and geo-data. The farmers are already responding to climate change through indigenous means with remarkable results; these results can be improved through the adoption of space based CSA practices. Finally, the existing spaced based tools for promoting CSA in Nigeria and their efficacy are assessed and potential tools discussed.

The first two potential space based tools discussed in this paper can be achieved through the collaborations of government research agencies like the NIMET, NARSDA and national and states' Ministries of Agriculture. NIMET will provide the meteorological data, NARSDA will provide the remote sensing images and geo-data, GIS officers will convert the data to useful agricultural information and agricultural agencies will disseminate the information through extension workers to the rural farmers. This project will not cost the government any extra fund except for mobilization. The last two potential space applications will cost the government some funds, but they are very reasonable cost. Policy makers can formulate relevant policies along this initiative that will ensure relevant remote sensing and climatic data are available real-time for rural farmers.

The study has opportunities for future research by assessing the efficacy of indigenous adaptation strategies and comparing them with technology based adaptation strategies and their efficacy. The cost implication of these strategies and affordability by farmers can be evaluated. Also, the level and cost of relationships between NIMET, NASRDA, agricultural agencies and farmers and how it can be improved can be assessed.

References

Ali G (2013) Assessment of small scale irrigation system of some selected farms in Ogun-Osun river basin. Unpublished M.Sc. Thesis of the department of Agricultural and Environmental Engineering, Obafemi Awolowo University, Ile-Ife

Annon D (2006) National gender policy; Federal ministry of women affairs and social development. Amana Printing, Kaduna

Box J, Yang L, Bromwich D, Bau L (2009) Greenland ice sheet surface air temperature variability: 1840–2007. J Clim 22:4029–4049

CARE (2010) Participatory and inclusive planning for adaptation to climate change in Northern Ghana. http://www.careclimatechange.org/. Accessed 12 Aug 2014

Cattaneo A, Arslan A (2013) Climate- smart agriculture project. FAO EPIC, Rome

Farmers' Academy Ede (FAE) (2013) Our crops are failing. FAE, Ede

Food and Agriculture Organization (FAO) (1997) Irrigation potential in Africa: a basin approach. FAO, Rome

Food and Agriculture Organization (FAO) (2002) The state of food insecurity in the world: Corporate document repository, 4th edn. FAO, Rome

Intergovernmental Pannel on Climate Change (IPCC) (2007) Summary for Policymakers. In: Climate change 2007: impacts, adaptation and vulnerability. Contribution of working group II to the 4th assessment report of the intergovernmental panel on climate change. Cambridge University Press, Cambridge

International Food Policy Research Institute (IFPRI) (2002) Reaching sustainable food security for all by 2020; Getting the priorities and responsibilities right. IFPRI, Washington, DC

Lybbert T, Sumner D (2010) Agricultural technologies for climate change mitigation and adaptation in developing countries: policy options for innovation and technology diffusion. International Centre for Trade and Sustainable Development (ICTSD), Geneva

Mohammed B, Abdulquadri A (2012) Comparative analysis of gender involvement in agricultural production in Nigeria. J Dev Agric Econ 4(8):240–244

National Aeronautics and Space Administration (NASA) (1999) Applying space technology to help farmers diagnose fields: remote sensing applied to precision agriculture. http://science.nasa.gov/science-news/science-at-nasa/1999/essd08nov99_1/. Accessed 18 Mar 2015

Ojo E, Adebayo P (2012) Food security in Nigeria: an overview. Eur J Sustain Dev 1(2):199–222

Oni K (2008) Agricultural growth and development in Kwara State: the way forward. Maiden lecture delivered at the fund raising for the building of agriculture house, organized by the Kwara State Council of the Agricultural and Allied Employees' Union of Nigeria (AAEUN) in Ilorin

Orefi A (2012) Food security in Nigeria and South Africa: policies and challenges. J Hum Ecol 38 (1):31–35

Parry M, Rosenzweig C, Iglesias A (2001) UNEP handbook on methods for climate change impact assessment and adaptation strategies. UNEP, Nairobi

Schalatek L (2009) Gender and climate finance: double mainstreaming for sustainable development. UNDP, New York

United States Environmental Protection Agency (USEPA) (2014) Climate change impacts and adapting to change. http://www.epa.gov/climatechange/impacts-adaptation/. Accessed 31 Mar 2015

Van Marrewijk L (2011) Climate smart agriculture in the Mutale Basin, South Africa. Sustainable livelihoods and biodiversity in developing countries. M.Sc.-thesis IVMVU, Amsterdam. www.livediverse.eu

Yang P, Bi L, Baum B, Liou K, Kattawar G, Mishchenko M, Cole B (2013) Spectrally consistent scattering, absorption, and polarization properties of atmospheric ice crystals at wavelengths from 0.2 to 100 μm. J Atmos Sci 70:330–347

Chapter 9
Trade-Offs Between Climate Change Adaptation and Mitigation Options for Resilient Cities: Thermal Comfort in Households

Vera Gregório, Sofia Simões, and Júlia Seixas

Abstract Sustainable development within the broader context of climate change is a fairly recent research topic that underpins positive trade-offs between mitigation, adaptation and societal objectives. Many decisions regarding energy and transportation infrastructures, buildings, sanitary and storm-water management, flood risk assessment and also biodiversity protection take place at the city level. In this context, climate change issues have been recognized as fundamental for urban planning, requiring an integrated response with combined mitigation and adaptation strategies.

This paper develops an approach to assess some of the trade-offs between climate change vulnerability and mitigation options for the residential building stock of 29 Portuguese Municipalities, within the context of a comprehensive Municipal Adaptation Strategies Project, *ClimAdaPT.Local*. The paper presents a methodology to evaluate climate change vulnerability of the residential building stock regarding thermal comfort of the occupants, based on expected impacts from climate change measures and adaptive capacity. Moreover, the impact on climate change vulnerability is assessed as a function of a set of mitigation options taken at the building level aiming to explore whether it would be cost-effective to invest on mitigation or adaptation measures or both.

Results are presented for the municipality of Cascais and major findings show the interplay between mitigation and adaptation measures which can be synergetic or antagonistic. The approach and methodology are being validated with the 29 municipalities covering varied climatic zones, construction materials and socio-economic contexts, which will result in a comprehensive range of trade-offs between mitigation options and adaptation needs at the city level.

V. Gregório (✉) • S. Simões • J. Seixas
Faculdade de Ciências e Tecnologia, CENSE, Center for Environmental and Sustainability Research, Universidade Nova de Lisboa, DCEA – Campus de Caparica, 2829-516 Caparica, Portugal
e-mail: v.gregorio@campus.fct.unl.pt; sgcs@fct.unl.pt

W. Leal Filho (ed.), *Innovation in Climate Change Adaptation*, Climate Change Management, DOI 10.1007/978-3-319-25814-0_9

Keywords Climate change • Mitigation • Adaptation • Thermal comfort • Households • Resilient cities

Introduction

The linkage between adaptation and mitigation of climate change impacts is a fairly recent research topic (Barker et al. 2007), with several studies considering that integrated assessment may result in a more effective response to climate change effects (Dang et al. 2003; Klein et al. 2005). Traditionally, mitigation and adaptation policies have been dichotomized due to different perceptions addressing climate change issues, which has led to different scientific approaches (mitigation more economic and technologic oriented and adaptation more social and ecologically oriented) (Shaw et al. 2014), as well as to different spatial, temporal and stakeholder dimensions (Biesbroek et al. 2010). Notwithstanding, the IPCC fifth report recognizes sustainable development as the broader context of concerns with climate change, and therefore identifies many opportunities for integrated responses between mitigation, adaptation and societal objectives (IPCC 2014).

The community and city level are scales where many decisions take place regarding energy and transportation infrastructures, buildings, forestry and biodiversity protection, sanitary and storm-water management and flood risk assessment (Rosenzweig et al. 2011). Following these trends, cities all over the world are pursuing resilient solutions to integrate mitigation and adaptation policies. Examples of these solutions for the built environment may be found at (US GBC, WGBC, C40 Cities 2015) and span a broad range of actions from 66 cities, including incentives targeting private sector to reduce intensity energy CO_2 emissions in residential sector, improve building codes, municipal green building policies and actions to built city climate resilience to minimize the impact of climate change. In this context, the need to improve thermal comfort of households assessing synergies between mitigation and adaptation options is quite relevant to achieve climate resilient cities, since there is a dual implication (Adger et al. 2007): on one hand human settlements and buildings are vulnerable to climate change (Gething 2010), while on the other hand buildings have huge potential to mitigate greenhouse gas (GHG) emissions (Urge-Vorsatz et al. 2009). However there is still a need to improve the awareness to spread integrated approaches to climate change mitigation and adaptation to increase synergies between options and to help local planners (Zimmerman and Faris 2011).

The impacts of climate change on the built environment can be grouped into three main categories: (1) impact on comfort and energy performance; (2) impact on buildings' structures, buildings' constructions and systems, and (3) impact on water systems management (Gething 2010). This paper addresses the first topic within the background of the project ClimAdaPT.Local.

ClimAdaPT.Local is an ongoing comprehensive Municipal Adaptation Strategies Project, aiming to build capacity in 29 municipalities in Portugal towards the elaboration of Municipal Strategies for Adapting to Climate Change (MSACC). This paper is integrated in the research work developed under the *ClimAdaPT.Local* project and it develops an approach to assess some of the trade-offs between climate change vulnerability and mitigation options for the residential building stock. A case study for the municipality of Cascais illustrates the approach.

The paper is structured into four sections: **introduction**, followed by a descriptions of the **approach** proposed to assess the trade-offs between adaptation and mitigation including how to assess the climate change vulnerability of the thermal comfort in dwellings, considering the calculation of potential impact index, adaptive capacity index, vulnerability index and the trade-offs assessment between mitigation and adaptation options. The third section discusses the **results for the case study** and the last section presents the **conclusion**.

Approach to Assess Trade-Offs Between Adaptation and Mitigation Options for Thermal Comfort in Dwellings

The approach presented in this paper assesses the trade-offs between adaptation needs and mitigation options, focused on thermal comfort (i.e. space heating and cooling) for residential households. Our approach is divided in two main parts: (A) assess the climate change vulnerabilities of dwellings regarding thermal comfort, and (B) assess trade-offs between adaptation and GHG mitigation options of these dwellings. The ultimate end-users of this approach are city planners and, as such, the whole approach was discussed and validated with representatives of three Portuguese municipalities.

By "thermal comfort", we mean that a dwelling should maintain a pre-defined indoor temperature during the whole year. In the case of Portugal, this is a temperature of 20 °C during the heating season and of 25 °C during the cooling season, as stated in the Portuguese regulation on the thermal characteristics of buildings (RCCTE - Decree - Law n.° 80/2006).

Our approach was developed for 29 municipalities in Portugal and exemplified here with results for Cascais (Fig. 9.1). Cascais was selected as the first municipality to be studied given the robust background of its city planners on energy and

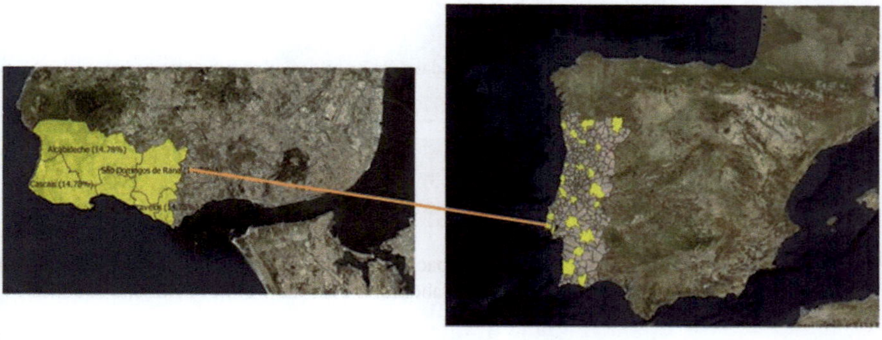

Fig. 9.1 Municipality of Cascais—case study (*left*) and location of the 29 Municipalities participating in the project *ClimAdaPT.Local* (right). *Source*: authors

adaptation strategies. This is a municipality of approximately 97 km^2 located 30 km west of Lisbon with a population of 206,470 inhabitants. It is divided in six civil parishes which are very different in terms of socio-economic indicators.

Assess the Climate Change Vulnerabilities of Dwellings Regarding Thermal Comfort

Regarding the first component, we developed our approach along the framework described in (Fritzsche et al. 2014) (Fig. 9.2) which entails addressing complementarily the potential impact and the adaptive capacity. The Fritzsche's framework includes four key components: (1) *exposure*—variables directly linked to climate parameters (e.g. temperature, precipitation), (2) *sensitivity*—the degree a system is affected by exposure (e.g. buildings typologies, heating and cooling technologies), (3) *potential impact*—measured by the combination between exposure and sensitivity (e.g. the potential impact on thermal comfort), (4) *adaptive capacity*—the ability of a system to adjust to climate changes (Adger et al. 2007), mostly related with societal environment (e.g. demography, literacy, socio-economic conditions).

Regarding the **adaptive capacity**, we have built an index to quantify the capacity of each civil parish in the municipality, varying from 0 (minimum capacity) to 5 (maximum capacity), based on detailed statistics data from the National Statistics Institute combining the following variables:

- Age of resident population, particularly the share of resident population with 4 years old or less and with 65 years old or more, reflecting the underlying

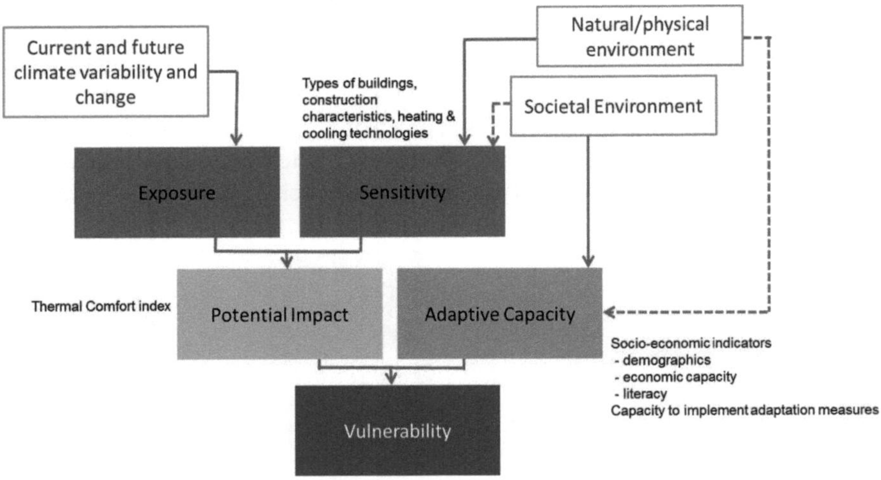

Fig. 9.2 Schematic representation of the approach to assess thermal comfort vulnerabilities of dwellings to climate change. *Source*: authors' elaboration over the work of Fritzsche et al. (2014)

assumption that very young or older people have more difficulties in adapting to climate change.

- Average monthly gain in euros, only available at the municipality level, to translate the capacity to implement adaptation measures, such as acquiring and using heating and cooling technologies.
- Property *vs*. rental of the dwellings, since we have assumed that persons living in rented dwellings have a more limited capacity to implement adaptation measures, as insulation options.
- Level of formal education of the resident population, in particular the share of population with a university degree, assuming that persons with higher education have more access to information on climate change issues and adaptation needs and measures, including access to funding opportunities, such as subsidies for retrofitting or for renewable heating and cooling technologies. Unemployment rate, reflecting that, in general, unemployed persons will have more economic difficulties and motivation to implement adaptation measures.

We have defined five classes of adaptive capacity ranging from zero (minimum capacity) to five, where five represents maximum capacity, according to the interval classes of each variable, as illustrated in Table 9.1.

An adaptive capacity index, ranging from 0 to 20, was built by using a weighted sum of these socio-economic indicators as follows: the share of young children weights 0.25, the share of senior citizens weights 0.5, the monthly gain 0.25 (because the data is only available at the municipality level), and the other indicators weigh 1. These weighs are naturally subjective but were found adequate by their end-users, the municipalities. We have experimented with different variations of the weighs between indicators from 0.1 to 1 and found that there were no substantial differences in the ranking of adaptive capacity index per civil parish.

Regarding estimation of the **potential impact** of climate changes in thermal comfort, we have developed our estimates based on the difference between the final energy consumed for space heating and cooling in reality (hereafter named as REAL, as reported by the DGEG-General Energy Directorate for the year 2012) and the final energy that would be needed to ensure the thermal comfort levels as stated in the above mentioned national buildings regulation (hereafter named as IDEAL), considering the RCCTE's required heating degree days (HDD) and cooling difference with outdoor temperature for space cooling for different climatic zones, and the different typologies of buildings. The higher the difference between the REAL and IDEAL energy consumed for heating and cooling, the higher the potential impact on thermal comfort (Fig. 9.3).

To estimate the REAL final energy consumption for space heating and cooling in the municipalities we used statistical historical data produced by (DGEG 2012) for each Portuguese municipality on the sales of electricity, LPG, natural gas and diesel to residential consumers. We have then assumed a share of these energy carriers to be consumed in space heating and cooling following the survey of (DGEG 2012) made to residential dwellings in Portugal. For energy consumption of biomass, which is quite significant in space heating in Portugal, we have assumed a value per

Table 9.1 Five level classes of adaptive capacity for each of the socio-economic indicators

Population with 4 years old or less		Population with 65 years old or more		Average monthly gain in euros		Ownership of the dwelling		Share of population with university degree		Unemployment rate	
Attribute range	Classes indicator (0–5)	Attribute range	Classes (0–5)	Attribute range	Classes (0–5)	Attribute range	Classes (0–5)	Attribute range	Classes (0–5)	Attribute range	Classes (0–5)
>9 %	0	>50 %	0	>1801 €	5	>50 %	5	>30 %	5	>30 %	0
7–9 %	1	40–49 %	1	1501–1800 €	4	40–49 %	4	26–30 %	4	26–30 %	1
5–7 %	2	30–39 %	2	1001–1500 €	3	30–39 %	3	20–25 %	3	20–25 %	2
3–5 %	3	20–29 %	3	901–1000 €	2	20–29 %	2	11–19 %	2	11–19 %	3
1–3 %	4	10–19 %	4	684–900 €	1	10–19 %	1	6–10 %	1	6–10 %	4
<1 %	5	<10 %	5	<683 €	0	<10 %	0	<5 %	0	<5 %	5

Source: authors

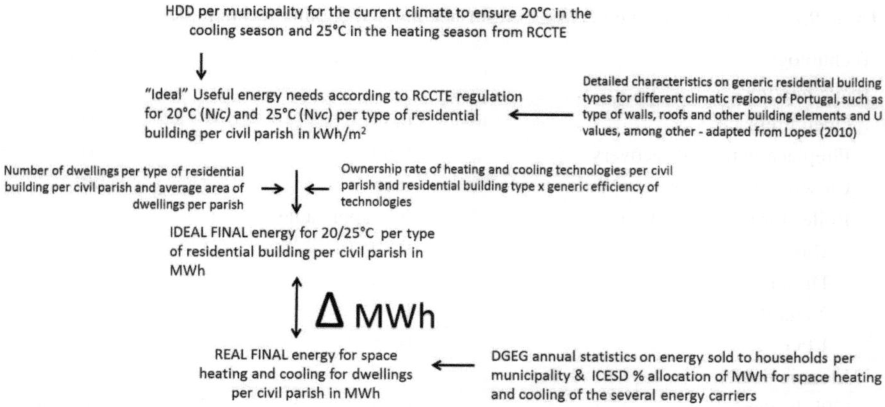

Fig. 9.3 Methodology to estimate the potential impact of climate change on thermal comfort of dwellings. *Source*: authors. ICESD refers to (DGEG and INE 2010)

household estimated by (DGEG and INE 2010), since there are no robust statistics available.

As in Fig. 9.3, the final energy consumption associated to the "IDEAL" thermal comfort is using the RCCTE requirements for a set of pre/defined typologies of the residential building stock in Portugal, based on the work of (Lopes 2010) and considering their age and construction types. Using the RCCTE's requirements we estimate the IDEAL useful energy needs in kWh/m^2 for space heating and cooling per dwelling and per type of residential building. We then convert the useful energy in final energy using average energy efficiencies of space heating and cooling technologies as presented in Table 9.2, as well as data from the last available population census (INE 2011) on: ownership rate of heating and cooling technologies per civil parish and building type, average area of dwellings per parish, and type of energy carrier consumed for space heating and cooling per residential building type.

The potential impact per each civil parish, estimated at this stage only for the current climatic conditions, is classified in a 20 level index, ranging from 1 (minimum impact) to 20 (maximum impact).

Finally, the **vulnerability** of households to climate change regarding thermal comfort is derived through a simple average from the combination of the adaptive capacity index with the potential impact index, ranging from a minimum of 0 to a maximum of 20.

The **georeferencing of the vulnerability indexes** was based on the geographic statistical referenced information extracted from Census 2011—BGRI (INE 2011) which contains different detailed levels of statistical information. We used georeferenced data at the level of civil parish since this was considered the most compatible scale in order to harmonize and make coherent different levels of data used in this methodology. While for adaptive capacity, we derived data at civil

Table 9.2 Average efficiency considered for heating and cooling technologies

Technology	Efficiency	References
Space heating		
Open fireplace	0.35	(a)
Fireplace with heat recovery	0.60	(a)
Closed biomass stove	0.55	(a)
Boiler for central heating	Not applicable	
Biomass	0.70	(b)
Diesel	0.75	(b)
Natural gas	0.75	(b)
LPG	0.75	(b)
Electric heater	1.00	(a)
GPL heater	0.85	Best guess
Air conditioning (heat pump)	2.20	(a)
Space cooling		
Air conditioning	2.38	(a)
Fan	1.00	Best guess
Heat pump	2.30	(b)

Source: (a) Gouveia et al. (2011), (b) ETSAP (2012)

parish scale, for the calculation of the potential impact we used energy consumption data at the municipality level.

All the vulnerability indicators calculated in this project have a geospatial identification (id) that assures the linkage to a GIS. Therefore each final index, including potential impact, adaptive capacity and vulnerability, was mapping according suitable class ranges. Given the design of this methodology, the intermediate geospatial data used for the calculation of the final indexes are also available in a GIS format. For example, regarding adaptive capacity index, the municipal technician is able to access disaggregated geospatially the variables behind the index (eg. share of population with 65 years old or more) and better understand where and why a civil parish has a certain adaptive capacity. It should be underlined that this methodology enables a dynamic interaction and updating process between the tool to calculate the vulnerability index, based on excel files, and the GIS mapping, allowing for a continuous spatial visualization of the vulnerability patterns.

Assess Trade-Offs Between Adaptation and GHG Mitigation Options Regarding Thermal Comfort of Dwellings

To assess the **trade-offs between mitigation and adaptation** options we propose an approach that quantifies the effect of measures to improve thermal comfort in terms of both GHG emissions reductions (i.e. mitigation) and variation of the vulnerability index (i.e. adaptation). The implementation of such measures has different motivations and it is not straightforward to allocate them to either mitigation or to adaptation. Some of the measures contribute for both adaptation and mitigation (i.e. they have a positive trade-off). Some other measures mitigate GHG emissions but do not lower or increase vulnerability (no trade-off), or they do lower vulnerability but increase GHG emissions (i.e. negative trade-off). We use our previously described approach for assessing vulnerabilities to estimate GHG emissions associated to heating and cooling in dwellings, change the input conditions to explore these trade-offs for the three examples in Table 9.2.

We considered three exploratory measures: improved efficiency (20 %) of heating and cooling technologies (*Efficiency* scenario), refurbishment interventions with 20 % of improvements on U-values of buildings' envelope (*Refurbishment*), and increasing the number of cooling appliances (20 %) to maintain the same level of thermal comfort (*Adjust Comfort*). We have quantified the magnitude of the trade-offs for each of these for Cascais.

Results and Discussion for the Case-Study of Cascais

The **adaptive capacity** of the civil parishes of Cascais, is presented in Table 9.3 and Fig. 9.4. The adaptive capacity of the civil parishes in Cascais ranges from 9 to 12 (in a scale of 0–20), and Alcabideche is the parish with lower adaptive capacity, due to the higher unemployment rate and the lower share of residents with a university degree (Tables 9.4 and 9.5).

As previously described, regarding **potential impacts,** we only were able to estimate figures at the municipality level (Table 9.6). We estimated, for 2012, a consumption of 63.8 GWh for REAL energy consumption associated with space heating, whereas 815.2 GWh would be required for an IDEAL comfort ensuring 20 and 25 °C indoors (Table 9.6). This represents a difference of approximately 8 % between the IDEAL and REAL final energy consumed. We have also estimated the

Table 9.3 Examples of measures to quantity trade-offs between mitigation and adaptation

Name	Measures to improve thermal comfort in dwellings	Type of potential trade-offs between mitigation and adaptation
Efficiency	Increase efficiency of heating technologies in 20 %	Neutral
Refurbishment	20 % improvements in the transmission coefficient of buildings envelope via insulated materials	Positive
AdjustComfort	Increase by 20 % the ownership of air conditioning appliances	Negative

Source: authors

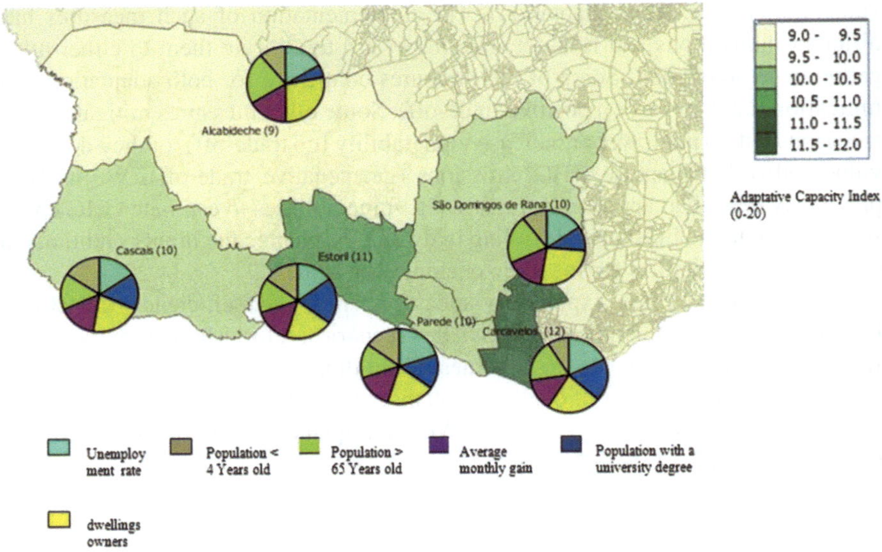

Fig. 9.4 Adaptive capacity index by civil parish and indicators' decomposition. *Source*: authors

CO_2 emissions associated to the space heating and cooling in Cascais using the emission factors from the (Decree - Law n.° 118/2013) (Table 9.7).

This discrepancy of 92 % between IDEAL and REAL energy consumption for thermal comfort is in line with the findings of (Gouveia et al. 2011) for Portugal and translates the rather poor thermal comfort currently experienced in most of the Portuguese households. Nonetheless it should be mentioned that there are other limitations with the used approach that influence this gap. The RCCTE IDEAL conditions translate into a whole dwelling heated to 20 °C for the whole of the heating season, although most people do not heat equally the whole of their homes. Additionally, there are quite some data gaps in the DGEG statistics as they do not report consumption per end-use and we have had to assume a single generic value for Portugal. Moreover, we have only compared IDEAL with REAL for the 2012 energy sales. These values fluctuate from 1 year to the other, for example due to the on-going economic crisis.

As depicted in Table 9.8 and Fig. 9.4, the final **vulnerability index** for the civil parishes of Cascais ranges from 13.1 to 14.6.

The **georeferenced indicators** for adaptive capacity and vulnerability are presented in Figs. 9.4 and 9.5, allowing a better understand of spatial disparities.

Table 9.4 Characterization of the indicators used to estimate the adaptive capacity for the civil parishes of Cascais

Civil parish	Share of population with 4 years old or less (%)	Share of population with 65 years old or more (%)	Average monthly gain in euros (€)	Share of dwellings owned (%)	Share of population with a university degree (%)	Unemployment rate (%)	Population (no. Persons)
Alcabideche	5.20	15.76	1161.20	50.28	9.61	12.69	42,162
Carcavelos	5.07	17.87	1161.20	57.04	27.34	10.41	23,347
Cascais	4.60	20.42	1161.20	40.28	24.68	12.06	35,409
Estoril	4.52	21.47	1161.20	41.99	26.92	12.72	26,399
Parede	4.56	22.46	1161.20	49.19	23.08	10.85	21,660
São Domingos de Rana	5.95	14.15	1161.20	59.70	11.39	12.38	57,502

Source: authors

Table 9.5 Adaptive capacity index estimated for the civil parishes of Cascais

Civil parish	Share of population with 4 years old or less	Share of population with 65 years old or more	Average monthly gain in euros	Share of dwellings owners	Share of population with a university degree	Unemployment rate	Total adaptive capacity index
	Classes (0–5)						Index (0–20)
Alcabideche	2	4	3	5	1	3	9
Carcavelos	2	4	3	5	4	4	12
Cascais	3	3	3	4	3	3	10
Estoril	3	3	3	4	4	3	11
Parede	3	3	3	4	3	4	10
São Domingos de Rana	2	4	3	5	2	3	10

Source: authors

Table 9.6 Estimated potential impacts for the municipality of Cascais

Energy consumption	GWh	Difference between ideal and real (%)
Space heating		
Real consumption in 2012	63.8	Circa 92 %
Estimated consumption for ideal thermal comfort	815.2	
Space cooling		
Real consumption in 2012	5.1	Circa 85 %
Estimated consumption for ideal thermal comfort	34.5	

Source: authors

Table 9.7 Energy consumption for space heating and cooling per fuel estimated for the municipality of Cascais and associated CO_2 emissions

Energy and emissions associated to space heating and cooling	Natural Gas	Electricity	Diesel	Butane	Propane	Biomass	Solar	Total
REAL consumption in 2012 (GWh)	8.0	23.6	0.4	0.00	0.025	36.3	0.5	68.9
Estimated consumption for IDEAL RCCTE Thermal Comfort (GWh)	102.3	270.7	5.5	0.00	0.3	464.5	6.3	849.7
Emission factor ($kgCO_2$/kWh)	0.20	0.14	0.27	0.17	0.17	0.00	0.00	n.a.
Emissions associated to REAL consumption (kt CO_2)	1.6	3.4	0.1	0.00	0.004	0.00	0.00	5.1
Emissions associated to thermal comfort consumption (kt CO_2)	20.6	40.0	1.4	0.00	0.054	0.00	0.00	61.1

n.a.—not applicable
Source: authors

Table 9.8 Vulnerability index estimated for the civil parishes of the municipality of Cascais

Civil Parish	Adaptive Capacity (Index 0–20)	Impact (Index 0–20)	Vulnerability (Index 0–20)
Alcabideche	9	18.5	14.6
Carcavelos	12		13.1
Cascais	10		14.3
Estoril	11		13.8
Parede	10		14.3
São Domingos de Rana	10		14.1

Source: authors

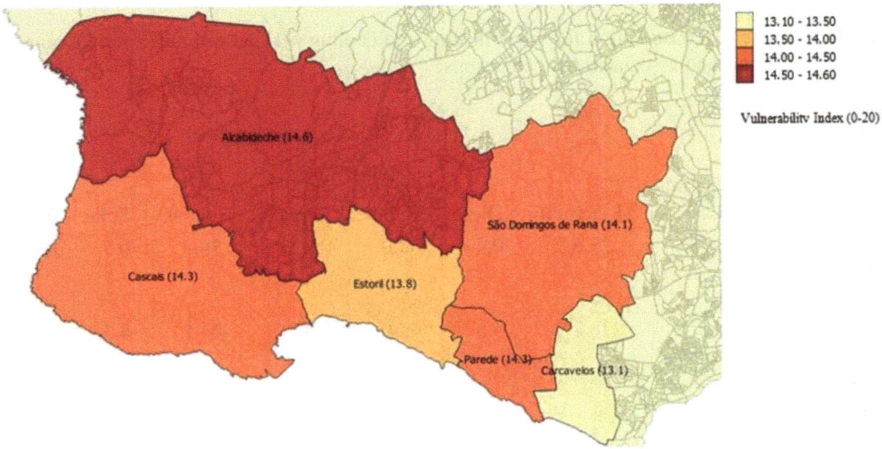

Fig. 9.5 Vulnerability index by civil parish. *Source*: authors

Trade-Offs Between Adaptation Needs and Mitigation Options

To assess trade-offs between mitigation and adaptation regarding thermal comfort of dwellings for Cascais we have replicated the previous calculations for the three measures identified in the previous section. The results in Table 9.9 show that when implementing insulation (Refurbishment) or replacing existing space heating and cooling technologies for more efficient ones (Efficiency), although there are lower emissions, there are no changes in vulnerability compared to BASE. When we increase the number of space cooling appliances by household (Adjust Comfort) the emissions increase as energy consumptions is higher but the vulnerability decreases from 14.6 to 13.9.

The trade-offs obtained from the *Efficiency* and *Refurbishment* measures are neutral situations for mitigation and adaptation options, since we did not reduce the vulnerability.

Table 9.9 Trade-off results between mitigation and adaptation for three scenarios

	CO_2 emissions IDEAL (kt)	Energy consumption IDEAL GWh	Impact heating (%)	Impact cooling (%)	Potential impact index (0–20)	Vulnerability (index 0–20)
BASE	61.1	849.8	7.8	14.8	18.5	14.6
Efficiency 20 % more efficiency	56.2	778.1	8.6	14.9	18.5	14.6
Refurbishment 20 % better building envelope	60.5	840.4	7.9	14.8	18.5	14.6
AdjustComfort 20 % more cooling	62.2	856.7	7.8	24.6	17.0	13.9

Source: authors

Conclusions

The conceptual methodology was conceived to assess the trade-offs between adaptation and mitigation options for residential buildings considering space heating and cooling necessary to ensure thermal comfort. The results reveal that when we adopt mitigation options to refurbish residential buildings or to use more efficient appliances, considering efficiency improvements of 20 %, the trade-offs are neutral. This situation means that to harvest benefits with mitigation options that may reduce the vulnerability and the need to apply extended adaptation measures, it will be necessary to increment the efficiency level for mitigation measures.

The system approach used in this research paper implied to integrate mitigation related data (e.g. energy consumption) and adaptation related data (eg. socio-economic variables) available at different administrative scales which resulted in different levels of accuracy for the indicators used to assess the **Vulnerability**.

The **Potential Impact** was derived from national and municipal data for the calculation of REAL energy while the IDEAL energy was estimated based on building typologies at the civil parish level. The REAL energy data for urban studies should be complemented in the future with local surveys adjusted to the local building typologies. We identified also the need to improve the national statistic databases with information at civil parish level. Despite the IDEAL energy is locked in very demanding cooling season (25 °C) and heating season (20 °C) temperatures defined in Portuguese regulation for indoors, the **Potential Impact** values achieved for Cascais demonstrate the poor thermal comfort of households.

The **Adaptive Capacity** index has a high local expression at civil parish scale since the majority of their socio-economic variables are available at that scale with exception of the revenue indicator. This index has potential and data availability from National Statistics to be created at block building level with geospatial representation, which is relevant for the integration of the adaption options with urban planning.

However to yield these synergies between mitigation and adaptation at urban level, it is crucial to deepen the interactions with urban spatial planning. In first place to understand which population is more vulnerable and where is their location inside the city, in second place to create mechanisms and policies that may account for non-economic costs and benefits that may reduce social disparities. Further research is suggested extending trade-off approach mitigation-adaptation to the global project including the 29 municipalities and deepening the interactions with urban planning at a neighbourhood scale.

Acknowledgements The work supporting this paper was funded by European Economic Area Financial Mechanism EEA – Grants through the project *ClimAdaPT.Local*, promoted by the Environment Portuguese Agency.

References

Adger WN, Agrawala S, Mirza MMQ, Conde C, O'Brien K, Pulhin J, Pulwarty R, Smit B, Takahashi K (2007) Assessment of adaptation practices, options, constraints and capacity. In: Parry ML, Canziani OF, Palutikof JP, van der Linden PJ, Hanson CE (eds) Climate change 2007: impacts, adaptation and vulnerability. Contribution of working group II to the 4th assessment report of the intergovernmental panel on climate change. Cambridge University Press, Cambridge, pp 717–743

Barker T, Bashmakov I, Alharthi A, Amann M, Cifuentes L, Drexhage J, Duan M, Edenhofer O, Flannery B, Grubb M, Hoogwijk M, Ibitoye FI, Jepma CJ, Pizer WA, Yamaji K (2007) Mitigation from a cross-sectoral perspective. In: Metz B, Davidson OR, Bosch PR, Dave R, Meyer LA (eds) Climate change 2007: mitigation. Contribution of working group III to the fourth assessment report of the intergovernmental panel on climate change. Cambridge University Press, Cambridge

Biesbroek GR, Swart RJ, Carter TR, Cowan C, Henrichs T, Mela H, Morecroft MD, Rey D (2010) Europe adapts to climate change: comparing national adaptation strategies. Glob Environ Chang 20:440–450

Dang HH, Michaelowa A, Tuan DD (2003) Synergy of adaptation and mitigation strategies in the context of sustainable development: the case of Vietnam. Clim Pol 3(1):S81–S96

Decree - Law n.° 80/2006, Diário da República, 1.ª série.67 (04-04-06)

Decree - Law n.° 118/2013, Diário da República, 1.ª série. (20-08-13)

DGEG (2012) Energy consumption per municipality and per activity sector for 2012. Lisbon. Retrieved from: http://www.dgeg.pt/

DGEG and INE (2010) ICESD – 2010 survey on energy consumption in households. Lisbon. Retrieved from: http://www.ine.pt/xportal/xmain?xpid=INE&xpgid=ine_publicacoes&PUBLICACOESpub_boui=127226704&PUBLICACOESmodo=2

ETSAP (2012) Space heating and cooling – IEA ETSAP – technology Brief R02. Retrieved from: www.etsap.org

Fritzsche K, Schneiderbauer S, Bubeck P, Kienberger S, Buth M, Zebisch M, Kahlenborn W (2014) The vulnerability sourcebook – concept and guidelines for standardised vulnerability assessments. GIZ GmbH, Adelphi, EURAC Research, pp 177. Retrieved from: http://www.adelphi.de/files/uploads/andere/pdf/application/pdf/vulnerability_sourcebook_-_guidelines_for_assessments_-_giz_2014.pdf

Gething B (2010) Design for future climate – opportunities for adaptation in the built environment. Technology Strategy Board, UK. Available at: http://www.arcc-network.org.uk/wordpress/wp-content/D4FC/01_Design-for-Future-Climate-Bill-Gething-report.pdf

Gouveia JP, Fortes P, Seixas J (2011) Projections of energy services demand for residential buildings: insights from a bottom-up methodology. Energy 47(1):430–442. doi:10.1016/j.energy.2012.09.042

INE (2011) Census 2011. National Institute of Statistics. Available at www.ine.pt

IPCC (2014) Climate change 2014: synthesis report. In: Core Writing Team, Pachauri RK, Meyer LA (eds) Contribution of working groups I, II and III to the fifth assessment report of the intergovernmental panel on climate change. IPCC, Geneva, 151 pp

Klein RJT, Schipper EL, Dessai S (2005) Integrating mitigation and adaptation into climate and development policy: three research questions. Environ Sci Pol 8:579–588

Lopes TP (2010) Potencial de poupança de energia na climatização de edifícios habitacionais. Dissertação apresentada na Faculdade de Ciências e Tecnologia da Universidade Nova de Lisboa para obtenção do Grau de Mestre em Engenharia do Ambiente, perfil Gestao e Sistemas A (p 163). Lisbon. Retrieved from: http://run.unl.pt/handle/10362/5014

Rosenzweig C, Solecki WD, Hammer SA, Mehrohtra S (2011) Climate change and cities first assessment report of the urban climate change research network. Cambridge University Press, Cambridge. Executive Summary available at: http://uccrn.org/documents/ARC3_Executive_Summary.pdf

Shaw A, Burch S, Kristensen F, Robinson J, Dale A (2014) Accelerating the sustainability transition: exploring synergies between adaptation and mitigation in British Columbian communities. Glob Environ Chang 25:41–51

Urge-Vorsatz D, Novikova A, Koeppel S, Boza-Kiss B (2009) Bottom-up assessment of potentials and costs of CO_2 emission mitigation in the buildings sector: insights into the missing elements. Energy Effic 2:293–316

US Green Building Council, World Green Building Council, C40 Cities (2015) Green building city market briefs. US GBC, WGBC, C40 Cities, Washington, DC. Retrieved from: http://c40-production-images.s3.amazonaws.com/researches/images/35_USGBC_C40_WGBC_City_Market_Brief_Compendium.original.pdf?1427985432

Zimmerman R, Faris C (2011) Climate change mitigation and adaptation in North American cities. Curr Opin Environ Sustain 3:181–187

Chapter 10
Adapting to Climate Change: Getting More from Spatial Planning

Jeremy Carter and Graeme Sherriff

Abstract Given the severity of future climate change projections and associated risks to human and natural systems, societies are now faced with a strong imperative to develop adaptation policies and actions in response. Spatial planning, which is the process through which the development and use of land is visualised, negotiated and regulated, has an important role to play in adapting to the changing climate. Despite positive steps forward in some locations, there remains a gap between spatial planning's potential capacity to support the achievement of adaptation goals and the realisation of this role in practice. This paper reports on the findings of an online Delphi survey undertaken to build understanding of the relationship between spatial planning and climate change adaptation. The survey secured the input of over 70 academics, planners and policy makers working across these fields in ten different countries. Its results offer insights on barriers inhibiting spatial planning's contribution to adaptation, which range from overarching systemic issues through to those concerning the detailed workings of the planning system. The Delphi survey also identified solutions that could help build the capacity of spatial planning to progress the adaptation agenda. Approaches include enhancing the adaptation knowledge, skills and technical capacity of planners and applying different concepts and methods to align spatial planning more closely with adaptation goals. In presenting and analysing the results of the Delphi survey, the aim of this paper is to help build the capacity of policy makers, practitioners and researchers to adapt spaces and places for the changing climate.

Keywords Climate change • Climate change adaptation • Spatial planning

---

J. Carter (✉)
Centre for Urban Resilience and Energy, University of Manchester, Oxford Road, Manchester M139PL, UK
e-mail: jeremy.carter@manchester.ac.uk

G. Sherriff
Sustainable Housing and Urban Studies Unit, The University of Salford, The Crescent, Salford M54WT, UK

W. Leal Filho (ed.), *Innovation in Climate Change Adaptation*, Climate Change Management, DOI 10.1007/978-3-319-25814-0_10

Introduction

There is now widespread acknowledgement that the climate is changing, that human activity is the principal driver of recent changes in climate variables, and that the coming decades are likely to be characterised by an intensification of extreme weather events and changes to the climate (IPCC 2014; The Royal Society and The US National Academy of Sciences 2014). There is growing awareness at the highest levels of the need to develop adaptation responses to climate change impacts due to the risks they pose to broader policy priorities from social welfare to economic competitiveness (EEA 2012; IPCC 2014; Stern 2007).

The spatial planning process, through which the development and use of land is regulated, has the potential to play an important role in adapting to the changing climate. Planning legislation, regulations and urban climate change strategies from countries and cities across the world are increasingly incorporating adaptation objectives (Reckien et al. 2014). Although practice is increasing it is not endemic, and this remains a relatively new agenda for urban areas and the organisations with the responsibility for guiding their future growth and development (Birkmann et al. 2010; Blanco et al. 2011; Carter 2011). Indeed, Blanco et al. (2011: 238) describe adaptation planning as a 'novelty', with Birkmann et al. (2010: 185) adding that 'the discourse on urban climate change adaptation has only recently gained momentum in the political and scientific arena.' Given the increasing prominence of climate change adaptation, and the emerging recognition of spatial planning's contribution to this agenda, the goal of this paper is to help build the knowledge and capacity of policy makers, practitioners and researchers in planning for the changing climate. This has been achieved via an international Delphi survey which engaged 71 people and identified barriers limiting the effectiveness of spatial planning in this context, in addition to solutions that offer the potential to strengthen its role within the adaptation agenda.

This paper begins with an overview of existing literature connecting to the contribution that spatial planning has to play as an adaptation response (section "Spatial Planning and Climate Change Adaptation: Background and Context"). The Delphi survey approach is then explained (section "Research Methodology: The Delphi Approach"). This covers reasons for using a Deplhi survey within this research, a description of how the study was carried out and details of who was engaged in the process. The Delphi results are then presented and analysed. The research findings focus on a discussion of barriers constraining the effectiveness of spatial planning in achieving adaptation goals (section "Barriers to Planning for a Changing Climate") and on solutions that could strengthen the role of planning in this respect (section "Strengthening the Contribution of Spatial Planning to Climate Change Adaptation"). Finally, the broader implications of the research findings outlined within this paper are identified and discussed, including opportunities for future research (section "Discussion and Conclusions: Getting More from Spatial Planning").

Spatial Planning and Climate Change Adaptation: Background and Context

Climate change impacts cross sectors, spatial scales and timescales. Addressing extreme weather and climate change risks will therefore involve a variety of responses (EEA 2012; IPCC 2014; Shaw et al. 2007). Sitting amongst this suite of responses, spatial planning has, in principle, a strong role to play in reducing current weather and climate risks. There are counter-arguments, however. Indeed, critics have argued that it is important to challenge the perception of spatial planning as a progressive force, and that instead spatial planning should be analysed '. . .as a form of neoliberal spatial governance' (Allmendinger and Haughton 2011: 89).

Over recent years, the profile of spatial planning as a flood risk management approach has been raised (White and Alarcon 2009; Wilson 2006), and there is now an emerging recognition of the contribution that spatial planning can make to climate change adaptation goals more broadly (Blanco et al. 2011; Carter et al. 2014; Davoudi et al. 2010). Within this paper, Davoudi et al.'s (2010: 14) definition of spatial planning is followed, namely: '. . .the processes through which options for the development of places are envisioned, assessed, negotiated, agreed and expressed in policy, regulatory and investment terms.'

The 'blue light' emergency services provide a reactive response to extreme weather hazards such as floods and heat waves. This can help residents and businesses deal with these events when they occur and in their immediate aftermath, but it does little to reduce their underlying exposure and vulnerability to these hazards. Here, spatial planning offers an important route into the development of anticipatory proactive adaptation responses that manipulate buildings and landscapes to reduce weather and climate risk (Blanco et al. 2011; Carter et al. 2014; Davoudi et al. 2010; Wilson 2006). The development (or lack of) spatial planning responses to adapt locations to changes in climate variables will influence the frequency and severity of climate change impacts within towns and cities. Corfee-Morlot et al. (2011: 170) acknowledge this point, stating that:

> Urban development patterns as we currently understand them today are a main driver of vulnerability to climate change. However, these same characteristics suggest that local authorities and other decision-makers that shape urban development may also play a central role in what is becoming an urgent agenda to adapt to climate change.

There are a number of reasons why spatial planning has the potential to make a significant contribution to climate change adaptation. The following themes are briefly introduced below:

- The cross-boundary nature of spatial planning.
- The cross-sectoral remit of spatial planning.
- The forum for engagement offered by spatial planning.
- The influence of spatial planning over urban form and building design.
- The long-term nature of spatial plans.

Natural landscapes bordering cities and urban areas can act as rainwater 'sinks', reducing the volume of water flowing through urban rivers and streams hence reducing flood risk. Similarly, green corridors running into cities, such as those flanking rivers, can help to transfer cool air into urban centres under heat wave conditions as they do in Stuttgart (Germany) (Kaźmierczak and Carter 2010). These examples point towards the need for cross-boundary and spatially integrated plans that encompass strategies to reduce weather and climate risks. In principle, strategic spatial planning can influence the development and use of land across a range of spatial scales to support adaptation goals by, for example, conserving and expanding absorptive surfaces and existing green spaces that offer valuable adaptation functions (Forest Research 2010).

In addition to exhibiting connections and interdependencies across spatial scales, urban systems are thematically complex and characterised by a multitude of linkages and feedbacks between natural, human and technological systems (Ravetz 2000). As a result, urban climate impacts cross themes including human health and well-being, the built and natural environment and different types of infrastructure—critical, emergency and social (Carter et al. 2014; EEA 2012). Spatial planning has a cross-sectoral remit, encompassing themes including land use and building design, housing, transportation, public facilities and services, biodiversity, open space and recreation, with some cities also including economic development and urban design (Blanco et al. 2011: 230). The capacity of planning to integrate and coordinate activity across different sectors places it in a frontline position for addressing climate change risks.

Due to the connections in urban systems that cross spatial boundaries and thematic areas, climate change impacts affect a wide range of stakeholder groups. Adger (2003: 387) describes adaptation as '. . .a dynamic social process. . .', and highlights that the effectiveness of adaptation strategies will depend on factors including the social acceptability of adaptation options. Public participation is an important element of many spatial planning systems (Rydin 2010), offering different opportunities for engagement from creating spatial plans through to decisions on development design and location.

The design of new developments can be influenced through spatial planning measures including building regulations and planning policies. In the context of adaptation, this may include design interventions such as requiring green roofs and shading canopies to reduce rainwater runoff and exposure to high temperatures respectively. It is estimated that by 2050, around 80 % of the UK's current domestic building stock will remain (Boardman et al. 2005). Retrofitting existing buildings to increase their climate resilience will therefore be necessary (Gething 2010). However, the planning system has little influence over existing buildings. The principal contribution of spatial planning in this context is to influence land use around buildings to reduce their exposure to climate hazards, for example by enhancing green spaces that provide cooling and storm water absorption capacity.

Modifying urban landscapes is a long term process, yet is ultimately necessary in order to adapt cities and urban areas to the changing climate. This emphasises that long-term strategic spatial planning is essential to the growth and development of cities. This is especially the case for critical infrastructure due to its long life span. Building the resilience of infrastructure to extreme weather and climate change pressures will be crucial over the coming decades as it is central to the effective functioning, productivity, and competitiveness of cities and urban areas (RAE 2011; Corporation Limited 2009). Spatial plans often extend 15–20 years into the future, providing scope for taking a forward looking view on the design and operation of urban areas and infrastructures. For this reason, and the other issues introduced above, spatial planning has an important role to play in climate change adaptation and resilience building.

Research Methodology: The Delphi Approach

The research findings that inform this paper were developed using a Delphi approach. Developed in the 1950s, Delphi techniques have been widely used across different research fields and is accepted as a method for achieving convergence of opinion concerning real-world knowledge and for forecasting future developments (Hsu and Sandford 2007). Examples of the previous use of Delphi include bringing together expert views on Finnish environmental policy (Wilenius and Tirkkonen 1997); surveying expert opinion on climate change (Doria et al. 2009); and evaluating options for climate change adaptation in Ontario (Lemieux and Scott 2011). The distinguishing feature of the Delphi approach is that it is an iterative process, consisting of two or more rounds. Analysis and feedback occurs after each round, with the outcomes of the analysis informing the approach taken to subsequent round (s) where the initial findings are consolidated and refined. Whilst a classical Delphi is intended to be a 'forum for facts' in which unbiased experts form a consensus, a policy Delphi can be understood as a 'forum of ideas' and does not necessarily seek consensus (Crisp et al. 1997). A policy Delphi may be used as a tool for analysis rather than decision making, aiding the development of potential solutions to identified problems and helping researchers target the roots of consensus, or disagreement (Miller 2001). The research findings presented in this paper emerged from a policy Delphi.

Literature and Internet searches were carried out to identify potential participants to involve in the Delphi survey. The research team had been working in the climate change field for a number of years, and was able to draw on established contacts. Coverage was enhanced by asking established contacts to suggest others to engage in the survey. A total of 71 people participated in the Delphi survey. The sample contained government employees ($n = 28$, 39 % of participants), academics ($n = 27$, 38 %), consultants ($n = 11$, 15 %) and representatives of the non-governmental sector ($n = 5$, 7 %). The participants were from organisations based in Australia, Austria, Belgium, Canada, Finland, Germany, Italy, Lebanon, Netherlands, Norway, Sicily, Spain, Sweden, Switzerland, UK, and the USA, with others working in EU or international level organisations. The majority of participants, nearly two-thirds, had researched related policy areas, and just over half had experience of developing policy on climate change adaptation. Fewer, one quarter, were involved in research on the science of climate change. Just under half worked in related areas of policy.

The study comprised two stages. Three relatively open questions were asked at Stage 1. These concerned; (1) identifying existing spatial planning initiatives linking to adaptation; (2) identifying significant factors reducing the effectiveness of spatial planning in contributing to adaptation; and (3) suggesting solutions that could strengthen the contribution of spatial planning to adaptation. Stage 1 answers were analysed and grouped. With the aim of keeping Stage 2 concise to maintain participation, questions were focused on analysing the solutions identified at Stage 1. Following this approach, the iterative nature of the Delphi approach was effectively realised, with open, qualitative questions in Stage 1 being subject to prioritisation via closed questions in Stage 2. The Delphi approach offered numerous advantages as a research method including the ability to involve a large number of participants whilst facilitating a degree of interaction (without the budgetary and time constraints of a workshop); the potential to capture new ideas in an emerging field; and the guarantee of anonymity for the participants. The key results of the Delphi are presented and discussed in sections "Barriers to Planning for a Changing Climate" and "Strengthening the Contribution of Spatial Planning to Climate Change Adaptation" below.

Barriers to Planning for a Changing Climate

A key goal of the research reported within this paper was to identify barriers constraining spatial planning's effectiveness in contributing to climate change adaptation. This was achieved by using a Delphi survey to gather information and insights on this issue. More knowledge about why planners are not taking action and the factors that are limiting them in this respect enhances capacity to devise effective strategies in response. Three broad groups of barriers were identified from the Delphi survey results, which differ according to their level of abstraction from the day-to-day workings of the spatial planning system. They are outlined below.

Systemic Issues External to the Planning System

High level systemic issues exert a significant influence over the way the planning system works, and ultimately on its potential effectiveness in supporting the achievement of adaptation goals. Related barriers identified through the Delphi survey included limited knowledge and awareness of climate change amongst different groups of society, insufficient data and tools to assess climate change risks, fragmented urban governance structures, a lack of political commitment to climate change and the influence of market pressures on land use and development.

Although these systemic issues lie beyond the direct influence of spatial planning systems, they affect their overarching structure and operation via their impact on legislative frameworks and the nature of dominant approaches and methods applied by planners. For example, market pressures and can lead to adaptation measures being marginalised within planning decision making due to pressure from developers. Similarly, a lack of political commitment to climate change can limit the inclusion of adaptation themes in planning legislation and regulations.

The Overarching Structure and Operation of the Planning System

The Delphi survey identified several barriers related to the overarching structure and operation of spatial planning systems themselves, suggesting that they encompass certain characteristics that work against the delivery of adaptation responses. Delphi participants identified that planning legislation and regulations do not always specify that adaptation issues must be addressed, which removes an important incentive for spatial plans to include adaptation policies. In the UK, for example, adaptation remains a largely voluntary exercise with little action prescribed within either the statute book or regulations guiding the development and use of land. The survey also identified that where adaptation features within planning frameworks, or where planning authorities progress adaptation responses voluntarily, useful guidance to support action is sometimes lacking. In addition, the Delphi identified that some planning authorities suffer from organisational weaknesses including a lack of policy integration across sectors and poor communication between departments. This in turn limits their capacity of planners to integrate adaptation activity across sectors and scales.

The Detailed Workings of the Planning System

Looking below the overarching structure of the planning system reveals barriers that relate to the day-to-day workings of planners and planning authorities. Here, the key issue identified by the Delphi participants was that planners do not have access to sufficient resources (including time, money and human resources) to effectively deliver adaptation responses. For example, funding shortages reduce opportunities to commission research or dedicate staff time to adaptation activities. This is compounded by the fact that, at the local level, planning authorities do not always have staff with a climate change remit and therefore lack the capacity to undertake their own analysis of impacts and adaptation responses.

At each step down this hierarchy of barriers, moving from high level systemic issues to the structure and operation of the planning system and down to the day-today workings of the system itself, the number of barriers identified via the Delphi survey declined. This emphasises that broader systemic issues are particularly significant when considering barriers limiting the contribution of spatial planning to the delivery of adaptation goals.

Strengthening the Contribution of Spatial Planning to Climate Change Adaptation

A key goal of this paper is to help move the urban adaptation planning agenda forwards. This section presents insights gathered via the Delphi survey on approaches to strengthen the adaptation contribution of spatial planning. At Stage 1 of the Delphi process, participants were requested to identify up to three changes or new initiatives that could support this outcome. A large number of 'solutions' were suggested. They were subsequently analysed and grouped under seven headline solutions, which are listed below. The number of Delphi participants that suggested a solution falling under each headline is provided in brackets. This figure is also given as a percentage of the total number of participants who provided a response to this question.

- *Solution 1:* Increase planner's climate change adaptation knowledge, skills and technical capacity through guidance, education, training and good practice exchange (n = 33, 54 %)
- *Solution 2:* Apply different concepts and methods to align spatial planning more closely with the adaptation agenda. For example extend planning time horizons, link adaptation and mitigation and promote the creation of multifunctional green spaces (n = 32, 52 %)
- *Solution 3:* Build an overarching framework of legislation and regulation to strengthen the capacity of spatial planning to deliver climate change adaptation (n = 27, 44 %)
- *Solution 4:* Strengthen institutional and organisational governance structure to facilitate connections, across sectors and spatial scales, between planning authorities and other organisations with a role to play in adaptation (n = 12, 20 %)
- *Solution 5:* Enhance awareness of climate change impacts and adaptation, and the role of spatial planning in delivering responses, amongst politicians and the public (n = 11, 18 %)
- *Solution 6:* Increase resources (financial and human) to enhance the capacity of planning authorities to develop and implement adaptation responses (n = 6, 10 %)
- *Solution 7:* Develop responses to counteract the market pressures and economic drivers that influence planning and development, in some cases to the detriment of adaptation goals (n = 6, 10 %)

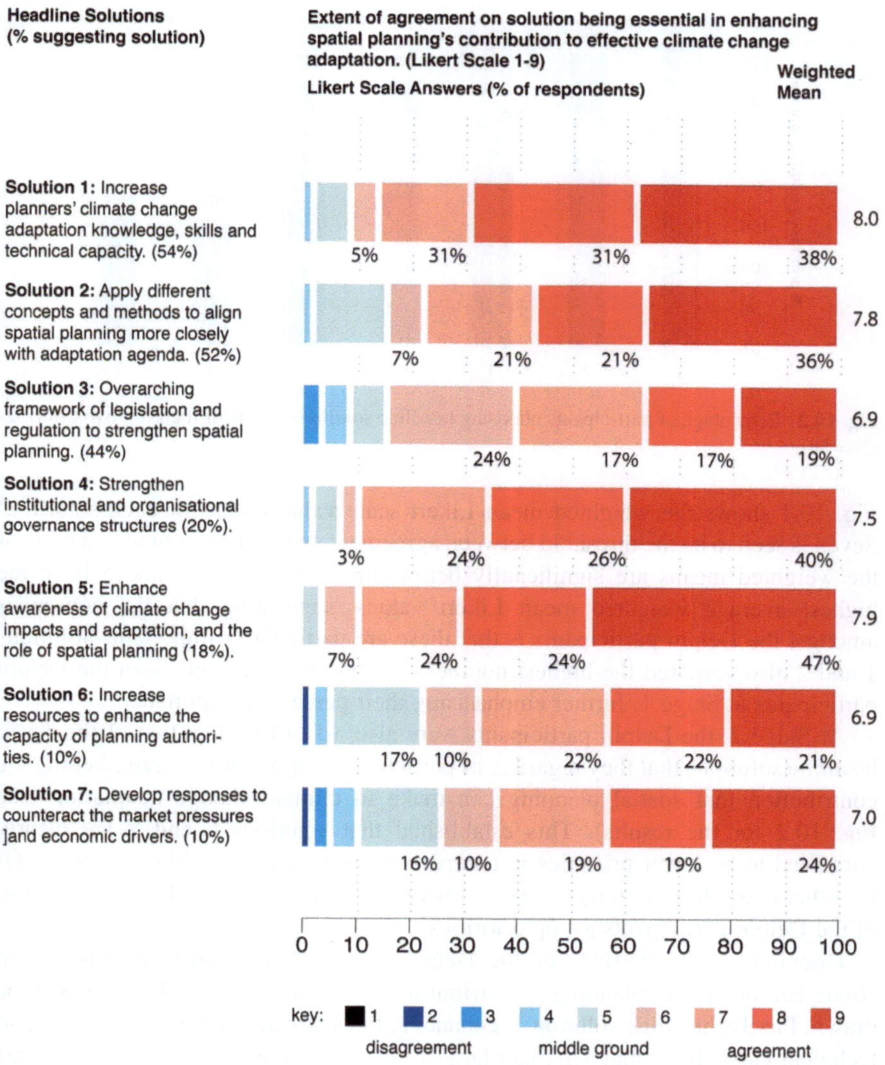

Fig. 10.1 Likert scale answers, with weighted mean, from stage 2 respondents (N = 58)

Stage 2 of the Delphi was dedicated to consolidating and verifying the analysis of solutions identified during Stage 1. The seven headline solutions were presented to the participants for evaluation. They were asked to use a numerical Likert scale to indicate to what extent they agreed that each solution is essential for enhancing spatial planning's contribution to effective climate change adaptation. The results of this question are presented in Fig. 10.1.

As a group, the Delphi participants, broadly agreed that each of the solutions has an important role to play in strengthening spatial planning's contribution to the adaptation agenda. The number at the right hand end of each bar displayed in

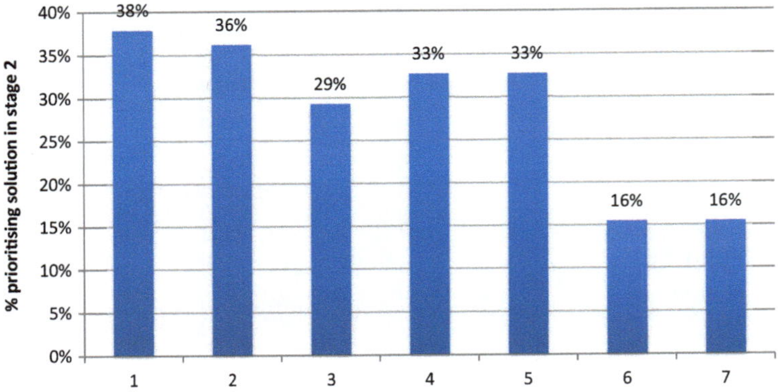

Fig. 10.2 Percentage of participants choosing headline solutions as one of their top two priorities (N = 58)

Fig. 10.1 shows the weighted mean Likert scale value given for each solution.[1] Seven is seen to be the threshold between agreement and middle ground and none of the weighted means are significantly below this. Solutions 1, 2 and 5 have the highest average weighted mean Likert values, suggesting that the consensus amongst the Delphi participants is that these are particularly important. Solutions 1 and 2 also garnered the highest number of suggested solutions from the Delphi participants at Stage 1, further emphasising their perceived importance.

At Stage 2, the Delphi participants were also asked to select two of the seven headline solutions that they regarded as particularly important for strengthening the contribution that spatial planning can make to climate change adaptation (see Fig. 10.2 for the results). This established that solutions 6 and 7 are clearly perceived to be lower priorities in comparison to the other headline solutions. Of the other five solutions, only two (solutions 1 and 2) were selected by over one third of the Delphi participants as top priorities.

Drawing on the findings of the Delphi survey, three broad conclusions on strengthening spatial planning's contribution to climate change adaptation can be drawn. Firstly, headline solution 1, enhancing the adaptation knowledge, skills and technical capacity of planners, had largest number of participants suggesting this solution at Stage 1. At Stage 2, it had the highest weighted mean Likert value (see Fig. 10.1) and also the highest percentage of participants selecting the solution as a top priority (see Fig. 10.2). The 'up-skilling' of planners is clearly regarded as a vital element of enhancing the connection between spatial planning and climate change adaptation. If this was to occur, other solutions such as applying different concepts and methods to align spatial planning more closely to the adaptation agenda (solution 2) would become easier to achieve. Also, the allocation of

[1] The weighted mean is calculated by summing the product of each Likert number and the number of times it was selected (1*a + 2*b + 3*c...) and dividing by the total number of answers.

additional resources to adaptation may follow from increasing knowledge and awareness of the agenda amongst planners.

Secondly, headline solution 2, applying different concepts and methods to align spatial planning more closely with the adaptation agenda, also consistently featured highly across the different Delphi metrics. This suggests that the planning system is not yet properly equipped to make an effective contribution to the adaptation agenda, and that new tools and approaches are needed. Some options suggested by the Delphi participants are already applied in some contexts, including scenario development and participatory planning, providing a platform for them to be practiced more widely.

Finally, the Delphi results show that headline solutions 6 and 7, increasing resources available to planners to target at adaptation issues and developing responses to counteract the market pressures and economic drivers that influence planning and development, were consistently perceived by the Delphi participants as lower priorities relative to the other headline solutions. Market pressures and economic drivers are systemic issues who's origin lies beyond the sphere of influence of planners working within local and regional authorities. Similarly, resourcing decisions are generally taken at the highest levels of planning authorities, often in order to meet national level legislation and regulations. Hence, the lower priority afforded to solutions 6 and 7 by the Delphi participants may reflect the challenges faced by planning officers in progressing related issues locally. However, their intractability does not mean that these issues should be sidelined as they remain important elements of strengthening spatial planning's contribution to the adaptation agenda.

Discussion and Conclusions: Getting More from Spatial Planning

Adapting to the risks associated with extreme weather and climate change is an important element of securing the future of urban areas, and cities from across the world are now starting to incorporate related objectives into urban spatial plans (Bulkeley and Broto 2013; Carter 2011; EEA 2012; Reckien et al. 2014). This is a valuable development as the planning profession can play a vital role in adaptation to climate change impacts. However, adaptation is not yet a mainstream planning issue. This paper points towards a number of barriers that help to explain this situation. The Delphi survey indicated that the system within which planners operate, which is influenced by issues including dominant free market ideologies, limitations in data, knowledge and awareness of climate change and low political commitment to climate change, puts a real break on realising spatial planning's potential as an adaptation response. This chimes with critiques of spatial planning that emphasise its politicisation and capture by neoliberal agendas (RCEP 2002; Allmendinger and Haughton 2011). Ultimately, as Evans (1997: 4) noted: 'Landuse policy is simply a public-policy process or mechanism, in principle no different to tax collection or waste management, through which public or government policy is enacted.' Given these systemic constraints, it seems appropriate to be cautious about over-promoting the degree to which spatial planning can deliver on climate change adaptation goals. This has been necessary when considering the role of planning within other major sustainability agendas, for example relating to the 'compact city' (Gleeson 2012).

Despite this, in principle, spatial planning does have an important role to play in adapting to climate change. This is due to synergies between the two agendas, which relate to issues including their long time frames, cross-sectoral reach and connections across spatial scales. The solutions identified by this study, linked to strengthening spatial planning's adaptation contribution, offer valuable insights into options for progressing this agenda. Here, enhancing the adaptation knowledge, skills and technical capacity of planners and applying different concepts and methods to align spatial planning more closely with adaptation goals are particularly important. Broadly, these solutions connect to the 'up-skilling' of planners and relate to building capacity within the planning profession to develop and implement climate change adaptation responses. Although the realisation of these solutions in practice is likely to require central government intervention to bring widespread consistent benefits, via the development of supportive legislation and guidance for example, progress can nevertheless be made at the local level where there is the political will to do so.

This study has opened several opportunities for further research. Given their breadth and scope, each of the barriers and solution identified by the Delphi survey represents a worthy research opportunity in its own right. However, it appears that the systemic barriers place the most problematic barriers on spatial planning's adaptation potential, and it would therefore be valuable to understand these in more detail. The Delphi survey has also help to prioritise the solutions that may offer the greatest scope for future research; effectively those afforded higher priority by the survey participants. It would also be useful to extend this assessment, perhaps by increasing the sample size or by looking at themes linked to different adaptation approaches such as flood defence or community resilience building. This study has also demonstrated the value of Delphi surveys, although it is important to recognise associated limitations of this research approach. These include that the interaction is not as rich as that achieved during a face to face workshop. Further, the issues identified via the Delphi process represent a consensus perspective of the issue which tends to marginalise less prominent issues which may nevertheless be significant. Also the findings of the Delphi are aspatial; that is they do not account for local context. The barriers and solutions identified are crosscutting and transferable, although prioritising which are most prominent or likely to succeed will depend on local context, with relevant factors including the current position of the adaptation agenda in policy hierarchies and the availability of information on climate change impacts and vulnerabilities.

This paper has addressed the deficit between spatial planning's potential role in adaptation and the realisation of this role in practice. The range of barriers and solutions highlighted within this paper offer spatial planners, and decision-makers with an influence over this sector, a better insight into possible strategies to connect the two crucial agendas of spatial planning and climate change adaptation. Although particular attention is paid to the planning community, the findings have broader relevance and highlight issues that lie at the heart of adapting to the changing climate. The reality is that adaptation buffers up against constraining factors, such as the dominant growth imperative and the short term nature of political priorities, which challenge the achievement of related objectives. However, given the severity of climate change projections and the related risk to urban areas, progress across a broad range of adaptation approaches is needed. Adaptation should not be seen as an exercise in its own right, but as a constituent element of designing and governing urban areas. This will help to broaden the adaptation debate and move it closer to key urban agendas such as supporting healthy communities and enhancing the economic competitiveness of cities.

Acknowledgements The funding to undertake this research was gratefully received from Bruntwood and the Oglesby Charitable Trust (via the EcoCities project) and the European Union (via the INTERREG IVC Green and Blue Space Adaptation for Urban Areas and Eco Towns project – GraBS). Finally, our thanks go to Graham Haughton for his valuable comments on an earlier version of this chapter.

References

Adger N (2003) Social capital, collective action and adaptation to climate change. Econ Geogr 79 (4):387–404

Allmendinger P, Haughton G (2011) Post-political spatial planning in England: a crisis of consensus? Trans Inst Br Geogr 37(1):89–103

Birkmann J, Garschagen M, Krass F, Quang N (2010) Adaptive urban governance: new challenges for the second generation of urban adaptation strategies to climate change. Sustain Sci 5 (2):185–206

Blanco H, McCarney P, Parnell S, Schmidt M, Seto K (2011) The role of urban land in climate change. In: Rozensweig C, Solecki W, Hammer S (eds) First urban climate change research network (UCCRN) assessment report on climate change in cities. Cambridge University Press, Cambridge

Boardman B, Darby S, Killip G, Hinnells M, Jardine C, Palmer J, Sinden G (2005) 40 % house. Environmental Change Institute, The University of Oxford, Oxford

Bulkeley H, Broto V (2013) Governing by experiment? Global cities and the governing of climate change. Trans Inst Br Geogr 38(3):361–375

Carter JG (2011) Climate change adaptation in European cities. Curr Opin Environ Sustain 3 (3):193–198

Carter JG, Cavan G, Connelly A, Guy S, Handley J, Kazmierczak A (2014) Climate change and the city: building capacity for urban adaptation. Prog Plan 95:1–66

Corfee-Morlot J, Cochran I, Hallegatte S, Teasdale PJ (2011) Multilevel risk governance and urban adaptation policy. Clim Chang 104(1):169–197

URS Corporation Limited (2009) Adapting energy, transport and water infrastructure to the long-term impacts of climate change. RMP/5456. Defra, London

Crisp J, Pelletier D, Duffield C, Adams A, Nagy S (1997) The Delphi method? Nurs Res 46:116

Davoudi S, Crawford J, Mehmood A (2010) Planning for climate change: strategies for mitigation and adaptation for spatial planners. Earthscan, London

Doria MF, Boyd E, Tompkins EL, Adger WN (2009) Using expert elicitation to define successful adaptation to climate change. Environ Sci Pol 12(7):810–819

European Environment Agency (EEA) (2012) Urban adaptation to climate change in Europe: challenges and opportunities for cities together with supportive national and European policies. EEA report, no. 2/2012. Office for Official Publications of the European Union, Luxembourg

Evans B (1997) From town planning to environmental planning. In: Blowers A, Evans B (eds) Town planning into the 21st century. Routledge, London

Forest Research (2010) Benefits of green infrastructure. Report by Forest Research. Forest Research, Farnham

Gething B (2010) Design for future climate: opportunities for adaptation in the built environment. Technology Strategy Board, Swindon

Gleeson B (2012) 'Make no little plans': anatomy of planning ambition and prospect. Geogr Res 50(3):242–255

Hsu CC, Sandford BA (2007) The Delphi technique: making sense of consensus. Pract Assess Res Eval 12(10):1–8

Intergovernmental Panel on Climate Change (IPCC) (2014) Climate change 2014: impacts, adaptation and vulnerability summary for policy makers. IPCC, Geneva

Kaźmierczak A, Carter JG (2010) Adaptation to climate change using green infrastructure, a database of case studies. The Town and Country Planning Association, London

Lemieux C, Scott D (2011) Changing climate, challenging choices: identifying and evaluating climate change adaptation options for protected areas management in Ontario, Canada. Environ Manag 48(4):675–690

Miller G (2001) The development of indicators for sustainable tourism: results of a Delphi survey of tourism researchers. Tour Manag 22:351–362

Ravetz J (2000) City-region 2020: integrated planning for a sustainable environment. Earthscan, London

Reckien D, Flacke J, Dawson RJ, Heidrich O, Olazabal M, Foley A, Hamann JJ-P, Orru H, Salvia M, Hurtado SDG, Geneletti D, Pietrapertosa F (2014) Climate change response in Europe: what's the reality? Analysis of adaptation and mitigation plans from 200 urban areas in 11 countries. Clim Chang 122:331–340

Royal Academy of Engineering (RAE) (2011) Engineering the future. Infrastructure, engineering and climate change adaptation—ensuring services in an uncertain future. The Royal Academy of Engineering, London

Royal Commission on Environmental Pollution (RCEP) (2002) Environmental Planning. HMSO, London

Rydin Y (2010) Governing for sustainable urban development. Earthscan, London

Shaw R, Colley M, Connell R (2007) Climate change adaptation by design: a guide for sustainable communities. Town and Country Planning Association, London

Stern NH (2007) The economics of climate change: the stern review. Cambridge University Press, Cambridge

The Royal Society, The US National Academy of Sciences (2014) Climate change evidence and causes. The Royal Society, The US National Academy of Sciences, Washington, DC

White I, Alarcon A (2009) Planning policy, sustainable drainage and surface water management: a case study of greater Manchester. Built Environ 35(4):516–530

Wilenius M, Tirkkonen J (1997) Climate in the making: using Delphi for Finnish climate policy. Futures 29(9):845–862

Wilson E (2006) Adapting to climate change at the local level: the spatial planning response. Local Environ 11(6):609–625

Chapter 11
Adaptations to Possible Climate Change Impacts: Problem Structuring Based on VFT Methodology

Luiz Priori, Marcelo Hazin Alencar, and Adiel Teixeira de Almeida

Abstract This paper presents a research study, the objective of which is to generate alternatives to improve the urban infrastructure systems of Recife, capital of the Brazilian state of Pernambuco, and to develop the city's adaptability to possible impacts of climate change. This city is always at risk from flooding and landslides caused by erosion of hillsides when heavy and/or prolonged rains occur. These cause destruction and deaths.

The research method applied was a case study and the data were collected through direct observation in an attempt to identify vulnerabilities in the infrastructure of the city. The data collected were analysed by 20 experts on urban infrastructure systems and the adaptation problem was structured around Value-Focused Thinking (VFT) methodology.

Among other findings, it became clear that electricity poles and wires are poorly maintained; there are urban drainage problems and substantial losses from the water distribution network; there is a lack of maintenance of the sewage collection network; and the municipal waste is collected irregularly. These are some of the problems which have led to putting forward a set of alternatives to improve the infrastructure of urban systems. It was concluded that the city is not in a state of readiness to counteract the adverse impacts of possible climate changes.

Keywords Climate change • Adaptation • Urban infrastructure • Value-focused thinking

---

L. Priori, Ph. D. (✉) • M.H. Alencar, Ph. D • A.T. de Almeida, Ph. D
Department of Management Engineering, Universidade Federal de Pernambuco, Av. Prof. Moraes Rego, 1235—Cidade Universitária, Recife, Pernambuco, Brazil
e-mail: luizpriori@gmail.com

W. Leal Filho (ed.), *Innovation in Climate Change Adaptation*, Climate Change Management, DOI 10.1007/978-3-319-25814-0_11

Introduction: Climate Change and Its Effects on the City of Recife

According to the Report of the United Nations Secretary-General's High-level Panel on Global Sustainability, as part of the United Nations Environment Programme (UNEP 2012), climate change is a risk to people in all countries and is particularly serious for the world's poorest populations. The IPCC (Intergovernmental Panel on Climate Change), defines climate change as a modification in the state of the atmosphere that can be identified by alterations in the average temperature and/or in the variability of climatic properties which persist for a long period of time i.e. decades or longer (IPCC 2012).

Urban areas have unique characteristics that make their residents particularly vulnerable to climate change (Beatley 2009; Gasper et al. 2011; UNISRD 2012). Several phenomena related to possible global climate change can be listed (IPCC 2012; Pilkey and Young 2009; Ribeiro 2008; Satterhwaite et al. 2007). However, this study limits itself to giving priority to two of these which are expected to occur in the coastal cities of the Northeast of Brazil: rising temperatures, greater frequency of rainfall and their effects on the urban environment.

The Brazilian Panel on Climate Change—PBMC—indicates that all of Brazil is likely to be at least 3 °C warmer by the end of the twenty-first century and that precipitation may increase by 30 % in some regions of the country and diminish by 40 % in others. The northeast region of Brazil, Pernambuco in particular, is notably vulnerable to the occurrence of extreme hydrological events, like droughts and storms (PBMC 2013).

This is due to the irregularity in the distribution and the high interannual variability of rainfall in the Northeast region, which facilitate the occurrence of prolonged droughts and heavy rains, alternately. Northeastern Brazil has an average cumulative rainfall which can vary between less than 500 mm/year and more than 2000 mm/year, which makes the region vulnerable to floods in the rainiest years as was highlighted in 1964, 1967, 1974, 1985, 1986, 1988, 1989, 1994, 2004, 2009 and 2010 (Lacerda et al. 2010; Nobre et al. 2013).

The City of Recife (Fig. 11.1) has an area of 210 km^2, which corresponds to 0.2 % of the territory of the state of Pernambuco. According to the 2010 Population Census, conducted by IBGE—the Brazilian Institute of Geography and Statistics (IBGE 2015)—Recife has a population of 1,537,704 inhabitants spread over 94 districts. Two serious effects of increased intense rainfall in some seasons of the year are part of everyday life of the city: the flooding and landslides from steep terrains.

The urban area of Recife is always at risk from flooding after heavy and/or prolonged rains. These produce large volumes of surface water. According to Satterhwaite et al. (2007), buildings, roads, infrastructure and other paved areas prevent rain water from seeping into the soil, thus reducing the natural drainage and increasing the flow of water. As a result, urban drainage systems are overwhelmed.

Besides the flooding problem, another risk factor which is difficult to mitigate in Recife and is likely to increase with climate change is the erosion of hill-sides. As

Fig. 11.1 View of Recife (*Source*: Authors' files)

the rains become more intense in some regions, run-off water will gain in velocity and thus will have the power to gouge channels and transport sediment, which will cause and/or accelerate erosion and endanger homes, or worse, during a heavy downpour, will wash away what is on the surface, including people and their houses. In addition, more intense erosion adds to the siltation of water bodies, which increases the possibility of flooding in the bottoms of valleys (Priori 2013).

Furthermore, Recife was constructed in a valley. That is to say, in an area which slows down the flow of waterways by reducing the river channels and accentuates the storage of excess water on floodplains. Cities in coastal areas are usually located in low-lying places where drainage is difficult without pumping. High tides or storms can hinder water draining into the sea and cause prolonged flooding with polluted water, and therefore aggravate health problems in cities. Effects of climate change are likely to increase the occurrence of heavier rains and more frequent and severe flooding, which is more difficult to predict (Tingsanchali 2012).

Infrastructure systems are highly vulnerable in disasters and should they fail, this leads to losses that are widely felt. The generation and distribution of electric power, transportation and other infrastructure systems are highly vulnerable in disaster situation and are interdependent. For this reason, a few issues must be addressed, such as multiple sources of disasters, multiple pathways for system failure, multiple and cascading interdependencies among the infrastructure systems, as well as giving consideration to many potential alternative measures to reduce the risk from failures that may occur within and across systems (Chang et al. 2014).

This research study aims to identify (in situ) and document corroborating aspects of increased urban vulnerability in relation to the resilience and adaptive capacity of Recife to climate change, regarding the six key elements that make up the urban

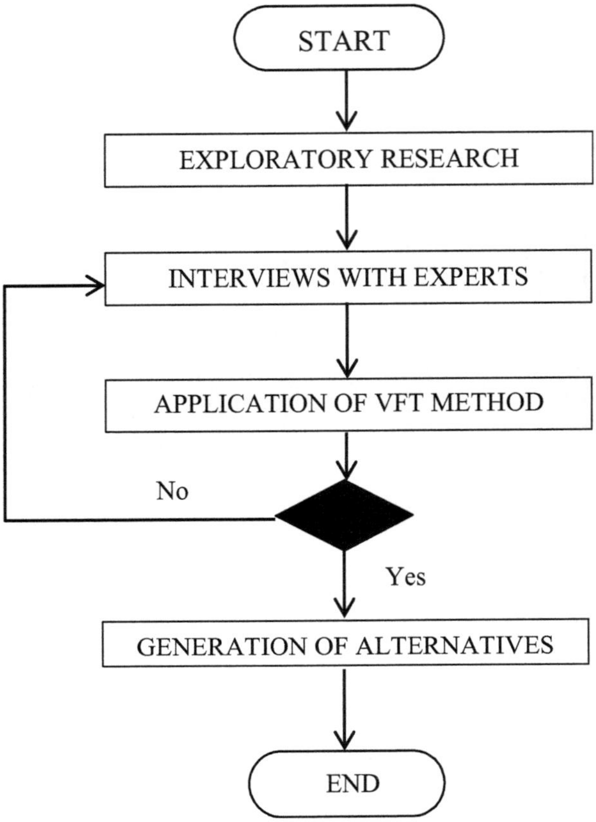

Fig. 11.2 Research steps (*Source*: The authors)

infrastructure, namely, the urban systems of energy, sanitation, waste, transport and communication, and the set of buildings (Priori 2014).

The research method applied was a case study and data were collected through direct observation, in an attempt to identify vulnerabilities in the infrastructure of the neighborhoods of Recife. These data were analysed and discussed with specialists in urban infrastructure systems and the results were organized in accordance with the VFT methodology, with a view to generating alternatives for improving urban city systems (Fig. 11.2).

Adaptive Capacity of the Urban Infrastructure

Adaptation can be defined as: "The process of adjustment to the actual or expected climate and its effects. In human systems, adaptation seeks to moderate or avoid harm or exploit beneficial opportunities. In some natural systems, human intervention may facilitate adjustment to expected climate and its effects" (IPCC 2014: 5).

According to Smit and Wandel (2006) adaptation is still a novel concept when related to the subject of climate change subject and their research work has shown that "adaptations in human communities are closely associated with, and reflective of, adaptive capacity and vulnerability". The debate on climate change adaptation is guided by four questions: Adapt to what? Whom or what adapt? How does adaptation happen? What are the limits to adaptation? (Pelling 2011).

The relationship between adaptation with vulnerability and resilience can be evidenced in the definition of vulnerability as: "The propensity or predisposition to be adversely affected. Vulnerability encompasses a variety of concepts and elements including sensitivity or susceptibility to harm and lack of capacity to cope and adapt" (IPCC 2014: 5); and resilience: "The capacity of social, economic, and environmental systems to cope with a hazardous event or trend or disturbance, responding or reorganizing in ways that maintain their essential function, identity, and structure, while also maintaining the capacity for adaptation, learning, and transformation" (IPCC 2014: 5). Resilience is a dominant theme as to assessing adaptation to climate change, while vulnerability has roots in disaster planning on a shorter and local time-scale (Thomalla et al. 2006). Adaptability is also defined as "the capacity of actors in a system to influence resilience" (Walker et al. 2004).

Adaptation to climate change also has a connection with reduction the risk from a disaster as there is a common goal: to diminish the impacts of extreme events and increasing urban resilience to disasters, particularly among vulnerable urban populations (Baker 2012; Bartlett et al. 2012; Priori 2014; Roaf et al. 2009; Solecki et al. 2011). However, local knowledge, local expertise and local capacity within local government bodies are necessary to a successful adaptation to climate change. Only in prosperous and well-governed cities is adaptation to the probable risks of climate change for the coming decades more likely to be successful. The real problem is that buildings and infrastructure cannot be adapted to these risks if the infrastructure systems are not there. Millions of people live in urban centres devoid of weather-resistant roads, where there is no supply of piped water, no sewage system and no electricity supply. The home of many of these people are poorly constructed and often in areas occupied illegally, a situation that inhibits investment in improving buildings and prevents appropriate infrastructure and services being establish. Many urban centres which need to adapt to avoid potentially catastrophic impacts have major shortcomings in all these prerequisites, which remain noted their lack of adaptability (Adger et al. 2005; PROVIA 2013; Satterhwaite et al. 2007).

Having effective urban infrastructure systems is of fundamental importance for responding to emergency situations and for the quick recovery of a community and its economy (Procyk 2010a, b; Yu 2010). The design of these systems, which includes water, sanitation, energy, communications and transportation, has to encompass possible failures, the impacts of which can be reduced by using adaptability measures, such as redundancy, diversity and modularity (Fig. 11.3), thus reducing the vulnerability of building structures and therefore mitigating the consequences of failure such that these are less damaging to society (Word Bank 2013).

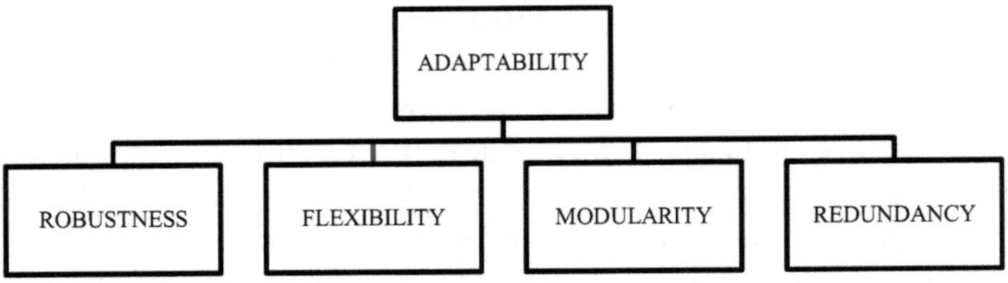

Fig. 11.3 Elements of the adaptability of urban systems (*Source*: The authors)

VFT Methodology

The VFT methodology (Keeny 1992) has been applied to a wide range of contexts in order to identify what the objectives of decision-makers should be. These contexts include for areas such as the military sector, mobile technology, tourist management, anticipating terrorists' objectives, environmental aspects and telecommunications management (Alencar et al. 2011; Hassan 2004; Kajanus et al. 2004; Keeney 2001; Keeney and Mcdaniels 2001; Keeney and Winterfeldt 2010; León 1999; Merrick and Garcia 2004; Merrick and Grabowski 2014; Morais et al. 2013; Sheng et al. 2005).

The adaptation problem was structured around VFT methodology, which contributes to a more consistent in the decision-making process proposed by Keeny (1992). The methodological objective was to seek creative solutions and not to be limited only to the known alternatives, since the first alternative that comes to mind is always similar to those already used (Alencar et al. 2011). This required recruiting help from other specialists in the research subject.

The VFT method is based on the adoption of values to improve the decision-making process, in which three aspects characterize the objectives: (1) the decision context; (2) an object; (3) a preference direction (De Almeida et al. 2015). According to Keeney (1996), values must be the driving force for this process, since they are fundamental in everyday decisions. Thinking that is focused on value is a way to aim at a critical resource in order to lead to better decisions, as this process generally deals with the choosing between known alternatives. It is common to limit a decision problem to the alternatives available. Keeny (1992) argues that the values are more fundamental than the alternatives in decision-making.

The alternatives are the means by which the fundamental value is found. Therefore, it is important to identify the structure, to scrutinize and gain an in-depth understanding of the objectives, by using appropriate devices to make them as complete as possible (Hammond et al. 1999). Following this approach, 20 experts in urban systems from university research staff and highly experienced public and privet sector professionals were interviewed on the subject under study. They answered a semi-structured questionnaire with open questions formulated to stimulate creative ideas by analysing each of the urban infrastructure systems and generating alternatives.

Structuring the Urban Adaptation Problem

The VFT methodology starts by identifying the objectives and then separating them into fundamental and means objectives, thereby constructing a hierarchy of the fundamental objectives and a network of means-ends objectives, which elucidates the relationships among all the objectives (Morais et al. 2013). The first step was to identify of the decision context and to define of the strategic objective: "Alternatives for improving the adaptation of the urban infrastructure to possible impacts from extreme hydrological events in the city of Recife".

Having defined the strategic objective, the second step was to identify the fundamental objectives of the research. For this purpose, the urban infrastructure was broken down into the six key urban systems: energy, sanitation, waste, transport, communication and the set of buildings. These were analysed by assessing their level of robustness, flexibility, modularity and redundancy in the event of being impacted by an extreme hydrological event.

Thus, the fundamental objectives, designed to meet the strategic research problem, were defined as the form of evaluating of the urban infrastructure elements by means of examining the adaptive capacity of their components (Table 11.1).

As a third step, the means-end objective network was set. The final configuration of the means-end network was constructed based on the results of the discussions with the experts interviewed. According to Keeny (1992) the means-end network expresses what is of interest in the decision-making context by indicating which objectives should be considered when developing a model that relates the alternatives to its consequences (Fig. 11.4).

The fulfilment of a fundamental objective is measured in terms of an attribute. Thus, the resulting structure which consists of objectives and attributes provides the decision-maker with an overview of the problem situation, hence allowing him to focus on the highest-value alternatives (Morais et al. 2013). The fourth step was to set the attributes based on the means-end network. Some of the attributes defined are shown in Table 11.2.

Table 11.1 Hierarchy of fundamental objectives

Alternatives for improving the adaptation of the urban infrastructure to possible impacts from extreme hydrological events in the city of Recife (strategic objective)	
1.	Evaluate the adaptability of the physical structure for power generation and transmission system
2.	Evaluate the adaptability of the structure of the sanitation system
3.	Evaluate the adaptability of the structure for managing municipal waste
4.	Evaluate the adaptability of systems and means of transport
5.	Evaluate the adaptability of the communication system
6.	Evaluate the structure of the set of buildings and the availability of shelters to house people in case of disaster

Source: The authors

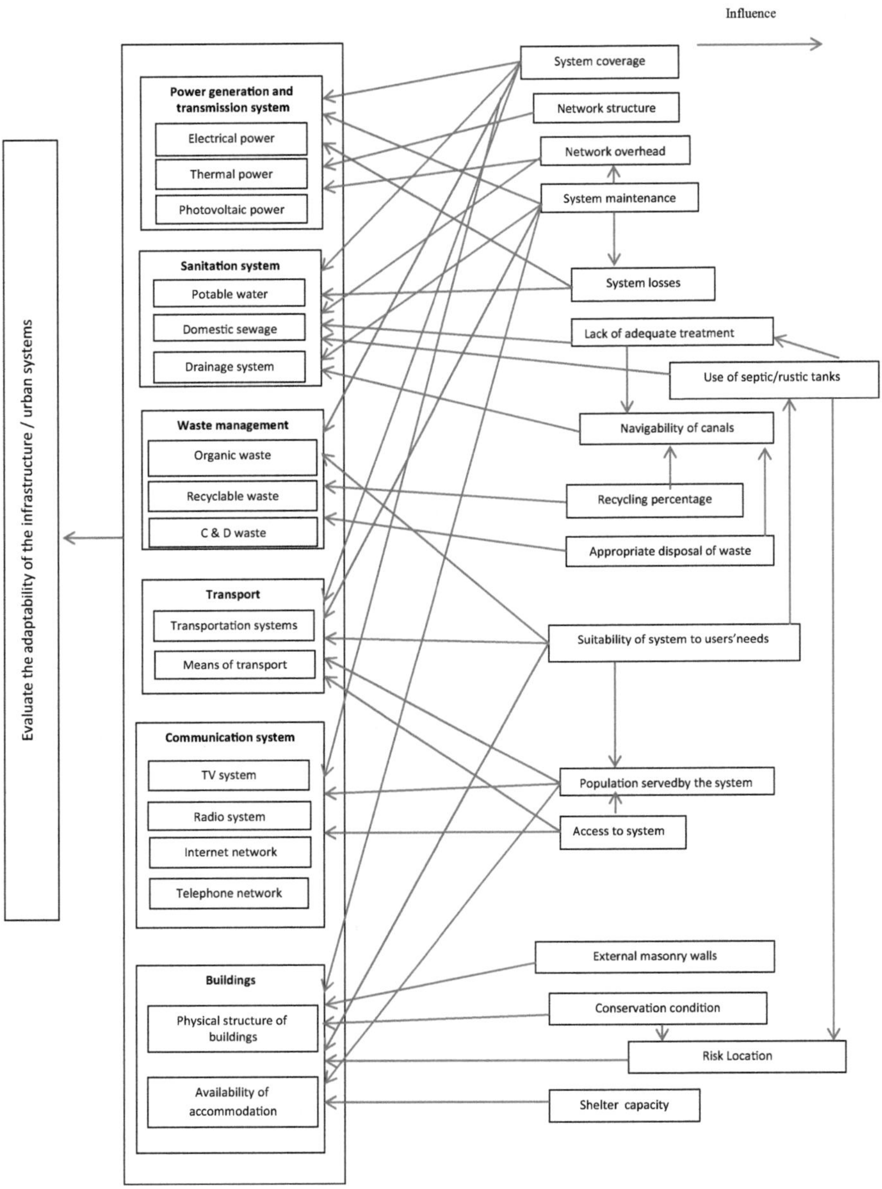

Fig. 11.4 Network of means-end objectives (*Source*: The authors)

Results Achieved

The application of the VFT, through interviews with experts, was used to good effect towards understanding the issues underlying the problem and inducing creative thinking, thereby stimulating the creation of alternatives as a starting

Table 11.2 Attributes of the fundamental objectives

Physical structure for power generation and transmission system	
1.1	*Generation and transmission capacity of electrical power*
1.1.1	Percentage of the network which is underground
1.2	*Capacity of thermal power generation system which uses gas as fuel*
1.2.1	Percentage of households served by the gas network
1.3	*Capacity of photovoltaic power generation*
1.3.1	Percentage of properties using photovoltaic panels for their power supply
Structure of sanitation system	
2.1	*Distribution and treatment of potable water*
2.1.1	Percentage of water losses in the pipes
2.2	*Distribution and treatment of domestic sewage*
2.2.1	Percentage of households with rudimentary drains and ditches
2.3	*Macro drainage system*
2.3.1	Percentage of canals which are navigable
2.4	*Micro drainage system*
2.4.1	Percentage of the city area covered by the micro drainage network
Structure for the management of municipal waste	
3.1	*Evaluation of organic waste management*
3.1.1	Percentage of households with regular garbage collection
3.2	*Evaluation of the management recyclable waste*
3.2.1	Percentage of waste which is recycled
3.3	*Evaluation of the management of construction and demolition (C&D) waste*
3.3.1	Percentage of C&D waste which is deposited in appropriate locations
Systems and means of transport	
4.1	*Evaluation of transportation systems*
4.1.1	Percentage of properties with sidewalks considered appropriate as laid down in legal requirements
4.2	*Evaluation of the means of transport*
4.2.1	Percentage of the population serviced by the bus system.
Communication system	
5.1	*Evaluation of the TV system*
5.1.1	Percentage of households with a TV set
5.2	*Evaluation of the Radio system*
5.2.1	Percentage of households with a radio set
5.3	*Evaluation of the Internet network*
5.3.1	Percentage of households with computers connected to broadband internet
5.4	*Evaluation of the telephone network*
5.4.1	Percentage of households with conventional or mobile phones
Set of buildings and availability of shelters to house people in case of disaster	
6.1	*Evaluation of the physical structure of the buildings*
6.1.1	Percentage of buildings located in barrier fall-risk sectors
6.2	*Evaluation of the level of available accommodation*
6.2.1	Percentage of resident in risk areas who can be sheltered

Source: The authors

Table 11.3 Alternatives to improving the adaptability capacity of infrastructure systems

Energy
• Improve the physical structure of the power grid by isolating the overhead cables and conductors, and placing the pipe lines underground
• Invest in the use of other energy sources that can be generated in the city, such as photovoltaic and thermal gas
Sanitation
• Develop a maintenance plan for cleaning the urban drainage system
• Adapt the city canals so that they are made navigable
Waste
• Establish a cleansing plan for the city, to diminish the amount of trash on public roads and sidewalks
• Prohibit the household waste from being deposited in plastic bags on the sidewalks, which encourages the spread of garbage on site and contributes to the spread of worms
Transport
• Implement urgently a plan for sidewalks maintenance and restoration
• Invest in the diversification of the mobility modals like Bus Rapid Transport (BRT) and Light Rail Transport (LRT)
Communication
• Register the cell phones of the population so people can be informed about what to do should extreme hydrological events occur
• Implant Wi-Fi network for free internet transmission to the population in areas at risk
Buildings
• Check the real situation of irregular buildings which makes it more difficult for civil defense forces to respond in case of an accident
• Adjust the number of dwellings and shelters needed for the population at risk

Source: The authors

point for understanding the decision opportunities. Table 11.3 shows some of the alternatives recommended for improving the infrastructure systems and the city's adaptability to possible impacts from extreme hydrological events, derived from the objectives identified on applying VFT methodology.

Conclusions

Throughout the analysis of the exploratory survey conducted in Recife, urban vulnerabilities were observed to be concentrated on: the poor maintenance of electricity poles and wires; urban drainage problems; losses from the water distribution network; the shortfall in and lack of maintenance of the sewage collection network; and the irregular collection of the municipal waste.

The mobility of vehicles and pedestrians, as well as inappropriate or lack of maintenance of the sidewalks are also important urban weaknesses. Irregular buildings occupying roads and sidewalks were also found in all districts. Hence, it can be concluded that the vulnerability of urban infrastructure systems throughout the entire city of Recife can be generalized and therefore the city is not in a state of readiness to face the adverse impacts of possible climate changes.

In order to organize the experts' ideas, opinions and proposals, the VFT methodology was fundamental for structuring the problem, thereby enabling a better evaluation of the decision-making context to be made. Based on the decision context analysed, the method also enable a structured scrutiny of objectives, verification of possible redundancies between objectives and the identification of fundamental objectives and means objectives. Moreover, this analysis also led to previously hidden objectives being considered. As a result, attributes have been defined to measure the fundamental objectives. The decision opportunities have been identified as well as the aspects that served as input for the development of alternatives.

Is worth underlining that the methodology used to structure the problem analysed, even though it was specifically applied to the infrastructure of the city of Recife and the results are limited to the situation investigated, can be used to structure similar problems in other cities or regions. It is very important to demonstrate that other places which have analogous situation can use the same methodology to learn more about their urban problems. It is suggested that to analyse and structure other problems the point of view already known should be set aside and all stakeholders must be involved in finding a solution for the problem being considered.

Further studies should consider the application of multicriteria methods of aiding decision-making as a way of selecting and prioritizing the alternatives of major relevance to decision-makers as regards evaluating the adaptability capacity of the urban infrastructure when confronted with extreme hydrological events.

Acknowledgments This study was partially sponsored by the Brazilian Research Council (CNPq) for which the authors are most grateful. The authors also gratefully acknowledge the valuable suggestions made by the experts who collaborated with this research study.

References

Adger WN, Arnella NW, Tompkinsa EL (2005) Successful adaptation to climate change across scales. Glob Environ Chang 15:77–86

Alencar LH, Mota MCM, Alencar MH (2011) The problem of disposing of plaster waste from building sites: problem structuring based on value focus thinking methodology. Waste Manag 31:2512–2521

Baker JL (ed) (2012) Climate change, disaster risk, and the urban poor—cities building resilience for a changing world. The International Bank for Reconstruction and Development, World Bank Publications, Washington, DC

Bartlett S, Dodman D, Hardoy J, Satterthwaite D, Tacoli C (2012) Social aspects of climate change in urban areas in low and middle income nations. International Institute for Environment and Development (IIED), World Bank Publications, Washington, DC

Beatley T (2009) Planning for coastal resilience: best practices for calamitous times. Island Press, Washington, DC

Chang S, Mcdaniels T, Fox J, DhariwaL R, Longstaff H (2014) Toward disaster-resilient cities: characterizing resilience of infrastructure systems with expert judgments. Risk Anal 34 (3):416–434

IPCC—Intergovernmental Panel on Climate Change (2012) Managing the risks of extreme events and disasters to advance climate change adaptation. http://www.ipcc.ch/pdf/special-reports/srex/SREX_Full_Report.pdf. Accessed on 28 Apr 2015

IPCC—Intergovernmental Panel on Climate Change (2014) Managing the risks of extreme events and disasters to advance climate change adaptation. http://www.ipcc.ch/pdf/special-reports/srex/SREX_Full_Report.pdf. Accessed on 28 Apr 2015

De Almeida AT, Cavalcante CAV, Alencar MH, Ferreira RJP, Almeida-Filho AT, Garcez TV (2015) Multicriteria and multiobjective models for risk, reliability and maintenance decision analysis, vol 231, International series in operations research and management science. Springer, New York

Gasper R, Blohm A, Ruth M (2011) Social and economic impacts of climate change on the urban environment. Curr Opin Environ Sustain 3:150–157

Hammond JS, Keeney RL, Raiffa H (1999) Smart choices: a practical guide to making better life decisions. Broadway Books, New York

Hassan O (2004) Application of value-focused thinking on the environmental selection of wall structures. J Environ Manag 70:181–187

IBGE—Brazilian Institute of Geography and Statistics (2015) Censo 2010. http://cidades.ibge.gov.br/xtras/perfil.php?lang=&codmun=261160&search=pernambuco|recife. Accessed on 28 Apr 2015

Kajanus M, Kangasb J, Kurttilac M (2004) The use of value focused thinking and the A'WOT hybrid method in tourism management. Tour Manag 25:499–506

Keeney RL (1996) Value-focused thinking: identifying decision opportunities and creating alternatives. Eur J Oper Res 92:537–549

Keeney RL (2001) Modeling values for telecommunications management. IEEE Trans Eng Manag 48(3):370–379

Keeney RL, Mcdaniels TL (2001) A framework to guide thinking and analysis regarding climate change policies. Risk Anal 21(6):989–1000

Keeney GL, Winterfeldt DV (2010) Identifying and structuring the objectives of terrorists. Risk Anal 30(12):1803–1816

Keeny RL (1992) Value-focused thinking: a path to creative decision making. Harvard University Press, Cambridge

Lacerda FF, da Silva HD Jr, Assad ED, Assis JMO, Moura MSB (2010) Extremos e variabilidade climática no Nordeste brasileiro e em Pernambuco. In: Galvíncio JD (ed) Mudanças climáticas e impactos ambientais. Ed. Universitária da UFPE, Recife, pp 1–23

León OG (1999) Value-focused thinking versus alternative-focused thinking: effects on generation of objectives. Organ Behav Hum Decis Process 80(3):213–227

Merrick JRW, Garcia MW (2004) Using value-focused thinking to improve watersheds. J Am Plan Assoc 70(3):313–327

Merrick JRW, Grabowski M (2014) Decision performance and safety performance: a value-focused thinking study in the oil industry. Decision analysis. Published online in Articles in Advance, pp 1–1219. http://dx.doi.org/10.1287/deca.2014.0291. Accessed on 15 Apr 2015

Morais DC, Alencar LH, Costa APCS, Keeney RL (2013) Using value-focused thinking in Brazil. Pesquisa Operacional 33(1):73–88

Nobre P, Siqueira LSP, de Almeida RAF, Malagutti M, Giarolla E, Castelão GP, Bottino MJ, Kubota P, Figueroa SN, Costa MC, Baptista M Jr, Irber L Jr, Marcondes GG (2013) Climate simulation and change in the Brazilian climate model. J Clim 26:6716–6732. doi:10.1175/JCLI-D-12-00580.1

PBMC—The Brazilian Panel on Climate Change (2013) Contribuição do Grupo de Trabalho 1 ao Primeiro Relatório de Avaliação Nacional do Painel Brasileiro de Mudanças Climáticas. Sumário Executivo GT1, Rio de Janeiro

Pelling M (2011) Adaptation to climate change. Routledge, London

Pilkey OH, YOUNG R (2009) The rising sea. Island Press, Washington, DC

Priori L Jr (2013) The possible impacts of climate change in the city of Recife, Brazil. IAPS—International Association People—Environment Studies Symposium 2013, La Coruña, Spain

Priori L Jr (2014) Resiliência urbana de cidades costeiras: Um recurso para enfrentar as mudanças climáticas. Research report. CAPES/PNPD/UFPE, Recife

Procyk A (2010a) Creating alternatives for improving infrastructure interdependency resilience. Analyzing infrastructures for disaster-resilient communities, practioner report #3. The University of British Columbia. http://www.chs.ubc.ca/dprc_koa/pdf_files/PR_No3_Creating%20Alternatives.pdf. Accessed on 15 Apr 2015

Procyk A (2010b) Setting priorities for improving infrastructure interdependency resilience. infrastructures for disaster-resilient communities, practioner report #4. The University of British Columbia. http://www.chs.ubc.ca/dprc_koa/pdf_files/PR_No4_Setting%20Priorities.pdf. Accessed on 15 Apr 2015

PROVIA (2013) Guidance on assessing vulnerability, impacts and adaptation to climate change. Consultation document. United Nations Environment Programme, Nairobi

Ribeiro WC (2008) Impactos das mudanças climáticas em cidades no Brasil. Parcerias Estratégicas 27:297–321

Roaf S, Crichton D, Nicol F (2009) Adapting buildings and cities for climate change: a 21st century survival guide, 2nd edn. Elsevier, Oxford

Satterhwaite D, Huq S, Reid H, Pelling M, Lnkao PR (2007) Adapting to climate change in urban areas: the possibilities and constraints in low- and middle-income nations. International Institute for Environment and Development. www.iied.org/pubs/display.php?o=10549IIED. Accessed on 16 Apr 2015

Sheng H, Nah FF, Siau K (2005) Strategic implications of mobile technology—a case study using value-focused thinking. J Strateg Inf Syst 14:269–290

Smit B, Wandel J (2006) Adaptation, adaptive capacity and vulnerability. Glob Environ Chang 16:282–292

Solecki W, Leichenko R, O'Brien K (2011) Climate change adaptation strategies and disaster risk reduction in cities: connections, contentions, and synergies. Curr Opin Environ Sustain 3:135–141

Thomalla F, Downing T, Spanger-Siegfrie E, Han G, Rockström J (2006) Reducing hazard vulnerability: towards a common approach between disaster risk reduction and climate adaptation. Disasters Spec Issue: Clim Chang Disasters 30(1):39–48

Tingsanchali T (2012) Urban flood disaster management. Procedia Eng 32:25–37

UNEP—United Nations Secretary-General's High-level Panel on Global Sustainability (2012) Resilient people, resilient planet: a future worth choosing. United Nations, New York

UNISRD—United Nations International Strategy for Disaster Reduction (2012) Making cities resilient report 2012. http://www.unisdr.org/files/28240_rcreport.pdf. Accessed on 30 Apr 2015

Walker B, Holling CS, Carpenter SR, Kinzig A (2004) Resilience, adaptability and transformability in social–ecological systems. The Resilience Alliance 9(2): Art. 5. http://www.ecologyandsociety.org/vol9/iss2/art5. Accessed on 10 Mar 2015

Word Bank (2013) Turn down the heat: climate extremes, regional impacts, and the case for resilience. International Bank for Reconstruction andDevelopment, World Bank, Washington, DC

Yu A (2010) Infrastructure disruptions and interdependencies during floods events. Infrastructures for disaster-resilient communities, practioner report #5. The University of British Columbia. http://www.chs.ubc.ca/dprc_koa/pdf_files/PR_No5_Flood%20Events.pdf. Accessed on 10 Mar 2015

Chapter 12
Assessing Vulnerability to Support Promotion of Adaptive Agricultural Practices in the Sahel

Alex Apotsos, David Miller, and Brent Simpson

Abstract Across the Sahel, rural producers struggle to maintain viable livelihoods owing to the harsh and variable climate. Unfortunately, climate change is expected to exacerbate many of the challenges they currently face. While a number of adaptive agriculture practices exist, numerous studies have demonstrated that the reasons farmers adopt (or do not adopt) these practices are complex and context dependent.

This paper seeks to broaden our understanding of how assessments of vulnerability can be used to inform development programs designed to promote adaptive agricultural practices by summarizing the methodologies and findings from four vulnerability studies relevant to the Sahel conducted for the United States Agency for International Development (USAID) from 2012 to 2014. These studies took different approaches that have advantages and challenges depending on the spatial scale being assessed. Specific assessment findings include identification of a significant difference between the practices being promoted and those being adopted, a need to improve our understanding of the current and future effectiveness of agricultural practices in a changing climate, and that vulnerability in eastern Senegal varies both spatially and with livelihood practiced. Together these assessments demonstrate that there is neither a methodology to assess vulnerability nor an agricultural practice that is likely to be universally acceptable or effective across the Sahel. This suggests that different development programs will require different levels of specificity in analysis, and that assessments should consider carefully the

---

A. Apotsos, Ph.D. (✉)
Geoscience Department, Williams College, Williamstown, MA, USA

U.S. Agency for International Development, Bureau for Africa, Washington, DC, USA
e-mail: aa13@williams.edu

D. Miller, Ph.D.
ACDI/VOCA, Washington, DC, USA
e-mail: dmiller@acdivoca.org

B. Simpson, Ph.D.
Investment Center, Food and Agriculture Organization, Rome, Italy
e-mail: bsimpson@anr.msu.edu

W. Leal Filho (ed.), *Innovation in Climate Change Adaptation*, Climate Change Management, DOI 10.1007/978-3-319-25814-0_12

spatial scale over which a development program will act and which factors affect adoption of practices at those scales. The lessons and findings from these studies will be useful to anyone seeking to support climate change adaptation or assess vulnerability in the Sahel.

Keywords Climate change • Vulnerability assessment • Adaptive agricultural practices • Sahel

Introduction

Africa and Climate Change

The Fifth Assessment Report of the Intergovernmental Panel on Climate Change (IPCC) states "Africa as a whole is one of the most vulnerable continents due to its high exposure and low adaptive capacity" (Niang et al. 2014). In the Sahel, rural producers, both small holder farmers and pastoralists, are vulnerable to a changing climate because they are exposed to extremely variable rainfall, are sensitive due to their high dependence on natural resources, and possess low adaptive capacity owing to their poor access to markets, finance, and effective governance structures (e.g., Trench et al. 2007). Conversely, rural producers have inhabited the Sahel for millennia and have found ways of surviving under challenging environmental and socio-economic conditions (e.g., Mortimore and Adams 2001). However, previous adaptations have primarily been reactive, and current development efforts seek to help rural producers proactively anticipate and adapt to current and future changes in climate. The challenge is to support proactive adaptation without reducing productivity under current conditions (Simpson and Burpee 2014).

Given the harsh environment in the Sahel, which is only expected to be exacerbated by climate change, it is essential to understand how adaptive agricultural practices can build resilience and improve productivity. This paper seeks to broaden our understanding of how vulnerability assessments can be used to inform the promotion of these practices. It is first demonstrated that a variety of different assessment approaches exist, that no approach is universally appropriate, and that the optimal approach will likely depend on several factors, including the spatial scale of the assessment. Furthermore, given that a single universally acceptable and effective adaptive agricultural practice is unlikely to be found and that there remains a significant gap in our understanding of the social and biophysical aspects associated with many adaptive agricultural practices, assessments should seek to identify, and development programs promote, a robust basket of solutions that can be tailored by individual stakeholders to best suit their circumstances.

Defining Vulnerability

The concept of vulnerability to climate change is not new, and has been reviewed extensively (e.g., Adger 2006; Eakin and Luers 2006; Füssel and Klein 2006; Hinkel 2011; Preston et al. 2011). Previous studies provide both an excellent summary of how different fields have approached and defined vulnerability as well as a number of conceptual models for assessing vulnerability. These conceptual models have evolved over the past decades to the point where social, economic, and cultural factors are recognized as being as important as biophysical factors (e.g., Füssel and Klein 2006; Preston et al. 2011). For example, the IPCC has defined vulnerability as being composed of three components: exposure, sensitivity and adaptive capacity (Parry et al. 2007). Comprehensive assessments of vulnerability must, therefore, examine aspects from both the physical and social sciences.

While various conceptual frameworks for assessing vulnerability have been proposed, operationalizing these frameworks has proven difficult (e.g., Nelson et al. 2010; Hinkel 2011). For example, the relationships between the components of vulnerability are ambiguous and poorly defined (Janssen and Ostrom 2006; Hinkel 2011), and no objective methods for selecting and aggregating measurable indicators of these components have been found (e.g., Eakin and Bojorquez-Tapia 2008; Hinkel 2011). Furthermore, vulnerability can differ significantly between and among individuals, groups, and communities, as well as across both space and time, owing to socio-economic characteristics (e.g., wealth, gender, culture), livelihoods, political contexts, and perceptions of climate change (e.g., Adger et al. 2007; Nielsen and Reenberg 2010; Antwi-Agyei et al. 2013; Djoudi et al. 2013; Carr and Thompson 2014). Climate is also only one of many factors affecting communities, and the relative importance of climate is often less in poor and marginal communities (e.g., Adger 1999; Ziervogel et al. 2006; Tschakert 2007). For example, households surveyed in Senegal viewed economic, political, and social factors, rather than climate, as the main drivers for adopting changes in agricultural practice (Mertz et al. 2009). However, the relative importance of climate change may increase if the rate of change increases or critical thresholds are reached.

Adaptive Practices

A wide range of adaptive agricultural practices are available and promoted within the Sahel (e.g., Boro et al. 2014; Ray and Simpson 2014). For example, many development agencies have been promoting the adoption of low-cost innovations, including soil conservation, water harvesting, agroforestry, crop-livestock integration, crop storage, and small dams (Barbier et al. 2009). However, levels of adoption vary depending on a complex array of factors including wealth, education, gender, culture, and political contexts (e.g., Nielsen and Reenberg 2010; Djoudi et al. 2013). Understanding why practices are adopted or not is difficult as these decisions are often made at the household level (e.g., Antwi-Agyei et al. 2013) and can thus vary significantly not only across communities but also within communities. For example, access to resources, including agricultural extension services and improved weather forecasts, has been shown to be important for farmers' adaptation (e.g., Ziervogel et al. 2006), while culture can also act as a major barrier to embracing some of the most successful adaptive strategies including labor migration, gardening, and the engagement of women in economic activities (Nielsen and Reenberg 2010). According to many farmers, in some areas of the Sahel most new techniques have been adopted because of growing land scarcity and new market opportunities, rather than because of climate variability and change (Barbier et al. 2009). While adoption of adaptive practices by farmers has been uneven, some success stories have been reported, especially in Burkina Faso (e.g., Sendzimir et al. 2011). Unfortunately, understanding the complex and locally specific factors that dictate household decision-making in rural communities is a highly resource intensive process, especially as external dynamics, such as changes in subsidies and market conditions, can affect household decision-making regardless of climatic parameters (e.g., Mertz et al. 2009). In some circumstances the promotion and uptake of certain adaptive practices can lead to or enhance inequities, where some stakeholders adapt to the detriment of others, especially women and marginal groups (e.g., Djoudi et al. 2013), or can lead to increased climate risks (e.g., Folle and Mulla 2014).

Assessments Overview and Key Findings

From 2012 to 2014 USAID commissioned four complementary vulnerability studies in West Africa (Table 12.1). These studies, which include sector level studies as well as more comprehensive assessments, and their key findings are briefly summarized below.

West Africa

This set of studies was conducted to inform the development of a climate change program for USAID's regional West Africa mission and includes two foundational documents relevant to the Sahel: (1) a background document summarizing the climate change impacts projected to affect West Africa, including the Sahel (Baptista et al. 2013), and (2) an assessment of the sources of uncertainty in regional climate change projections (Blamey et al. 2013). Together these two documents laid the foundation for all future studies in West Africa. They corroborated anecdotal evidence gathered from Sahelian farmers who remarked "the climate has lost its memory" by re-emphasizing the understanding that the main climate change impacts in the Sahel are likely to be associated with increasing temperatures and

Table 12.1 Vulnerability studies

West Africa	Background paper for the ARCC West Africa regional climate change vulnerability assessment	Baptista et al. (2013)
	The status and possible evolution of climate projections in West Africa	Blamey et al. (2013)
Mali	Climate change in Mali: organizational survey and focus groups on adaptive practices	Boro et al. (2014)
	Key issues in water resources	Murray-Rust (2013)
	Agricultural adaptive practices impact modeling assessment	Folle and Mulla (2014)
	Institutional Analysis of L'Agence de L'Environnement et du Developpement Durable (AEDD) and L'Agence Nationale de la Météorologie (Mali-Météo)	Freudenberger et al. (2014)
	Climate change in Mali: country vulnerability map	de Sherbinin et al. (2014)
Senegal	Senegal climate change vulnerability assessment and options analysis	Miller et al. (2014)
Sahel	An approach to conducting phenological screening of the impact of climate change in the West African Sahel	Simpson (2014a)
	Agricultural adaptation to climate change in the Sahel: an approach to evaluating the performance of agricultural practices	Simpson (2014b)
	Agricultural adaptation to climate change in the Sahel: expected impacts on pests and diseases afflicting livestock	Younan and Simpson (2014)
	Agricultural adaptation to climate change in the Sahel: profiles of agricultural management practices	Ray and Simpson (2014)
	Agronomic profiles of fifteen crops important in the Sahel	del Rio and Simpson (2014a)
	Agricultural adaptation to climate change in the Sahel: expected impacts on pests and diseases afflicting selected crops	del Rio and Simpson (2014b)

more variable rainfall. These documents demonstrate further that owing to the complex factors that affect the climate in the Sahel, much of the uncertainty associated with future rainfall projections is likely to remain in the near future. Therefore, delaying decisions for a few years will not bring greater clarity.

Mali

This set of studies was conducted to inform a new climate change adaptation program in Mali, and was sequenced in a stepwise manner, where different studies were added as the need arose. The initial piece of the assessment involved constructing a vulnerability hotspot map to identify highly vulnerable areas within Mali [see de Sherbinin et al. (2014) for details].

Institutional Assessment

Given the increasing variability and unpredictability of precipitation in the Sahel, rural producers need forward looking information to complement their historical climate knowledge. One potential source of climate and weather information is the National Meteorological Service, commonly called Mali-Météo. An institutional assessment of Mali-Météo found that while it has some strong internal capacity to collect, analyze, and deliver climate information, it needs to develop a long-term business plan to capture the financial resources necessary to cover its operating expenses. In the meantime, continued government or international support will be necessary to improve Mali's network of observation stations as well as to enhance Mali-Météo's analytical capabilities through updated hardware, software, and skills training (Freudenberger et al. 2014).

Adaptive Practice Promotion and Adoption

To understand what adaptive agricultural practices are currently being promoted and adopted in Mali, two surveys, one of non-governmental organizations (NGOs) and one with groups of farmers, were conducted (Boro et al. 2014). NGOs generally take a conservative approach to climate change adaptation that builds on current agricultural practices and addresses immediate concerns (e.g., productivity and mitigating risk). Virtually all the practices promoted by the NGOs predate the recent increase of climate change funding. Farmers were found to make adjustments to adapt to a broad range of changes (i.e., degradation of natural resources, market demands, population pressure, climate conditions) around them. The factors affecting the adoption of specific practices appear to be complex, with different practices being adopted at different rates depending on location, livelihood, and gender. In general, the adoption of a field practice appears to decrease as the resources required by a practice increase. Although the focus groups were conducted in villages where the NGOs worked, many of the practices most commonly promoted by those NGOs were among those with the lowest levels of adoption.

Impact Modelling

To better understand the effectiveness of certain water harvesting practices under a changing climate, the impact of four practices (i.e., bunds, vegetated filter strips, contour ridges, and *zai* pits) on surface runoff, crop yield, soil organic carbon content, and soil loss was assessed for a hypothetical site in the Mopti region of Mali (Folle and Mulla 2014). In general, water harvesting practices improved crop yield and soil fertility under projected future climate conditions, but the effectiveness of the practices depends on a number of site specific parameters, including soil type, crop choice, and slope steepness. In a few cases, the water harvesting techniques negatively impacted yields of some crops in certain soils by inducing waterlogging.

Water Resources Assessment

Mali's relatively abundant water resources are concentrated along the Niger and Senegal Rivers, creating two distinct water regimes in Mali: abundant water for communities near the major rivers and significant water stress for communities away from the rivers (Murray-Rust 2013). Areas along the rivers will continue to have adequate access to water, and thus limited exposure to climate change, as long as there is sufficient capital to fund the necessary infrastructure. Areas away from the rivers face significant exposure to climate change owing to their dependence on the highly variable rainfall.

Senegal

This assessment was conducted to inform both a new climate change program as well as potentially the agricultural programming in Senegal. Of the studies discussed here, the Senegal assessment was envisioned and implemented as the most integrated and comprehensive, where all components addressed a single purpose and hypothesis (Miller et al. 2014). This assessment focused on four departments in eastern Senegal that stretch north–south across several agro-ecological zones and experience high levels of food insecurity. In general, livestock raising plays a greater role in livelihoods in the more arid north, while crop farming plays a greater role in the more humid south. This assessment hypothesized that household livelihood systems that include livestock are less vulnerable to climate change than household livelihood systems that depend primarily on crop farming. The assessment employed a number of methods including focus groups, household surveys, key informant interviews, climate data analysis, climate projections down-scaling, and crop and pasture modelling. The analyses revealed noteworthy differences in vulnerability between the different regions and livelihoods. Overall, livestock-dependent households in the northern communities appear to be the most vulnerable, while livestock-dependent households in the southern communities appear to be the least vulnerable. Between these two extremes were households with crop-dominant and mixed crop-livestock livelihoods. Therefore, within the limits of the methodology used, the hypothesis appears to hold true in the southern, but not the northern, portion of the study zone, suggesting that reality is more complex than initially hypothesized.

Sahel

The Sahel studies were conducted to inform USAID's resilience programming under the Resilience in the Sahel Enhanced (RISE) program being implemented in Niger and Burkina Faso. Owing to the complexity and size of the region, these studies focused on developing tools and approaches that could reduce the need for smaller organizations to expend limited resources developing their own methodologies.

Tools: Methodologies

The fact that the effectiveness of agricultural practices and the productivity of crops are site and parameter specific precludes a rigorous assessment of future impacts relevant to the entire Sahel. Therefore, two approaches were developed to facilitate and support localized assessments. The first outlines a phenological screening methodology for crops (Simpson 2014a), with a focus on critical environmental thresholds that affect crop productivity. The second outlines a methodology for evaluating the performance of agricultural management practices (Simpson 2014b). In both approaches, future climatic conditions must be defined and "adaptation profiles" for each practice or crop established. Both approaches stress the importance of moving beyond overly simplified analyses based on total annual rainfall, and focus instead on intra-annual variations in rainfall, which are particularly important to rural producers.

Tools: Information

Adaptive Practices and Crops

The approaches described above require an understanding of the environmental factors important to crops and agricultural practices. Therefore, agronomic profiles of 15 important Sahelian crops were developed. Each profile describes the geographic distribution and importance, life cycle, known rainfall and temperature requirements at different stages of physiological development, and thresholds related to tolerances of soil conditions, water availability, and temperature for each crop (del Rio and Simpson 2014a). Based on these factors, the crop's potential adaptability to changes in climate is discussed. Similarly, profiles of 32 agricultural management practices were developed (Ray and Simpson 2014). Each profile provides an overview of the practice, its technical specifications, the context in which it functions best, and the practice's limitations and weaknesses.

Pest and Diseases

To increase our understanding of how climate change might affect pests and diseases afflicting selected crops (del Rio and Simpson 2014b) and livestock (Younan and Simpson 2014), assessments were conducted that catalog the current prevalence and impact of more than 100 pests and diseases in the Sahel and discuss the important environmental factors that influence their prevalence. The potential effects of climate change on disease and pest prevalence was estimated for two simplified future climate scenarios (i.e., hot/wet and hot/dry) based on the environmental factors identified as important.

Discussion

Significant Gaps in Knowledge and Information

These studies clearly reveal that there remains a significant lack of information and understanding associated with the biophysical response of crops, agricultural practices, pests and diseases in the Sahel to climate change. This reflects both the size and complexity of the region and the extent to which further research is needed. The initial intent of the Sahel studies was to identify and evaluate a range of adaptive agricultural practices in order to provide general guidance concerning which would be the most effective under a changing climate. However, initial efforts highlighted the limits of and gaps in the available information on future climate conditions, crop responses, agricultural management practices, and the response of pests and diseases in the Sahel. For example, the available literature provides only a basic description of the technical specifications of field level adaptive management practices. Similarly, crop literature does not enable the development of detailed phenological profiles covering the likely range of climate change stressors for the major crops and varieties. Even less information is available concerning the effects of climate change on the prevalence of pests and diseases. Furthermore, most analyses conducted to date look primarily at bulk changes in total annual rainfall, instead of intra-annual variations, which have the greatest impact on Sahelian agriculture. While it is possible that some of the necessary information exists within local institutions, this information risks being lost if it is not published or shared more widely. The challenges associated with developing this fundamental biophysical information was clearly demonstrated by the impact assessment conducted in Mali, where several field level factors (e.g., slope, soil type, crop choice) were found to influence significantly the effectiveness of water harvesting practices in one small area of the Mopti Region.

Gap between Promoted and Adopted Practices

While there is a continuing emphasis on understanding why adaptive practices are (or are not) adopted, a significant gap exists between the practices that are being promoted and those that are being adopted. This is partly owing to the complex set of factors that rural producers must contend with in making decisions and to our incomplete understanding of the biophysical responses of these practices to climate change. For example, the diversity of contexts influencing household decision-making across the large spatial scales of the Sahel makes finding a single, universally acceptable practice almost impossible. Furthermore, practices that allow rural producers to proactively adapt to future conditions must not negatively affect productivity under current conditions (Simpson and Burpee 2014), and there is often a time lag between when a practice is promoted and when it is adopted by different community members (Simpson 2015).

While significant research has focused on why farmers adopt practices, less research appears to have been conducted concerning how NGOs and governments select practices to promote (e.g., Biermann 2009). During a meeting held in Dakar, Senegal where the Senegal assessment findings were discussed, the assembled stakeholders, representing the private sector, government, and civil society, were asked to rank 13 adaptive options both individually and then in groups. Interestingly, many of the options ranked highly by individuals were then ranked much lower by groups composed of mostly the same individuals. From the stakeholder responses recorded during the event, many of the options ranked highly by individuals were ranked lower by the groups owing to concerns associated with practicality. While this may suggest that individuals gravitate toward innovative and ideal solutions and groups tend to coalesce around more traditional and tested interventions, there are a number of other factors that dictate how groups make decisions, and more research is needed to understand how groups and organizations decide which practices and crops to promote.

Thresholds of Specificity in Analysis

As our understanding of how vulnerability varies among individuals, households and communities is still evolving, the research community will continue to probe the important nuances and underlying complexity of household decision-making. However, development programs are often designed at larger spatial scales (e.g., national or regional scale) where it is difficult to generalize across or account for household and community level variations. Therefore, while it is important to acknowledge these variations, conducting household-level assessments at the program design stage may not be cost-effective. For example, the Senegal assessment found that the reality in eastern Senegal is more complex than the initial hypothesis that owning livestock decreases vulnerability to climate change suggests. Similarly, the Mali studies identified two clear regimes of water stress: communities near and far from the major rivers. Both of these general findings are extremely important to ensure appropriate design of national and even subnational programs. However, the resources necessary to refine these general findings to the point where they could suggest specific interventions appropriate at the household or community level would be significant and would expand with increasing geographic area. Therefore, from the studies conducted here, it appears that thresholds of specificity exist beyond which further detail and refinement may have decreasing marginal benefit until the next threshold is reached. However, assessment of these smaller scale variations may be appropriate once individual communities have been selected for intervention.

Furthermore, these studies highlight the advantages and disadvantages of integrated, piece-meal, and tool-based assessments in a data poor region like the Sahel. The integrated Senegal assessment was able to provide a more complete, nuanced understanding of vulnerability, but the findings are less relevant to other regions and to questions beyond the original hypothesis. The piece-meal approach taken in Mali allowed for the flexibility to provide targeted information as new demands arose, but in the end this assessment did not produce the comprehensive understanding developed in Senegal. Finally, in the Sahel, owing to the size and complexity of the region, tools were created to assist future studies at more appropriate geographic scales. While none of these approaches is inherently better or worse, it does appear that each has strengths and weaknesses that make them more or less relevant at different spatial scales, and thus at providing different levels of specificity in analysis. In order to understand and achieve the necessary threshold of specificity it is important to clearly define the study objective and then maintain consistent communication with the relevant decision-makers.

Robust Solutions

Given the complex array of factors affecting adoption of adaptive practices, there is a need for robust interventions that are not only effective across a range of future climate scenarios but that can also be tailored by individual households to their circumstances and needs, which likely vary with time. For example, robust programs could be created that promote a suite of possible adaptive options (e.g., credit, seeds, innovative practices, improved information, capacity building). Individuals and households could then select the most appropriate and seek beneficial synergies across the different forms of assistance. Another option is to continue to strengthen the creation, dissemination, and use of climate and weather information. While challenges remain (e.g., Antwi-Agyei et al. 2013; Carr and Thompson 2014), the provision of more accurate and more spatially and temporally explicit information would allow individual households to make more informed decisions. The studies conducted here found a growing interest in, but continued hesitancy with, this type of information. However, even this information would need to be provided as part of a larger suite of support as technological solutions, such as seasonal forecasting, are not sufficient to address the underlying social drivers of vulnerability (Agrawala and Broad 2002).

Programs promoting robust solutions will need to allow for flexible implementation that ensures rural producers can address a variety of stresses, the resources necessary to complete implementation are available, and the interventions consider individuals farm different crops, receive information in different manners, and operate under different social, cultural and economic constraints. Finally, the success of these programs may have to be measured in a different manner as adoption of any one practice will be inherently uneven across various spatial and temporal scales. Success will thus not depend on the uptake of a specific practice, but on the effectiveness of the overall strategy (i.e., basket of practices) employed by different households.

Limitations

While these studies allow some general observations to be drawn, they were not developed and implemented with the intent to conduct a detailed cross-analysis nor do they cover the vast array of possible assessment approaches. Thus direct comparisons and general conclusions are illustrative, and further study is necessary. Furthermore, these assessments, like most assessments conducted to inform development programming, were conducted under significant time and resource constraints and were developed to inform specific information needs. Therefore, their intent was neither to produce rigorous data and information for analysis nor to assess the effectiveness of the approach. A more detailed and rigorous analysis of assessment approaches would have examined a set of nested assessments that covered the same area but at different spatial scales. Furthermore, while methodologies exist for assessing the impact of specific practices, future studies could explore ways to evaluate the effectiveness of vulnerability assessments to inform development programming.

Conclusions

In the Sahel, significant uncertainty and spatial variation exists in both future climatic conditions and how households make livelihood decisions, and this uncertainty and complexity are unlikely to decrease significantly in the near future. Understanding this complexity at large spatial scales is a resource intensive process, and studies of vulnerability meant to inform development programming will not be able to assess all the local contextual factors on which the uptake of specific agricultural practices depends. Therefore, this paper has sought to highlight a number of viable approaches to assess vulnerability, and to demonstrate that the optimal approach likely depends on a number of factors, including spatial scale, and that assessments should not necessarily seek to identify a single universally acceptable or effective adaptive agricultural practice. Assessments should be tailored to identify the socio-cultural and biophysical factors acting at the relevant spatial scales, and should seek to fill the still significant gaps in our understanding associated with promoting adaptive practices in the Sahel. Assessments will continue to be essential in informing development programming, and as our understanding of the underlying socio-cultural and biophysical aspects improves new and more refined assessment approaches will likely be developed and implemented.

Acknowledgements The authors would like to acknowledge the support of the U.S. Agency for International Development through the African and Latin American Resilience to Climate Change (ARCC) project. The authors would also like to thank all the people who helped conduct these assessments, who are, unfortunately, too numerous to list here.

Disclaimer Alex Apotsos was employed by USAID when the research for the current paper was conducted. The views and opinions expressed in this paper are those of the authors and not necessarily the views and opinions of the United States Agency for International Development.

References

Adger WN (1999) Social vulnerability to climate change and extremes in coastal Vietnam. World Dev 27:249–269

Adger WN (2006) Vulnerability. Glob Environ Change 16:268–281

Adger WN, Agrawala S, Mirza MMQ, Conde C, O'Brien K, Pulhin J, Pulwarty R, Smit B, Takahashi K (2007) Assessment of adaptation practices, options, constraints and capacity. In: Parry ML, Canziani OF, Palutikof JP, van der Linden PJ, Hanson CE (eds) Climate change 2007: impacts, adaptation and vulnerability. Contribution of Working Group II to the fourth assessment report of the intergovernmental panel on climate change. Cambridge University Press, Cambridge, pp 717–743

Agrawala S, Broad K (2002) Technology transfer perspectives on climate forecast applications. Res Sci Technol Stud 13:45–69

Antwi-Agyei P, Dougill AJ, Fraser EDG, Stringer LC (2013) Characterising the nature of household vulnerability to climate variability: empirical evidence from two regions of Ghana. Environ Dev Sustain 15(4):903–926

Baptista S, Brottem L, de Sherbinin A, Edquist M, Fischer A, Levy M, Schnarr E, Simon C, Sundareshwar PV, Trzaska S (2013) Background paper for the ARCC West Africa regional climate change vulnerability assessment. USAID, Washington, DC

Barbier B, Yacouba H, Karambiri H, Zoromé M, Some B (2009) Human vulnerability to climate variability in the Sahel: farmers' adaptation strategies in Northern Burkina Faso. Environ Manag 43:790–803

Biermann M (2009) The role of local NGOs in anticipating and responding to climate change. Department of Geography, Pennsylvania State University, Pennsylvania

Blamey R, Coop L, Jack C, Loveday B, Sutherland K (2013) The status and possible evolution of climate projections in West Africa. USAID, Washington, DC

Boro D, Bouaré D, Cissouma S, Diarra M, LaLumia C, Miller D, Niang M, Sangaré Y, Upton S (2014) Climate change in Mali: organizational survey and focus groups on adaptive practices. USAID, Washington, DC

Carr ER, Thompson MC (2014) Gender and climate change adaptation in agrarian settings: current thinking, new directions and research frontiers. Geogr Compass 8(3):182–197

de Sherbinin A, Chai-Onn T, Giannini A, Jaiteh M, Levy M, Mara V, Pistolesi L, Trzaska S (2014) Climate change in Mali: country vulnerability map. USAID, Washington, DC

del Rio A, Simpson BM (2014a) Agronomic profiles of fifteen crops important in the Sahel. USAID, Washington, DC

del Rio A, Simpson BM (2014b) Agricultural adaptation to climate change in the Sahel: expected impacts on pests and diseases afflicting selected crops. USAID, Washington, DC

Djoudi H, Brockhaus M, Locatelli B (2013) Once there was a lake: vulnerability to environmental changes in northern Mali. Reg Environ Change 13(3):493–508

Eakin H, Bojorquez-Tapia LA (2008) Insights into the composition of household vulnerability from multicriteria decision analysis. Glob Environ Change 18(1):112–127

Eakin H, Luers AL (2006) Assessing the vulnerability of social environmental systems. Annu Rev Environ Resour 31:365–394

Folle S, Mulla DJ (2014) Agricultural adaptive practices impact modeling assessment. USAID, Washington, DC

Freudenberger M, Lo H, Boulahya MS (2014) Institutional Analysis of L'Agence de L'Environnement et du Developpement Durable (AEDD) and L'Agence Nationale de la Meteorologie (Mali-Meteo). USAID, Washington, DC

Füssel H, Klein R (2006) Climate change vulnerability assessments: an evolution of conceptual thinking. Clim Change 75:301–329

Hinkel J (2011) Indicators of vulnerability and adaptive capacity: towards a clarification of the science–policy interface. Glob Environ Change 21:198–208

Janssen MA, Ostrom E (2006) Resilience, vulnerability, and adaptation: a crosscutting theme of the International Human Dimensions Programme on Global Environmental Change. Glob Environ Change 16:237–239

Mertz O, Mbow C, Reenberg A, Diouf A (2009) Farmers' perceptions of climate change and agricultural adaptation strategies in rural Sahel. Environ Manag 43:804–816

Miller D, Wasson M, Trazska S, Coxon S, Dinshaw A, McGray H, Baptista S, del Rio A, Brottem L, Finan T, Baro M, Lo HM (2014) Senegal climate change vulnerability assessment and options analysis. USAID, Washington, DC

Mortimore MJ, Adams WM (2001) Farmer adaptation, change and 'crisis' in the Sahel. Glob Environ Change 11:49–57

Murray-Rust H (2013) Key issues in water resources. USAID, Washington, DC

Nelson R, Kokic P, Crimp S, Meinke H, Howden SM (2010) The vulnerability of Australian rural communities to climate variability and change: Part I—Conceptualising and measuring vulnerability. Environ Sci Policy 13:8–17

Niang I, Ruppel OC, Abdrabo MA, Essel A, Lennard C, Padgham J, Urquhart P (2014) Africa. In: Barros VR, Field CB, Dokken DJ, Mastrandrea MD, Mach KJ, Bilir TE, Chatterjee M, Ebi KL, Estrada YO, Genova RC, Girma B, Kissel ES, Levy AN, MacCracken S, Mastrandrea PR, White LL (eds) Climate change 2014: impacts, adaptation, and vulnerability. Part B: regional aspects. Contribution of working group II to the fifth assessment report of the intergovernmental panel on climate change. Cambridge University Press, Cambridge, pp 1199–1265

Nielsen JØ, Reenberg A (2010) Cultural barriers to climate change adaptation: a case study from Northern Burkina Faso. Glob Environ Change 20:142–152

Parry ML, Canziani OF, Palutikof JP, van der Linden PJ, Hanson CE (2007) Climate change 2007: impacts, adaptation and vulnerability. In: Parry ML, Canziani OF, Palutikof JP, van der Linden PJ, Hanson CE (eds) Contribution of working group II to the fourth assessment report of the intergovernmental panel on climate change. IPCC, Cambridge, pp 23–78

Preston BL, Yuen EJ, Westaway RM (2011) Putting vulnerability to climate change on the map: a review of approaches, benefits, and risks. Sustain Sci 6:177–202

Ray M, Simpson BM (2014) Agricultural adaptation to climate change in the Sahel: profiles of agricultural management practices. USAID, Washington, DC

Sendzimir J, Reij C, Magnuszewski P (2011) Rebuilding resilience in the Sahel: regreening in the Maradi and Zinder regions of Niger. Ecol Soc 16:1

Simpson BM (2014a) An approach to conducting phenological screening of the impact of climate change in the West African Sahel. USAID, Washington, DC

Simpson BM (2014b) Agricultural adaptation to climate change in the Sahel: an approach to evaluating the performance of agricultural practices. USAID, Washington, DC

Simpson BM (2015) Planning for scale: using what we know about human behavior in the diffusion of agricultural innovation and the role or agricultural extension. MEAS Technical Note. USAID, Washington, DC

Simpson BM, Burpee CG (2014) Adaptation under the "New Normal" of climate change: the future of agricultural extension and advisory services. MEAS Discussion Paper, USAID, Washington, DC

Trench P, Rowley J, Diarra M, Sano F, Keita B (2007) Beyond any drought: root causes of chronic vulnerability in the Sahel. The Sahel Working Group, CARE International, Oxfam, London

Tschakert P (2007) Views from the vulnerable: understanding climatic and other stressors in the Sahel. Glob Environ Change 17:381–396

Younan M, Simpson BM (2014) Agricultural adaptation to climate change in the Sahel: expected impacts on pests and diseases afflicting livestock. USAID, Washington, DC

Ziervogel G, Bharwani S, Downing TE (2006) Adapting to climate variability: pumpkins, people and policy. Nat Res Forum 30:294–305

Part II
Innovation in Sectorial Approaches to Adaptation

Chapter 13
Adaptation of the Artisanal Fisher Folks to Climate Change in the Coastal Region of Ondo State, Nigeria

Mosunmola Lydia Adeleke and Matthias Wolff

Abstract Climate change is a global problem and has become an important agenda in both public and private discourse in recent times. Adaptation is more relevant for poorer nations because of their relative vulnerability to the impacts of climate change, [IPCC (Contribution of working group II to the 4th Assessment Report of the IPCC. University Press Cambridge, 2007a); Climate change: impacts, adaptation and vulnerability. Working group II contribution to the Intergovernmental Panel on Climate Change. Summary for policy makers. IPCC Secretariat, Geneva, Switzerland, 2007b]. However, this paper examined adaptation of the artisanal fisher folks to climate change in the coastal region of Ondo state, Nigeria. Data for climate variables (sea surface temperature and rainfall) were obtained from reanalyzed Satellite data and were further analyzed using sophisticated ecological models for further inferences. 5-Scale Likert scale was used to assess the current adaptation strategies adopted by artisanal fisher folks. The result revealed that rainfall pattern showed great variation over time in terms of volume and intensity; this observation is in line with the observation of FAO, 2008, that, climate change is modifying the distribution of marine and freshwater species. The trend of temperature pattern for about 30 years clearly showed high fluctuation overtime. It was observed that the fisher folks in the study area had various adaptation strategies adopted for maximum fish production/catch for livelihood and improve standard of living.

Keywords Adaptation • Artisanal fisher folks • Climate change • Coastal region • Nigeria

---

M.L. Adeleke (✉)
Federal University of Technology, Akure, Nigeria
e-mail: mosunmolalydia@yahoo.com

M. Wolff
Leibniz Center for Tropical Marine Ecology (ZMT), Fahrenheitstraße 8, 28359 Bremen, Germany
e-mail: mwolff@zmt-bremen.de

W. Leal Filho (ed.), *Innovation in Climate Change Adaptation*, Climate Change Management, DOI 10.1007/978-3-319-25814-0_13

Introduction: Climate Change a Global Problem

The Intergovernmental Panel on Climate Change (IPCC), predicted that during the next decades, billions of people, especially those in developing countries will face changes in rainfall patterns that will contribute to severe water shortages or flooding, rising temperatures that will cause shifts in crop growing season and aquatic organisms distribution (FAO 2008). Climate change is a global problem and has become an important agenda in both public and private discourse in recent times. It has brought about much anxiety and bewilderment in the wake of global disasters perpetrated by flood, storm and other natural hazards. Nigeria, like all the countries of sub-Saharan Africa, is highly vulnerable to the impacts of climate change (IPCC 2007a, b) and should be concern because of high vulnerability due to its long 800 km coastal line that is prone to sea-level rise and the risk of fierce storms (Apata et al. 2009; Adeleke and Balogun 2013).

Materials and Methods

Linux (ferret) was used get explore satellite or reanalyzed climate variables data from the Climate Research Unit (CRU). Ferret is an interactive computer visualization and analysis environment designed to meet the needs of oceanographers and meteorologists analyzing large and complex gridded data sets. The model data sets are generally multi-gigabyte in size with mixed 3 and 4-dimensional variables defined on staggered grids. Calculations may be applied over arbitrarily shaped regions. Fully documented graphics are produced with a single command. This model was used to establish the fact that climate change has occurred in the coastal region of Ondo State, Nigeria. Kruskal Wallis test was also used, to assess the current knowledge of the artisanal Fisher folks on climate change. It permits the analysis of decisions across more than two categories, allowing the determination of choice probabilities for different categories. This was done listing some perception statements. 5-Scale Likert scale was used to assess the current adaptation strategies adopted by artisanal fisher folks.

Result and Discussion

The fact that climate change has occurred in the coastal region of Ondo state, Nigeria was established by the use of temperature and rainfall data. The Temperature and rainfall data used for this study was obtained from the downscaled ensemble product and they collectively provide values of Temperature and Rainfall between 1901 and 2011 (Appendices I and II).

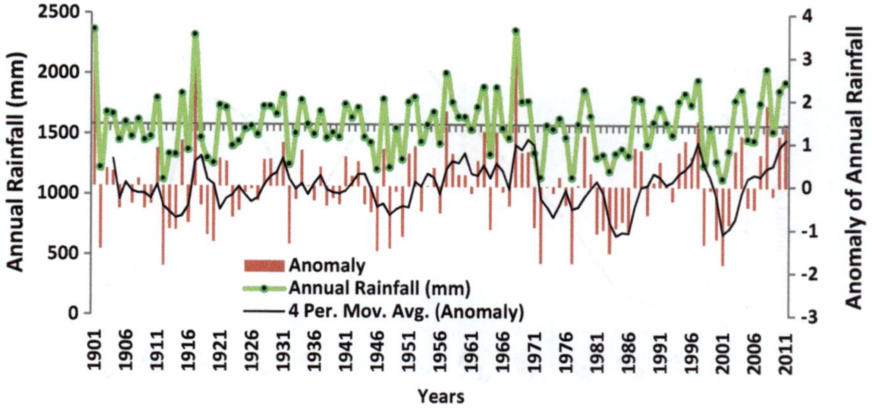

Fig. 13.1 Annual rainfall distributions in the study area (*Source*: Authors)

Annual Rainfall of the Study Area (1901–2011)

Figure 13.1 shows the time series plot of annual rainfall receipt extracted in the study area. The annual values were calculated from the monthly rainfall amounts. Other computations were done for the mean annual rainfall amount observed in the study area was 1579.283 mm and a standard deviation of 246.0073 mm. This was employed in the computation of the Normalized Anomaly of annual rainfall. A four percentile moving average anomaly was superimposed on the time series of the anomaly plot in order to show the flow of rainfall received and the cycle of rainfall distribution in the study area.

There was a noticeable cycle of annual rainfall amount but the steepness of such cycles increased beginning from 1979. Positive anomaly shows years of above long time average rainfall received while negative anomaly describes years of below average rainfall receipt.

Figure 13.2 shows the average monthly distribution of rainfall for January–December (2011), the graph above shows a double peak in monthly rainfall distribution with the first peak experienced in June and the second peak in October. A reduction in monthly rainfall receipt in often experienced in August, while months between March and October experience more than 100 mm monthly rainfall receipt.

Annual Temperature of the Study Area (1901–2011)

Anomaly plot of averaged annual temperature over the years is presented in Fig. 13.2. It was observed that Anomaly plot of averaged annual temperature shows years of below average and above average temperature. Between 1902 and 1910 were years below average temperature followed by above average

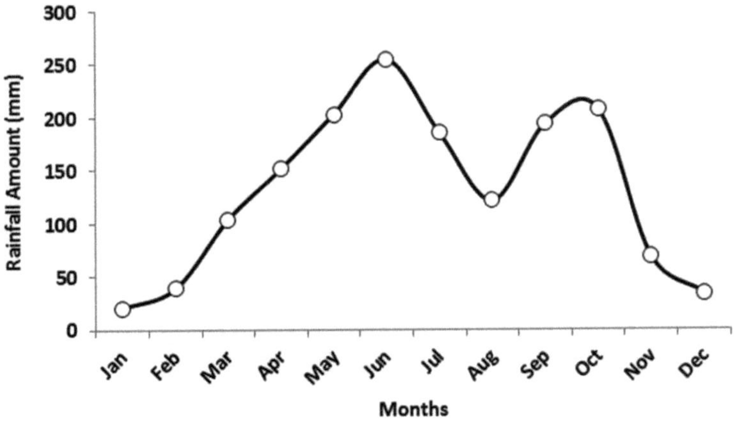

Fig. 13.2 Average monthly distributions of rainfall in the study area (*Source*: Authors)

temperature distribution between in 2003 and 2009. The four percentile moving average of the annual temperature anomaly curve implies that climate change has been taking place in the study area and this might had have both positive and negative effects on the fish distribution and production.

Assessment of the Current Knowledge of the Fisher Folks on Climatic Variables in the Study Area

The Kruskal Wallis (KW) test analysis was carried out to assess the current knowledge of the Fisher folks on increase and decrease effects of some climatic variables to ascertain the fact that climatic change had occurred in the study area. The result showed that the respondents in the study area have different perception and attitude towards increase or decrease in rainfall as a factor of climate change. There is also a significant difference in the responses of the Fisher folks in terms of increase or decrease in temperature, and highly significant at 0.01 level of significance.

The result of the test showed that there is significant difference in the responses of the artisanal Fisher folks to their current knowledge and perception in terms of increase or decrease in wind speeds, directions, duration and time. Also, increase or decrease in sea level was also significant and the respondents have different perception and attitude toward it occurrence. The Kruskal Wallis test for increase or decrease in solar radiation was highly significant at 0.01 level of significance. Kruskal Wallis value for increase or decrease in wave (ocean currents) and tides (high and low) were as well significant ($p < 0.01$) (Tables 13.1 and 13.2).

The implication of this is that any increase or decrease in any of these climatic variables would result in either positive or negative change which might have significantly effect on fish distribution and production. From the result presented

Table 13.1 Current knowledge of the Fisher folks on increase in climatic variables

Variables	Frequency	Mean rank	Equivalent	Corresponding group	Kruskal Wallis statistics
Increase in temperature leads to high yield/catch	37	87.95	2.377	Undecided	153.528***
	188	270.98	1.441	Disagreed	
	174	147.19	0.8459	Strongly disagreed	
Increase in rainfall leads to high yield/catch	37	101.00	2.729	Undecided	47.525***
	188	226.69	1.206	Disagreed	
	174	192.22	1.103	Disagreed	
Increase in wind speed leads to high yield/catch	37	78.53	2.122	Undecided	94.310***
	188	244.44	1.300	Disagreed	
	174	177.81	1.022	Strongly disagreed	
Increase in solar radiation leads to high yield/catch	37	83.99	2.270	Undecided	125.140***
	188	263.48	1.401	Disagreed	
	174	156.08	0.897	Strongly disagreed	
Increase in wave and tides leads to high yield/catch	37	98.16	2.653	Undecided	39.207***
	188	220.14	1.171	Disagreed	
	174	199.90	1.149	Disagreed	
Rise in sea level leads to high yield/catch	37	82.82	2.238	Undecided	94.900***
	188	249.75	1.328	Disagreed	
	174	171.16	0.983	Strongly disagreed	

Source: Authors
***Significant at 0.01 level significant (≤ 0.01)

in Figs. 13.1 and 13.3 for annual rainfall and temperature distributions and how it affects the Fisher folks day to day activities. It is therefore, expedient to assess the current inadvertent and intentional adaptation strategies adopted by the artisanal Fisher folks in the study area.

Adaptation Strategies Adopted by the Artisanal Fisher Folks in the Study Area

The distribution of these fish species are affected by the effects of climate variables such as rainfall, air and water temperatures, solar radiation, low and sea level rise, ocean currents, tides and wind (speeds, directions, duration and time). It was observed that the Fisher folks in the study area had various inadvertent and intentional adaptation strategies adopted for maximum fish production/catch i.e. coping strategies employed by Fisher folks for effective and efficient fish production Tables 13.3 and 13.4.

Table 13.2 Current knowledge of the Fisher folks on decrease in climatic variables

Variables	Frequency	Mean Rank	Equivalent	Corresponding group	Kruskal Wallis statistics
Decrease in temperature leads to high yield/catch	37	114.99	3.107	Agreed	32.950***
	188	224.36	1.182	Disagreed	
	174	191.76	1.102	Disagreed	
Decrease in rainfall leads to high yield/catch	37	90.28	2.44	Undecided	94.869***
	188	253.06	1.346	Disagreed	
	174	160.00	0.954	Strongly disagreed	
Decrease in wind speed leads to high yield/catch	37	118.81	3.211	Agreed	28.575***
	188	221.40	1.177	Disagreed	
	174	194.14	1.116	Disagreed	
Decrease in solar radiation leads to high yield/catch	37	89.70	2.424	Undecided	46.407***
	188	222.93	1.185	Disagreed	
	174	198.68	1.142	Disagreed	
Decrease in wave and tides leads to high yield/catch	37	101.14	2.734	Undecided	39.572***
	188	224.17	1.192	Disagreed	
	174	194.9	1.120	Disagreed	
Low sea level leads to high yield/catch	37	94.88	2.564	Undecided	37.972***
	188	218.39	1.1616	Disagreed	
	174	202.49	1.1637	Disagreed	

Source: Authors
***Significant at 0.01 level significant (≤0.01)

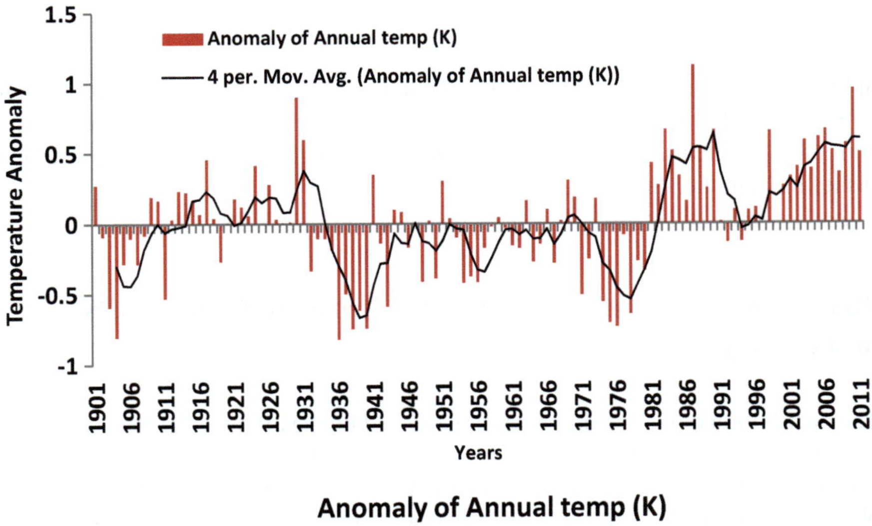

Fig. 13.3 Annual temperature distributions in the study area (*Source*: Authors)

Table 13.3 Adaptation strategies adopted by the Artisanal Fisher folks in the study area (increase in some climatic variables)

S/N	Variables	Effects on fish	Effects on fisher folks	Adaptive strategies	Characteristics of the adaptive strategies
1.	Increase in temperature (January–February) Harmattan	Make the fish to run deep inside the water for hiding and catch is low	Get fishes in the small streams	1. Adjusting fishing period to early morning or night 2. Changing the fishing gears to hook and lines 3. Setting traps	It is affordable and effective
2.	Increase in rainfall (March–May) (June–July)	Highest catch of fishes Fishes would come out for food	Fishermen/women go far to the sea and catch big fishes They go to fish at times 3 times daily	1. Setting of traps various types of traps –Igun/awa –Iyanma –Bamboo 2. Change the fishing gears to favour the season 3. Adjusting fishing period to morning 4. Construction of fish walls	1. Affordable 2. Effective 3. Available
3.	Increase in wind speed (March–September)	Fishes are disturbed and move in the direction of the wind	Fisher folks may stop fishing at this particular time. It will result in low fish catch	1. Setting of traps 2. No specific adaptative strategy	
4.	Increase in intensity of sun (November–February)	It will make fishes to run away down the stream/sea	Low catch	1. Fish very early or late in the night 2. No specific adaptative strategy	1. Effective
5.	Increase in wave and tides (March–November)	Fish will not come out for food and it will result in low catch of fish	Catches depend on season	1. Put stick in your "igun" to make it stronger	Relative advantage
6	Rise in sea level (May–September)	1. It makes fish to migrate far with the sea 2. Fishes run to cooler areas	1. Fisher folks may not enter the sea for fishing, instead, they may go fishing at nearby streams	1. No specific strategy 2. The use of traps 3. Use of cutlass to kill fishes that hide under the grass	1. Affordable

Source: Authors

Table 13.4 Adaptation strategies adopted by the Artisanal Fisher folks in the study area to (decrease in some climatic variables)

S/N	Variables	Effects on fish	Effects on fisher folks	Adaptive strategies	Characteristics of the adaptive strategies
1.	Decrease in temperature (November–March)	1. Allows fishes to come out for food	1. Normal fishing takes place. 2. Catch is high	1. Changing fishing methods by adjusting the fishing period and fishing gears	Environmentally friendly, affordable and effective
2.	Decrease in rainfall (November–March)	Fishes are exposed because the water that serves as habitat for them might have been dried off	1. It is seasonal 2. Catch is high 3. Normal fishing takes place	1. Night fishing operation 2. Changing fishing gears to hook and line	1. Affordable 2. Effective 3. Available
3.	Decrease in wind speed (November–March)	It dictate the movement of the fish	1. The sea dictates 2. Favourable for fishing 3. High fish catch	No specific strategies (it depends on the types of fishing gears to use, time to fish, etc.)	Effective
4.	Decrease in solar radiation/sun intensity (April–October)	It has no adverse effect on the fish	Favourable for fishing	No specific strategy	
5.	Decrease in wave and tide (November–March)	No effect on fish	1. The catch depends on the type of gear used, time out for fish (fishing time/period) 2. Favorable for fishing	Fortification of the fishing gears	Affordable
6.	Low sea level (December–March)	Fish move and come out during this period for food	1. Allow to catch more fish 2. High catch 3. High income	Use baits on the hooks and 'igun' to attract the fish	Effective

Source: Authors

Temperature, especially water temperature, exacts a major control over the distribution and activities of aquatic organism. It was observed from that increases in temperature which happens in the study area (January–February) Table 13.3, the respondents believe that the fishes would find means of hiding away from the high temperature and this may result in low catch.

The strategies used by Fisher folks include adjusting their fishing periods to either very early in the morning or late in the night. The nocturnal activity of some of the fishes, especially, the mullets (*Mugil* spp.) made some of Fisher folks to prefer fishing in the night because the chance of a good catch will be enhanced; this technique is known as *lamp fishing*. They also coped by setting traps and at times change their fishing gears depending on time, place, target species and types of fish. These methods are effective and affordable to Fisher folks.

On the other hand, Fisher folks observed decrease in temperature (November–March) Table 13.4. They believed it affect behavioral phenomenon of the fishes and influence the presence of fish and in the way they are caught. At this period, Fisher folks would have maximum fish production and normal fishing would take place as a result of the conducive and favorable temperature. The implication of this is that, more Fisher folks would like to fish at this period because catchability would be high with minimum unit effort and increase in revenue of the Fisher folks hence, increase in standard of living of the Fisher folks.

Increase in rainfall was observed (March–July) as presented in Table 13.3. Fisher folks believed that it was the best period to catch fish and they refer to it as the season of 'new moon'. At this period, fish production would be at the highest peak, because both the fishermen and women go far to the sea covering about 8–14 nautical miles for bigger fishes. Some of the respondents can go for fishing three or more times daily. Fisher folks made used of various fishing gears and adaptation measures for maximum fish production. The fishing strategies include: 'Siege fishing' this is usually performed by three fishermen, the strategy is very useful during low tides when the fishermen manage to visualize fishes. Another strategy was when the fishermen 'pay out the net', i.e. allowing the current to take it until a certain point where the fish get trapped in it. The nets used can be gillnets which captures species of the families Ariidae, Centropomidae, Gerreidae, Mugilidae and Sciaenidae.

Decrease in rainfall would take place between November and March in the study area Table 13.3. This season is also favorable for Fisher folks because the water might have dried off as a result of the dry season. At this period, normal fishing would take place and fish production would also increase. The fishing strategy was the attachment of lamp in the bow area of the stern of the canoe, this 'kerosene lamp' attracted the fish to jump as a reaction to the light reflection. The result implies that, both increase and decrease in rainfall would lead to increase in fish production.

Some Fisher folks in the coastal region of the state, were not familiar with the cardinal directions of the winds, however, they were able to name these winds based on their environmental characteristics. Increase in wind speed usually occurred between March and November, in the study area. Redistribution of fish take place at this period and fishes try to hide from their predators. During windy weather, Fisher folks were discouraged from fishing because of the low fish catch. There are no specific adaptation strategies to combat the effect of increase wind in the study area (Table 13.3).

Conversely, decrease in wind occurred between November and March as presented in Table 13.4. The result showed that decrease in wind speed is of greater advantages to Fisher folks because it aid in the voyage of the fish, facilitates easy sailing that had no adverse effect on the fish, and it would invariably lead to high fish production. At this period, the income of Fisher folks may increase, if there are market and good sales for their products. There are no specific strategies for adaptation since Fisher folks already know which species of fish will be abundant according to the period of the year. Fish production would depend on time of fishing and type of fishing gears used. It implies that, there would be increase production at this season compared to increase wind.

Solar Radiation or light penetration is one of the major factors controlling the distribution of aquatic plants which are restricted to areas where there is enough light energy for photosynthesis. Therefore, increase or decrease in solar radiation would have significant effect in population of fishes in the water. The result showed that increase in solar radiation took place between November and February in the study area has presented in Table 13.3. Fisher folks in the study area witnessed low catch at this period because fishes try to move down deep the water due to high intensity. Fisher folks adjust their fishing period to very early in the morning, late in the evening or night when the sun might have set. Fisher folks preferred to fish when there was decrease in sun intensity/solar radiation because it was believed to be favourable for fishing. Increase in fish production is the implication (Table 13.4).

In the study area, increase waves/ocean current and tides occurred between months of (March–November) Table 13.3. These would disrupt the activities of the fish and prevent the fish by changing their feeding and movement behaviour. Fish production and catchability would be low at this time. Fortification of fishing gears was the adaptative measure Fisher folks adopted to increase fish production/catchability. On the contrary, Fisher folks experienced increase in fish production when there were decreases in ocean current and tides, because, the fishes would not be disrupted and would come to the surface of the water for food (Table 13.4). The techniques and strategies of fishing used by Fisher folks have a close relation with the moon phases and consequently with the variation of the tides. The tides consist of periodic changes of short duration in the height of the ocean surface in a certain place, caused by a combination of the gravitational forces of the moon and sun with and the movement of the Earth.).

Notwithstanding, some Fisher folks prefer to fish in the new moon phase, because the nights are much darker and this facilitates the catch, especially of mullets (*Mugil* sp.). The period of low sea level that occurred between December and March (Table 13.3). Fisher folks preferred to fish at this period because, the fishes would come out for food. The used of more bait on hooks would be an added advantage. It is therefore, advisable for Fisher folks to fish during this period because increase in production would lead to increase in income and the standard of living of the people would also improve.

Conclusion

This paper examined adaptation of the artisanal fisher folks to climate change in the coastal region of Ondo state, Nigeria. Primary data were collected through administering 400 structured questionnaires, to the Fisher folks. Secondary data were obtained from reanalyzed Satellite data. There was a noticeable cycle of annual rainfall amount but the steepness of such cycles increased beginning from 1979. Between 1902 and 1910 were years below average temperature followed by above average temperature distribution between in 2003 and 2009. This confirmed the fact that climate change has occurred in the coastal region of Ondo state, Nigeria. The result of the assessment of the current knowledge of the Fisher folks on climatic variables showed that, the Fisher folks had different perception/attitude as well adaptation strategies towards increase or decreases the climatic variables. Based on the results of the findings, it is therefore recommended that; Fisher folks should be trained by the extension agents, on how to keep climatic and fishing production activities records. Also, livelihood diversification should be encouraged in the study area.

Appendix I: Annual rainfalls in the coastal region of Ondo State, Nigeria (January 1901–December 2011)

	Annual rainfall (mm)	Anomaly
1901	2363.6	3.18777
1902	1220.6	−1.45843
1903	1677.8	0.400048
1904	1662.8	0.339074
1905	1450	−0.52594
1906	1596.3	0.068757
1907	1477.1	−0.41578
1908	1619.4	0.162656
1909	1449.9	−0.52635
1910	1478.4	−0.4105
1911	1795	0.876457
1912	1124.5	−1.84907
1913	1334	−0.99747
1914	1329.4	−1.01617
1915	1836.5	1.045151
1916	1370.6	−0.8487
1917	2322.6	3.021108
1918	1470.4	−0.44302
1919	1301.4	−1.12999
1920	1261.9	−1.29055

(continued)

	Annual rainfall (mm)	Anomaly
1921	1736.5	0.638659
1922	1718.9	0.567116
1923	1402.3	−0.71984
1924	1440.6	−0.56415
1925	1544.4	−0.14221
1926	1565.4	−0.05685
1927	1497.6	−0.33245
1928	1730.4	0.613863
1929	1732.1	0.620773
1930	1667.3	0.357366
1931	1828.2	1.011412
1932	1250.6	−1.33649
1933	1507.2	−0.29343
1934	1783.9	0.831336
1935	1582.1	0.011035
1936	1498.5	−0.32879
1937	1689.4	0.447201
1938	1469.4	−0.44708
1939	1511.6	−0.27554
1940	1476.6	−0.41781
1941	1749.4	0.691096
1942	1637.8	0.237451
1943	1720.5	0.57362
1944	1477.2	−0.41538
1945	1433.5	−0.59301
1946	1210.1	−1.50112
1947	1790.8	0.859384
1948	1224.8	−1.44136
1949	1549.2	−0.1227
1950	1292	−1.1682
1951	1765.7	0.757354
1952	1809.3	0.934585
1953	1439.3	−0.56944
1954	1579.7	0.001279
1955	1685.5	0.431348
1956	1427.2	−0.61862
1957	2008.4	1.743911
1958	1767.3	0.763858
1959	1646.6	0.273222
1960	1644.4	0.264279
1961	1541.7	−0.15319
1962	1726.1	0.596383

(continued)

	Annual rainfall (mm)	Anomaly
1963	1893.1	1.275225
1964	1337	−0.98528
1965	1889	1.258559
1966	1549.6	−0.12107
1967	1470.8	−0.44139
1968	2365.3	3.194681
1969	1773.6	0.789467
1970	1778.2	0.808166
1971	1349.4	−0.93487
1972	1143.3	−1.77265
1973	1577.1	−0.00929
1974	1544.5	−0.14181
1975	1632.7	0.21672
1976	1476.6	−0.41781
1977	1144.1	−1.7694
1978	1586.5	0.028921
1979	1868.6	1.175635
1980	1655.7	0.310213
1981	1313.1	−1.08243
1982	1334.4	−0.99585
1983	1201.4	−1.53648
1984	1345.6	−0.95032
1985	1383.1	−0.79788
1986	1327.5	−1.02389
1987	1803.2	0.909789
1988	1789.1	0.852473
1989	1419.7	−0.64911
1990	1602.9	0.095585
1991	1724	0.587847
1992	1600	0.083797
1993	1500.4	−0.32107
1994	1780	0.815483
1995	1843.1	1.071979
1996	1753.8	0.708982
1997	1956.3	1.532128
1998	1253.4	−1.3251
1999	1562.5	−0.06864
2000	1284.2	−1.1999
2001	1138.2	−1.79338
2002	1367.6	−0.86089
2003	1787.5	0.84597
2004	1873.8	1.196772

(continued)

	Annual rainfall (mm)	Anomaly
2005	1468.3	−0.45155
2006	1456.1	−0.50115
2007	1766.1	0.75898
2008	2047.9	1.904475
2009	1533	−0.18855
2010	1870.64	1.183927
2011	1942.76	1.477089
Averaged rainfall receipt		1579.385
Standard deviation		246.0073

Source: Computed from ensemble satellite products (November, 2012)

Appendix II: Temperature in the coastal region of Ondo State, Nigeria (January 1901–December 2011)

1901	0.275
1902	−0.08333
1903	−0.5875
1904	−0.8
1905	−0.275
1906	−0.09583
1907	−0.27917
1908	−0.075
1909	0.191667
1910	0.166667
1911	−0.525
1912	0.033333
1913	0.233333
1914	0.225
1915	0.166667
1916	0.070833
1917	0.458333
1918	0.041667
1919	−0.2625
1920	0.008333
1921	0.179167
1922	0.120833
1923	0.058333
1924	0.416667
1925	0

(continued)

1926	0.279167
1927	0.033333
1928	0.008333
1929	0.0125
1930	0.9
1931	0.595833
1932	−0.32917
1933	−0.1
1934	−0.1
1935	−0.17917
1936	−0.81667
1937	−0.49583
1938	−0.74167
1939	−0.6125
1940	−0.7375
1941	0.35
1942	−0.13333
1943	−0.58333
1944	0.1
1945	0.08325
1946	−0.16408
1947	0.005583
1948	−0.40842
1949	0.02225
1950	−0.38475
1951	0.30425
1952	0.038
1953	−0.09375
1954	−0.4175
1955	−0.37167
1956	−0.4125
1957	−0.16667
1958	−0.01833
1959	0.045
1960	−0.0275
1961	−0.15292
1962	−0.17
1963	0.162083
1964	−0.26567
1965	−0.14042
1966	0.100083
1967	−0.27633
1968	0.021667

(continued)

1969	0.307833
1970	0.185583
1971	−0.50217
1972	−0.24725
1973	0.1745
1974	−0.55242
1975	−0.70275
1976	−0.73017
1977	−0.07775
1978	−0.63958
1979	−0.2625
1980	−0.32917
1981	0.428333
1982	0.27175
1983	0.6645
1984	0.515667
1985	0.336583
1986	0.155083
1987	1.118583
1988	0.533333
1989	0.246917
1990	0.6595
1991	0.014083
1992	−0.13108
1993	0.095167
1994	−0.12533
1995	0.08975
1996	0.1095
1997	0.005583
1998	0.650833
1999	−0.0005
2000	0.264583
2001	0.332455
2002	0.399083
2003	0.5855
2004	0.386833
2005	0.607583
2006	0.664583
2007	0.515917
2008	0.35925
2009	0.5645
2010	0.953667
2011	0.499583

Source: Computed from ensemble satellite products (November, 2012)

References

Adeleke ML, Balogun AA (2013) Occurrence of climate change and its effects on the fishing activities in the coastal region of Ondo State, Nigeria. Int J Dev Econ Sustain 1(1):1–13, European–American journals, IJDES-101

Apata TG, Samuel KD, Adeola AO (2009) Analysis of climate change perception and adaptation among arable food crop farmers in south western Nigeria. Contributed paper prepared for presentation at the international association of agricultural economists' 2009 conference, Beijing, China, 16–22 Aug 2009

Food and Agriculture Organization (FAO) (2008) Report of the FAO expert workshop on climate change implications for fisheries and aquaculture: Rome, Italy, 7–9 Apr 2008. FAO fisheries report no. 870 (Food and Agriculture Organization of the United Nations, 2009.2008). http://tiny.cc/fisheries7

Intergovernmental Panel on Climate Change (IPCC) (2007a) Climate change 2007: impacts, adaptations and vulnerability. In: Parry ML, Canzian OF, Palutikof JP, van der Linden PJ, Paul J, Hanson C (eds) Contribution of working group II to the 4th assessment report of the IPCC. Cambridge University Press, Cambridge

Intergovernmental Panel on Climate Change (IPCC) (2007b) Climate change: impacts, adaptation and vulnerability. Working group II contribution to the intergovernmental panel on climate change. Summary for policy makers. IPCC Secretariat, Geneva, Switzerland

References

Chapter 14
Comparative Analysis of Woody Composition of Farmlands and Forest Reserve Along Afram River in a Tropical Humid Savanna of Ghana: Implications to Climate Change Adaptation

Emmanuel Amoah Boakye, Dibi N'da Hyppolite, Victor Rex Barnes, Stefan Porembski, Michael Thiel, François N. Kouamé, and Daouda Kone

Abstract Riparian forests (RF) composition is important for moderating climate change impacts on agricultural watersheds. However, they are under threat from deforestation of catchment areas. The study used remote sensing techniques and field inventorying to assess woody species composition of RF on farmland (FA) and protected area (PA) along Afram rivercourse in the humid savanna of Ghana. Analysis of Landsat images revealed a reduction in forest cover from 1986 (50 %) to 2014 (31 %) in the river catchment. Ground survey of 60 randomly selected plots (500 m^2 per plot) equally divided between FA and PA along the river in a 50 m buffer zone showed a reduction in the number of woody species (diameter $\geq$5 cm) from PA (58) to FA (39). Shannon-Wiener Index for species diversity also reduced from PA (3.8 ± 0.05) to FA (3.1 ± 0.08). Diameter class distribution of species of both PA and FA showed a reversed J-shaped curve indicating successful regeneration. Reduction in species density per hectare from PA (545 ± 18) to FA (277 ± 13) is likely to increase the surface exposure of the riparian area in FA. This will heighten risks of climate disasters such as fires and flooding. Education of farmers on the importance of riparian forests may ensure their protection.

E.A. Boakye (✉) • D.N. Hyppolite • F.N. Kouamé • D. Kone
WASCAL Graduate Research Program in Climate Change and Biodiversity, Université Felix Houphouët Boigny, Abidjan, Côte d'Ivoire
e-mail: eaaboakye@yahoo.com

V.R. Barnes
Kwame Nkrumah University of Science and Technology, Kumasi, Ghana

S. Porembski
Institute of Biosciences, University of Rostock, Rostock, Germany

M. Thiel
Department for Geography and Geology, University of Wuerzburg, Wuerzburg, Germany

W. Leal Filho (ed.), *Innovation in Climate Change Adaptation*, Climate Change Management, DOI 10.1007/978-3-319-25814-0_14

Keywords Riparian buffer • Biodiversity • Humid savanna • Agricultural watershed

Introduction

In the savanna agricultural landscape, riparian forests are naturally resilient to climate change impacts as the continuous supplies of moisture coupled with the topographic heterogeneity and closed linear canopy limit grassy fuel loads, increase relative humidity, decrease temperature and wind speed to reduce fire risks (Sambare et al. 2011; Azihou et al. 2013). Ecologically, riparian forests (RF) are important as they protect farmlands from flooding, drying and sedimentation (Sambare et al. 2011; Gray et al. 2014). They also serve as habitat for fauna such as birds, insects and other organisms that are essential for crop pollination, seed dispersal and nutrient cycling (McCracken et al. 2012; Gray et al. 2014). Riparian forests have social benefits including provision of tourism, medicines, nutrition, firewood, and raw material for different crafts and construction (Ceperley et al. 2010; Gray et al. 2014). Culturally, riparian forests are sometimes designated as sacred groove (Ceperley et al. 2010). Due to these functions and many others, RF are protected by international conventions, national laws and policies (McCracken et al. 2012; Gray et al. 2014).

Within the water-limiting savanna environment, riparian catchments are hotspots for agricultural production (Natta et al. 2003; Goetze et al. 2006). As a result, riparian forests are under threat of deforestation which consequently changes their microclimatic conditions to increase climate change effects on species and associated functions (Callo-Concha et al. 2012). Globally, land areas dedicated to agricultural production are much greater than protected forest reserve areas (Traoré et al. 2012; Gray et al. 2014). This means that agricultural landscapes cannot be excluded from biodiversity conservation (Gray et al. 2014). With appropriate management, agricultural landscapes can contribute to the preservation of biodiversity and delivery of ecosystem services (McCracken et al. 2012; Gray et al. 2014). However, in spite of the global knowledge on the threat of agricultural production to riparian forests, our understanding in this area is limited in the tropical savannas of Ghana and West Africa in general (Natta et al. 2003; Ceperley et al. 2010; Sambare et al. 2011).

Several studies have demonstrated that farming activities cause deforestation and reduce the woody composition (diversity and structure) on agricultural landscapes (Ceperley et al. 2010; Okiror et al. 2012). In other studies, farmlands maintained high woody composition suggesting that not all farming practices have negative effects on biodiversity (Boakye et al. 2012; Traoré et al. 2012; Gray et al. 2014). To enhance the management of riparian forests in savanna agricultural landscape, this study compares woody vegetation composition in farmlands and that in protected forest reserve area by using the headwaters of the Afram river located in the humid savanna of Ghana as the case study. Two related hypothesis are tested in this study. Firstly, riparian forest in protected area hosts a higher diversity of woody species than on farmland (Ceperley et al. 2010; Okiror et al. 2012). Alternatively, no such diversity is observed in protected area (Boakye et al. 2012; Traoré et al. 2012). Secondly, the structure of the riparian woody species on farmland mimics that in protected area (Traoré et al. 2012). Alternatively, the riparian forest structure on farmland differs from the protected area (Boakye et al. 2012). The paper concludes on the findings of the research in the light of climate change adaptation on farmlands in the savanna zones of Ghana. It is anticipated that this will serve as an important baseline for the management of farmland biodiversity as well as the enforcement of the freshwater buffer zone policy of Ghana.

Materials and Methods

Study Area

The study was conducted in the headwaters of the Afram river catchment in the north-eastern part of the Ashanti region of Ghana (Kyerematen et al. 2014) (Fig. 14.1). The climate of the river catchment is characterized by distinct wet and dry seasons. The highest rainfall occurs between May and October and the annual average is 1400 mm. The hottest months occur from January to April and the annual average is 27 °C. The soil is composed of a well-drained sandy loam (Callo-Concha et al. 2012). The topography is flat to gently undulating with small areas of steep slopes occurring locally. The vegetation consists of Guinea savanna which forms a transition zone between closed forest and the Sudan savanna. Along the Afram river is located the Kogyae Strict Nature Reserve [Protected Forest Reserve

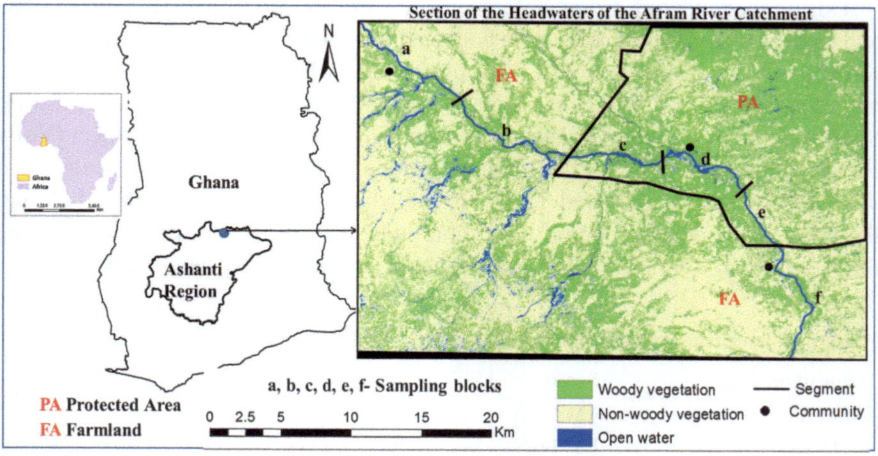

Fig. 14.1 The location map of a section of the research area in Ghana

Area (PA)], gazetted in 1971. Protected area (PA) is defined as clearly delimited area of natural vegetation that have been officially classified with appropriate legal status by public authorities with the aim of ensuring protection of natural resources as well as ecosystem functions and services (Traoré et al. 2012). The Kogyae reserve was thus established to protect the tributaries of the Afram river and also as refugeum for wildlife within the locality. The Kogyae reserve is under the management of the Forestry Commission of Ghana. The main land use activity of the communities in the vicinity of the reserve is farming (FA) and along the river, cereals are the widely cultivated food crops. The farmlands are affected by various anthropogenic activities including extensive livestock grazing, bush fires, and various harvestings of timber and non-timber forest products such as wood, leaves, bark, flowers and fruits (Kyerematen et al. 2014; Egyir et al. 2015).

Forest Cover Assessment (1986–2014)

Selection of Landsat Images and Ground Control Points

Satellite data inputs for multi-temporal studies of forest cover were obtained from the Landsat Thematic Mapper (TM), Landsat Enhanced Thematic Mapper Plus (ETM+), and Landsat Operational Land Imager (OLI). The images (Table 14.1) were downloaded from the United States Geological Survey National Center for Earth Resources Observation and Science via the GLOVIS data portal (http://glovis.usgs.gov/). Images with no cloud cover and which were available within the time frame in 1986, 2000 and 2014 were downloaded. All the dates of the selected images were within the dry season when the grassy layers have been scorched thereby increasing the detectability of forests. The Landsat images were geo-referenced to UTM projection WGS 84.

During the fieldwork from September to December, 2013, ground control points and forest canopy density data were collected using GPS and spherical densiometer respectively to classify the 2014 Landsat image. Ground control points for the classification of the 2000 Landsat image was collected with reference to a historic Landcover map prepared for the study area under the GLOWA-Volta Project (Volta Basin Authority Geoportal 2000). This was done by first identifying the features on the Landcover map prepared in 2000 and which could still be verified during fieldwork. This entailed identifying stable landcover in the protected area, along the Afram river, farms and settlements, which had been in existence since 2000.

Table 14.1 Attributes of the Landsat TM, ETM+ and OLI imagery used in the study

Acquisition date	Sensor	Spatial resolution (m)	Path/row
9/02/1986	TM	30	194/55
14/03/2000	ETM+	30	194/55
8/01/2014	OLI	30	194/55

Forest Classification (1986–2014)

Supervised classification procedures using ERDAS Imagine 2011 software were implemented to classify the Landsat images using the Maximum Likelihood Classification algorithm. Areas with tree canopy of 20 % and greater were located on the image and signature were selected and used as training set for classifying "forest areas". Areas with less than 20 % of canopy were classified as "non-forests". This procedure was undertaken with reference to Potapov et al. (2009). In the case of the 1986 map, reference was made to the original topographic map of the catchment published in the management plan of the Kogyae Nature Reserve (Wildlife Department of Ghana 1994) plus local knowledge from the field. Further, qualitative assessments of the classified images were done by examining the classified images visually and relating it to field knowledge. This ensured that the classified map output reflected reality. Protected forest reserve area boundary was obtained from the geodatabase of the Forestry Commission of Ghana. Analysis of forest cover in terms of area for 1986, 2000 and 2014 were carried out in ArcGIS 10.1.

Accuracy Assessment of Forest Classification

Fifty percent of the collected ground control points (test data set) were used for the accuracy assessment of the Landsat map of 2000 and 2014. The classified images were then crossed with the test data to generate confusion matrix. The confusion matrix was used to calculate the different accuracy measures i.e., producer's, user's accuracy, class mapping accuracy for each class and the overall accuracy. Kappa statistics were also calculated as additional information for evaluating the accuracies of the maps. It was not possible to carry out accuracy assessment for the 1986 map because of the lack of a satellite derived historical reference map. It is however, assumed that the accuracy assessments for the Landcover maps of 2000 and 2014 are sufficient to shed light on the overall classification procedures adopted for this study.

Woody Vegetation Inventory

Sampling

Owing to the narrow extent of riparian forest in a landscape, high resolution image, ALOS AVNIR (10 m) of 27 February, 2011 (Fig. 14.1) was utilized for mapping woody vegetation within the river catchment with maximum likelihood classification algorithm (Bagan et al. 2012) at accuracy of 89 % (Fig. 14.1) (confusion matrix not shown) to facilitate inventory with stratified randomized design in Farmland (FA) and protected area (PA). Whether in PA or FA, the rivercourse was divided into three segments, each of length, ranging 6–8 km at a buffer zone of 50 m on each side of the river channel. The inventory for species with diameter at breast height (DBH) ≥ 5 cm was conducted in 60 random rectangular plots (500 m^2 per plot), 30 each in PA and FA and 10 plots per segment. Tree caliper was used to measure the DBH of the species and the height was measured with Vertex IV and Transponder III, Haglof Sweden. Specimens of the species recorded were taken to the herbarium of the Forestry Research Institute of Ghana for confirmation of identification.

Analysis of Woody Species Richness and Diversity

Shannon-Wiener (SWI) (Shannon 1948) and Simpson (SI) (Simpson 1949) indices were calculated as measures of woody species diversity. Further Pielou Equitability index (Natta et al. 2003) was used to assess the evenness of the species distribution.

$$\text{Shannon-Wienerindex (SWI)} = -\sum P_i \ln P_i$$

where $P_i = n_i/N$ with n_i = number of individuals of species i and N = total number of individuals in a plot.

$$\text{Simpson index (SI)} = 1 - \sum P_i^2$$

$$\text{Pielou Equitability index (PEI)} = \text{SWI}/Hm$$

where $Hm = \ln S$ with S = number of species in a plot.

These indices are commonly used for forest and savanna diversity assessment in Ghana and West Africa in general (Boakye et al. 2012; Traoré et al. 2012; Tom-Dery et al. 2013). They were adopted in the study to facilitate comparison of the findings. Species richness (SR) used in this study refers to the number of different species recorded in a plot.

Structure and Size-Class Distribution of Species

For each landuse management regime (PA or FA), the following structural parameters were calculated:

1. Woody species density; the average of the number of individuals per hectare
2. Basal area; the average cross-sectional area of woody species per hectare was calculated from the DBH below:

$$\text{Basal area} = \sum \left(\text{DBH}^2 \pi 4^{-1}\right) \text{ where } \pi = 3.14$$

To establish the size-class distributions, diameters of all species were used to construct histogram with size classes of 5 cm interval. This was similarly done for the heights of species at 5 m interval classes.

Student's *t*-test was used to estimate the significance of the differences between the protected area and farmland after testing for normality using Statistical Package Software for the Social Sciences, Version 17. Results were considered significant at $P < 0.05$.

Results

Landcover Maps and Accuracy Assessment

The quantified forest cover trends mapped between 1986 and 2014 are depicted in Table 14.2 and Fig. 14.2. The trend generally shows increasing deforestation from 1986 to 2014. In 1986, the forest cover was estimated at 50 % of the entire

Table 14.2 Landcover proportions from 1986 to 2014 at the Afram catchment

Landcover	1986	%	2000	%	2014	%
Forest (ha)	143,453	50	105,595	37	87,897	31
Non-forest (ha)	142,497	50	180,355	63	198,053	69
Total	285,950		285,950		285,950	

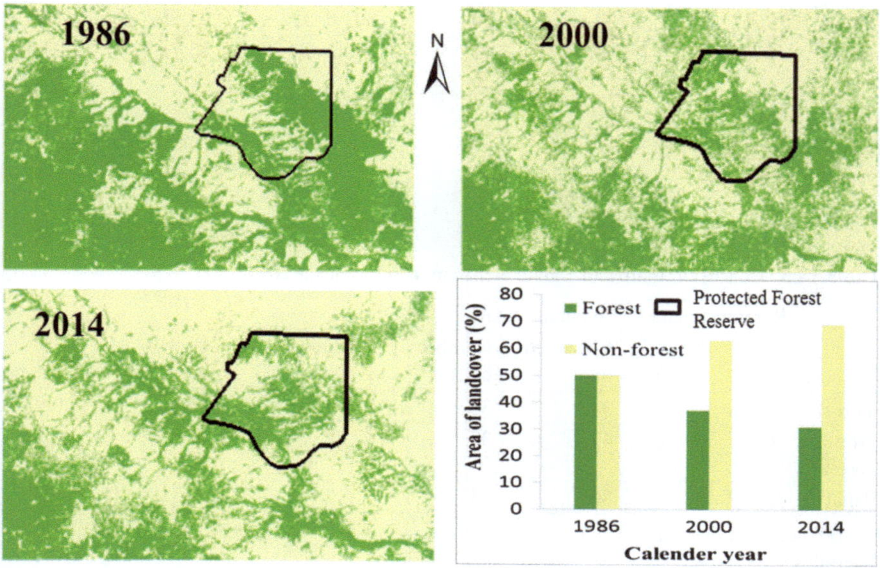

Fig. 14.2 Landcover in the headwaters of Afram river catchment for the studied years (1986–2014)

headwaters studied and was reduced to 37 % by 2000. Currently, the area of forest cover is only 31 % of the headwaters of the Afram river catchment. The "non-forest" which comprises primarily farmland and grassland have been increasing in area coverage since 1986 (50 %) through 2000 (63 %) to 2014 (69 %).

Within the protected area only (PA) (Table 14.3), the quantity of forest has been reducing steadily from 1986 (54 %) through 2000 (42 %) to 2014 (40 %). Analysis of forest cover on farmland (FA, outside the protected area) also shows decreasing forest area from 1986 (50 %) through 2000 (36 %) to 2014 (30 %) (Table 14.3).

The results of the classification accuracy assessment for 2000 and 2014 maps are presented in Tables 14.4 and 14.5 respectively. Overall accuracies of 89 % for the 2000 image and 91 % for the 2014 image. The Kappa coefficient for 2000 and 2014 were 0.77 and 0.83 respectively. The "forest" and "non-forest" classes of both 2000 and 2014 had producer's and user's accuracies over 80 %.

Table 14.3 Forest cover conversions from 1986 to 2014

Landuse	1986	%	2000	%	2014	%
Protected area (ha)	18,137	54	14,128	42	13,437	40
Farmland (ha)	125,316	50	91,468	36	74,461	30

Total area of protected forest reserve = 33,587 ha, Total area of farmland = 250,632 ha

Table 14.4 Confusion matrix of Landcover map using Landsat 2000

	Reference data		Accuracy total				
	Forest	Non-forests	Classified total	Number correct	Producers accuracy (%)	User accuracy (%)	Kappa
Forest	29	2	31	29	80.56	93.55	0.88
Non-forest	7	42	49	42	95.45	85.71	0.68
Total	36	44	80				

Overall accuracy = 89 %
Overall Kappa = 0.77

Table 14.5 Confusion matrix of Landcover map using Landsat 2014

	Reference data		Accuracy total				
	Forest	Non-forests	Classified total	Number correct	Producers accuracy (%)	User accuracy (%)	Kappa
Forest	36	1	37	36	85.71	97.3	0.94
Non-forest	6	37	43	37	97.37	86.05	0.73
Total	42	38	80	73			

Overall accuracy = 91 %
Overall Kappa = 0.83

Woody Species Richness and Diversity

A total of 63 woody species distributed within 24 families were recorded along the Afram river in both PA and FA. Three most species rich families in both protected area (PA) and farmland (FA) were Fabaceae (36 %), Rubiaceae (13 %) and Moraceae (8 %). The total number of specimen recorded was 1232 with 817 in PA and 415 in FA. The number of species decreased from PA (58) to FA (39). With this, 34 species were common to both PA and FA. Twenty-four and five species were found exclusively in PA and FA respectively. Examples of the species found only in the PA were *Albizia glaberrima* (2 %), *Albizia zygia* (1 %), *Alchornea cordifolia* (1 %) etc. The five species recorded exclusively in FA were *Acacia macrostachya* (1 %), *Anthocleista nobilis* (1 %), *Azadirachta indica* (1 %), *Canthium vulgare* (2 %) and *Raphia hookeri* (1 %). Some of the species common to the PA and FA were *Pterocarpus santalinoides* (11 %), *Mitragyna inermis* (11 %), *Cynometra megalophylla* (7 %) etc.

Furthermore, six dominant species in the PA were *Pterocarpus santalinoides* (12 %), *Mitragyna inermis* (11 %), *Cynometra megalophylla* (7 %), *Antiaris toxicaria* (6 %), *Ceiba pentandra* (5 %) and *Sterculia tragacantha* (4 %). Similarly, 6 dominant species on FA were *Mitragyna inermis* (12 %), *Cynometra megalophylla* (9 %), *Pterocarpus santalinoides* (8 %), *Antiaris toxicaria* (6 %), *Diospyros mespiliformis* (5 %) and *Anogeissus leiocarpa* (4 %). The diversity of the riparian species on FA was significantly lower than that in the PA in any of its measures (SWI, SI, PEI, SR) (Table 14.6).

Structure and Size-Class Distribution of Species

There was a significant reduction ($t = 12.4$, $df = 58$, $p < 0.0001$) in the mean density of woody species from the protected area (PA) (545 ± 18) to Farmland

Table 14.6 Diversity of woody species on farmland (n = 30) and protected aea (n = 30)

Diversity	Landuse	Mean	SE	t-value	p-value
SWI	PA	3.80	0.05	7.84	<0.0001*
	FA	3.10	0.08		
SI	PA	0.95	0.003	3.02	0.004*
	FA	0.93	0.007		
SR	PA	15.3	0.49	8.56	<0.0001*
	FA	9.7	0.44		
PEI	PA	0.96	0.003	2.19	0.032*
	FA	0.94	0.01		

SR, SWI, SI and PEI connote Species Richness, Shannon-Wiener, Simpson, and Pielou Equitability Indices respectively. *SE* Standard Error, Degrees of freedom (58)
*$p < 0.05$

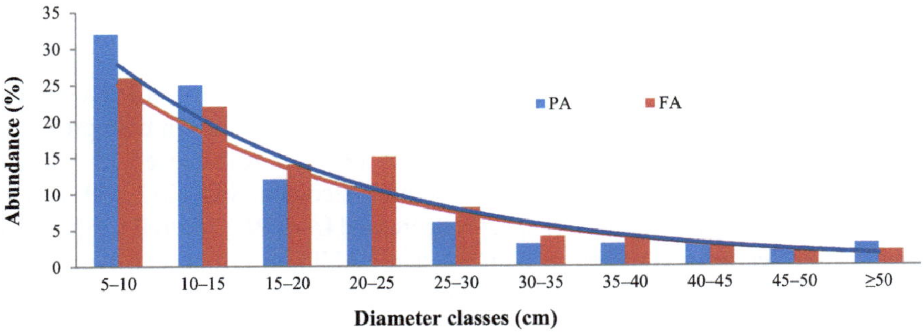

Fig. 14.3 Diameter class distribution of individuals' ≥5 cm DBH in riparian forests in protected area (PA) and farmland (FA)

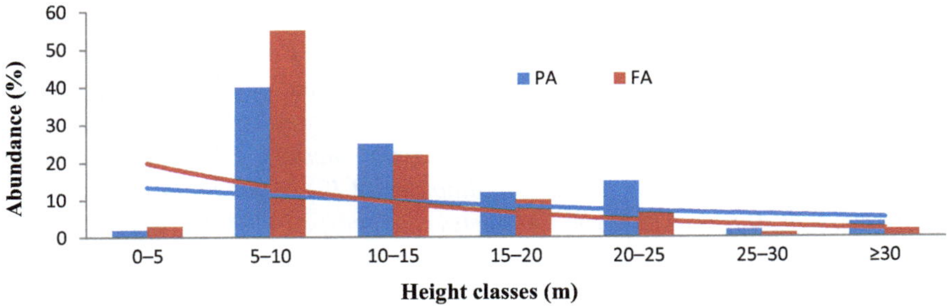

Fig. 14.4 Height class distribution of individuals' ≥5 cm DBH in riparian forests in protected area (PA) and farmland (FA)

(FA) (277 ± 13). The mean basal area of woody species per hectare also showed a significant reduction ($t = 2.603$, $df = 58$, $p = 0.01$) from the PA (19.63 ± 1.74) to FA (14.10 ± 1.22). The diameter class distribution of woody species of RF on FA was similar to that in PA as both showed a reversed J-shaped curve with high number of individuals of diameter less than 15 cm (Fig. 14.3). The height classes' distribution of the species also followed a similar trend as the diameter classes (Fig. 14.4). A high number of individuals had heights between 5 and 10 m for both PA and FA.

Discussion

Landcover Map Accuracies

The confusion matrix of the 2014 classification was an improvement over the 2000 classification product (Tables 14.4 and 14.5). This could be as a result of the use of current validation dataset as observed during fieldwork as opposed to the 2000 classification where reference was made to historic Landcover map and local knowledge. For both 2000 and 2014 classifications, errors were minimized by choosing only two landcover classes (forest/non-forest), with spectrally distinct signatures. The accuracy of the 2000 and 2014 classifications exceed the 85 % overall accuracy threshold used by the United States Geological Survey to determine acceptability (Chai et al. 2009).

Forest Cover Change

Forest cover change assessment is an important starting point for understanding degradation patterns in any ecosystem. The result of the landcover change analysis (section "Landcover Maps and Accuracy Assessment") shows the deforestation of the headwaters of the Afram river in both protected area and on farmlands between 1986 and 2014 (Fig. 14.2 and Table 14.3). This is an indication that the protected area has not been completely successful in preventing habitat destruction within its boundary. The finding is not peculiar to the study area. This is because evidence of deforestation has been reported in protected areas across the tropics (Chai et al. 2009; Traoré et al. 2012). Consequently, concerns have been raised on the effectiveness of protected area management for biodiversity conservation (Chai et al. 2009; Traoré et al. 2012). The effects of deforestation includes changes in elements such as light and wind which influence the microclimatic conditions of the forest remnants to exert a strong effect on biological diversity (Goetze et al. 2006). However, within the tropical environment, deforestation is as a result of humans striving to meet basic needs of food, energy and shelter. Farmers remove woody vegetation and in turn replace them with crops (Kyerematen et al. 2014; Egyir et al. 2015). Also rural households depend on fuelwood as their main source of energy, and the increasing demand by the populace contribute to the reduction in forest area (Kyerematen et al. 2014). Again due to illegal activities of hunters and poachers, wildfire is prevalent within the protected area and farmlands of the study area, which according to Callo-Concha et al. (2012) contributes to tremendous forest loss annually.

Changes in Riparian Woody Species Richness and Diversity

Although not all forest cover conversions have negative effects on biodiversity (Traoré et al. 2012), field inventory in this study have confirmed a reduction in the number of woody species in riparian forests (RF) on the farmland (FA) when compared to that in the protected area (PA). This finding in the FA was unexpected as riparian forest is designated as protected area in all landscapes under the freshwater buffer zone policy of Ghana (Government of Ghana 2011). Controlled human activities in protected areas play important role in reducing disturbance on biodiversity than lands outside protected areas (Okiror et al. 2012). Therefore, the reducing species on farmland could be as a result of the poor enforcement of the policy prescription prohibiting agricultural activities in the buffer zone (Government of Ghana 2011).

The result (Table 14.6) further showed that the riparian forest in PA is significantly diverse as oppose to the FA. This is also reported in other studies (Ceperley et al. 2010; Okiror et al. 2012) and on that basis, the first null hypothesis is accepted. In contrast to this research, it has been found in other studies that the species diversity value on agricultural watershed in the tropics is enhanced as a result of deliberate preservation of trees by farmers (Boakye et al. 2012; Traoré et al. 2012; Gray et al. 2014). In spite of the reducing farmland diversity in this study, the Shannon-Wiener diversity values of both FA and PA are within the range (2.4–5.4) reported in other savannas of West Africa (Natta and Porembski 2003; Natta et al. 2003) indicating that the RF in the study area (PA and FA) still has the potential to conserve high species diversity (Okiror et al. 2012). This could be as a result of the intensity and frequency of floods, small-scale variation in topography, soils and canopy structure of the riparian area that create a diversity of habitats for a wide variety of species to co-exist (Sambare et al. 2011).

The fact that the RF on farmland is less diverse reduces its resilience to disturbance (Scherer-Lorenzen et al. 2005). Studies have shown that less diverse ecosystems are prone to climatically induced catastrophes such as diseases and alien species invasions (Scherer-Lorenzen et al. 2005). Under this situation, the ability of the forests to provide ecosystem services and functions needed for sustainable food production is hampered (Eilu et al. 2003; Manning et al. 2006; Okiror et al. 2012). Loss of floral diversity results in poor habitats conditions which are detrimental to the survival of climate sensitive faunal species such as birds and insects that are essential for crop pollination, seed dispersal and nutrient fixation on farmlands (Sambare et al. 2011).

Structure and Size-Class Distribution of Species

The "J" shaped curve distribution (Figs. 14.3 and 14.4) of diameters of riparian woody species on farmland (FA) mimic that in the protected area (PA) as stated in the second null hypothesis. This means that the woody species on FA has the potential to regenerate naturally and face no danger of extinction (Sambare et al. 2011). According to Lykke (1998), for a population to maintain itself, it needs to have abundant juveniles which will recruit into adult size classes. This type of species distribution gives the ecosystem a stable population structure for the sustenance of the ecological succession of riparian forests (Sambare et al. 2011). Under this condition, the forest cover is perpetually maintained to regulate periodic events such as floods and drought that have damaging effects on crop production and greatly increases concerns over food supply (Okiror et al. 2012).

The decline in the density and basal area of the woody species (section "Structure and Size-Class Distribution of Species") on FA, may have been caused partly by the repeated manual weeding of the riparian area for crop cultivation (Ceperley et al. 2010). Also the use of agro-chemicals for weeding may have affected the regenerative capacity since some of these chemicals kill the seeds that are dispersed (Fischer et al. 2009; Ceperley et al. 2010). The fact that the riparian woody density is lower in FA is likely to result in a much drier forests due to the increase in the surface exposure of the riparian area for soil moisture loss. This increases the vulnerability and frequency of the riparian forests to wildfires (Sambare et al. 2011; Azihou et al. 2013). Such fires can break the resilience of the riparian ecosystem to intensify climate change impacts to such a degree that species physiological tolerances can be exceeded and the rates of biophysical forest processes altered (Scherer-Lorenzen et al. 2005).

Conclusion and Recommendation

Riparian forests in the headwaters of the Afram river in the humid savanna of Ghana contribute to the conservation of high woody species composition. That notwithstanding, reduction of forests cover from 1986 to 2014 coupled with the decline in the woody species composition in the riparian forests from protected area to farmland may be increasing the risk of the riparian area in the farmlands to climate change impacts such as fires and flooding. On that basis, the first null hypothesis of this study is accepted due to the high riparian diversity value in the protected area as oppose to that on the farmland. The research finding also supports the second null hypothesis partly in that the diameter distribution of the riparian woody species on farmland mimics that in the protected area. The basal area and density were lower for the riparian woody species on farmland. To ensure the continuous flow of benefits from the riparian forests for sustainable food production and farmlands' resilience to climate change, it is important that riparian biodiversity is sustainably managed. This may be achieved by enforcing the freshwater buffer zone policy of Ghana. There should also be the conscious effort to educate farmers to retain riparian woody species or replanted to enhance the composition. The study further recommends that the buffer zone is excluded from farming for the full recovery of the riparian forests.

Acknowledgements The authors wish to express their sincere gratitude to the West African Climate Change and Adapted Landuse programme of the German Federal Ministry for Education and Research for providing financial support to the corresponding author to carry-out this research as part of his postgraduate studies. Thanks to all colleagues for their insightful review of the manuscript.

References

Azihou FK, Glele KR, Ronald B, Brice S (2013) Distribution of tree species along a gallery forest-savanna gradient: patterns, overlaps and ecological thresholds. J Trop Ecol 29(1):25–37

Bagan H, Kinoshita T, Yamagata Y (2012) Combination of AVNIR-2, PALSAR, and polarimetric parameters for land cover classification. IEEE Trans Geosci Remote Sens 50(4):1318–1328

Boakye EA, Gils HV, Osei EM Jr, Asare VNA (2012) Does forest restoration using taungya foster tree species diversity? A case of the Afram Headwaters Forest Reserve in Ghana. Afr J Ecol 50 (3):319–325

Callo-Concha D, Gaiser T, Ewert F (2012) Farming and cropping systems in the West African Sudanian savanna. WASCAL research area: Northern Ghana, Southwest Burkina Faso and Northern Benin. ZEF working paper 100, Bonn

Ceperley N, Montagnini F, Natta A (2010) Significance of sacred sites for riparian forest conservation in Central Benin. Bois Et Forets Des Tropiques 303(1):5–23

Chai SL, Tanner E, McLaren K (2009) High rates of forest clearance and fragmentation pre- and post-National Park establishment: the case of a Jamaican montane rainforest. Biol Conserv 142:2484–2492

Egyir IS, Ofori K, Antwi G, Ntiamoa-Baidu Y (2015) Adaptive capacity and coping strategies in the face of climate change: a comparative study of communities around two protected areas in the coastal savanna and transitional zones of Ghana. J Sust Dev 8:1

Eilu G, Obua J, Tumuhairwe JK, Nkwine C (2003) Traditional farming and plant species diversity in agricultural landscapes of south-western Uganda. Agric Ecosyst Environ 99(1–3):125–134

Fischer J, Stott J, Zerger A, Warren G, Sherren K, Forrester RI (2009) Reversing a tree regeneration crisis in an endangered ecoregion. Proc Natl Acad Sci USA 106(25):10386–10391

Goetze DB, Hörsch B, Porembski S (2006) Dynamics of forest-savanna mosaics in north-eastern Ivory Coast from 1954 to 2002. J Biogeogr 33(4):653–664

Government of Ghana (2011) Ghana riparian buffer zone policy for managing freshwater bodies. Ministry of Water Resources, Works and Housing

Gray CL, Slade EM, Mann DJ, Lewis OT (2014) Do riparian reserves support dung beetle biodiversity and ecosystem services in oil palm-dominated tropical landscapes? Ecol Evol 4 (7):1049–1060

Kyerematen R, Owusu EH, Acquah-Lamptey D, Anderson RS, Ntiamoa-Baidu Y (2014) Species composition and diversity of insects of the Kogyae Strict Nature Reserve in Ghana. Open J Ecol 4:1061–1079

Lykke AM (1998) Assessment of species composition change in savanna vegetation by means of woody plants' size class distributions and local information. Biodivers Conserv 7 (10):1261–1275

Manning AD, Fischer J, Lindenmayer DB (2006) Scattered trees are keystone structures—implications for conservation. Biol Conserv 132(3):311–321

McCracken DI, Cole LJ, Harrison W, Robertson D (2012) Improving the farmland biodiversity value of riparian buffer strips: conflicts and compromises. J Environ Qual 41(2):355–363

Natta AK, Porembski S (2003) Ouémé and Comoé: forest-savanna border relationships in two riparian ecosystems in West Africa. Botanische Jahrbücher für Systematik 124(4):383–396

Natta AK, Sinsin B, van der Maesen LJG (2003) Riparian forests and biodiversity conservation in Benin (West Africa). Paper presented at the XII World Forestry Congress, organized by FAO in Quebec, Canada, 21–28 Sept 2003, pp 126–127

Okiror P, Chono J, Nyamukuru A, Lwanga JS, Sasira P, Diogo P (2012) Variation in woody species abundance and distribution in and around Kibale National Park, Uganda. International Scholarly Research Network Forestry 490461

Potapov P, Laestadius L, Yaroshenko A, Turubanova S (2009). Measuring and assessing forest degradation. Global mapping and monitoring the extent of forest alteration: the intact forest landscapes method: Food and Agriculture Organization of the United Nations (FAO). Forest resource working paper 166

Sambare O, Bognounou F, Wittig R, Thiombiano A (2011) Woody species composition, diversity and structure of riparian forests of four watercourses types in Burkina Faso. J For Res 22 (2):145–158
Scherer-Lorenzen M, Korner C, Schulze ED (eds) (2005) Forest diversity and function: temperate and boreal systems. Springer, Berlin
Shannon CE (1948) A mathematical theory of communication, part 1. Bell Syst Tech J 27 (3):379–423
Simpson EH (1949) Measurement of diversity. Nature 163:688
Tom-Dery D, Hinneh P, Asante WJ (2013) Biodiversity in Kenikeni forest reserve of Northern Ghana. Afr J Agric Res 8(46):5896–5904
Traoré L, Ouédraogo A, Thiombiano A (2012) To what extent do protected areas determine the conservation of native flora? A case study in the Sudanian zone of Burkina Faso. Int Sch Res Not Bot 168(196):10
Volta Basin Authority Geoportal (2000) Land cover in the Volta Basin area. http://131.220.109.2/geonetwork/srv/en/main.home. Accessed 21 Nov 2014
Wildlife Department of Ghana (1994) Kogyae Strict Nature Reserve development and management plan. http://gem.tropicalforests.ox.ac.uk/sites/kogyae. Accessed 5 Oct 2014

Chapter 15
A Stochastic Weather Generator Model for Hydroclimatic Prevision in Urban Floods Risk Assessment in Abidjan District (Cote d'Ivoire)

Jean Homian Danumah, Samuel Nii Odai, Mahaman Bachir Saley, Joerg Szarzynski, Kwaku Adjei, and Fernand Koffi Kouame

Abstract Flood risk occurrence is very often related to heavy precipitation; and available future weather data is a potential source for long term flood risk prediction. The aim of this paper was to determine and analyze trends in rainfall, temperature and PET under present and future climatic conditions using Long Ashton Research Science-Weather Generator (LARS-WG) software, in prediction of flood risk occurrence in Abidjan. This work was based on the integration of Hydro climatic daily data within LARS-WG software. The processing steps are: (1) calibrating and validating the model using 50 years measured data, (2) generating baseline data for 50 years, (3) processing future scenario data based on baseline already set using HADCM3 and (4) Comparing baseline and generated scenario data. The resulting statistics show that temperature will increase by 0.32, 1.36 and 2.54 °C for the periods 2011–2030, 2046–2065 and 2080–2099 respectively. Then rainfall in the same period will increase by 4 %, 6 % and 10 % respectively. The mean and high flooding risk will then increase in long term within this urban area. Thus this future large extension of flooding occurrence imposes to take future weather scenario into account in prediction and management of flooding risk in Abidjan District.

Keywords Climate change • Flood risk • LARS-WG • Abidjan • Cote d'Ivoire

J.H. Danumah (✉) • S.N. Odai • J. Szarzynski • K. Adjei
WASCAL PhD Programme (West African Science Service Center on Climate Change and Adapted Land Use), GRP on Climate Change and Land Use, Department of Civil Engineering, Kwame Nkrumah University of Science and Technology, Kumasi, Ghana
e-mail: danumahjean@yahoo.fr

M.B. Saley • F.K. Kouame
Centre Universitaire de Recherche et d'Application en Télédétection (CURAT), Laboratoire Associé Francophone (LAF n° 401), UFR des Sciences de la Terre et des Ressources Minières, Université de Cocody, 22 BP 801 Abidjan 22, Cote d'Ivoire

W. Leal Filho (ed.), *Innovation in Climate Change Adaptation*, Climate Change Management, DOI 10.1007/978-3-319-25814-0_15

Introduction

Climate change is considered to be the biggest challenge facing by the mankind in the twenty-first century (Osman et al. 2014). Climate change is a global issue and needs to be examined at global, regional and local scale. Population growth too with its demand and development has become a world issue on this century (Txomin et al. 2013). Population demand, green gas emission from industrialization and land use lead to severe climate variability (Yan et al. 2014) and change such as increases of extreme events (flood, drought, and tsunami) occurrence across the world and more specifically in West Africa.

More recently in Cote d'Ivoire, we have registered increasingly important phenomena of floods, with its effects such as death, properties damage and population exodus. Extreme rainfall is the main natural disaster which causes loss of many lives; destroy infrastructures, and displacement of people during the rainy season in Abidjan. Statistical analysis in 2013 shows that 26 % of the district of Abidjan is flood risk area and 21, 13, and 15 people died in 2009, 2010, and 2011 respectively due to floods. Besides, with a total of 80,000 people living in areas subject to risk of flooding in the district with 40,000 people in risk areas in Cocody, 12,500 people in Abobo, 10,000 in Adjame, 9500 in Yopougon and 8000 in Attecoube communes. However, the use of future scenario of weather data for understanding present condition and managing future hydrologic events in Cote d'Ivoire is practically non-existent. Weather data analysis for many years were based on determining break on the times series using some statically methods such as Pettit and Buishand test, application of Nicholson indices to bring out the wet and dry period in case on rainfall variability and shows general trend, inter-annual behavior (Brou 1997, 2005; Savané et al. 2001; Goula et al. 2006; Kouassi et al. 2008). Also assessing flood risk based on rainfall variability using remote sensing and GIS techniques in West, Southwest and South of Cote d'Ivoire (Savane et al. 2003; Saley et al. 2005, 2013).

Flood risk occurrence is very often related to heavy precipitation, which means that there is a need of knowledge about hydroclimatic variability for most reliable flood management and mitigation; then available future weather data becomes a potential source for short term and long term extreme rainfall prediction. For all that, general circulation models (GCM) has become an important tool to generate future scenarios and helpful to predict extreme rainfall under climate change. However, the output from Global Climate Models (GCMs) cannot be used directly at a site because of their very coarse spatial resolution (Semenov et al. 1998). Thus, weather generator can serve as a good computationally tool to produce at local scale (Abidjan site) specific climate change scenarios at the daily time-step with better resolution and fit reality.

Long Ashton Research Station-Weather Generator (LARS-WG) model which is a stochastic weather generator specially designed for climate change impact studies is one of the important general circulation models for future climate change. LARS-

WG is a numerical model which produces synthetic daily time series of a suite of climate variables, such as precipitation and temperature, with certain statistical properties (Semenov et al. 1998). LARS-WG has been tested for different sites worldwide and shown its ability to model rainfall extremes with reasonable skill (Semenov 2007). LARS-WG too has been used in various studies, including assessment of the impact of climate change (Goodarzi et al. 2014; Molanejad et al. 2014; Osman et al. 2014; Noori et al. 2013; Semenov 2007; Semenov 2014; Semenov and Brooks 1999; Trnka et al. 2014; Semenov et al. 1998; Rao 2014). According to Sobhany and Fateminiya (2014), high accuracy of climate data modelling in different climatic stations has been confirmed by many researchers using LARS-WG.

The significant lack of research gap identified by this study is that recent scientific work undertaken in the district of Abidjan concentrated on rainfall variability during past and current condition as flood risk drivers within two communes of Abidjan: Attecoube and Abobo (Savane et al. 2003 and Hauhouot 2008). This is a piece-meal approach and does not provide a solution to the problem of flood occurrence within the district. Other studies (Kouamé et al. 2013; Jourda et al. 2006; Ahoussi et al. 2013) in the district, not directly focus on extreme rainfall pointed out the inefficiency of the drainage network and impervious area which are part of the main drivers of floods. However, these studies are fragmented and did not consider the entire district and future scenario model to link climate change and natural disaster. Then, no studies have yet been undertaken to understand and analyze trends in weather data under present and future climatic conditions implications in urban flood risks at Abidjan.

The aim of this paper was to determine and analyze trends in hydroclimatic data (rainfall, temperature, solar radiation and PET) under present and future climatic conditions using Long Ashton Research Science-Weather Generator (LARS-WG) model, in prediction of flood risk occurrence in Abidjan. For this purpose, assessment process of observation and simulation data are conducted following three steps, calibration, validation and generating scenario using downscaling by SRA1B scenarios, HAdCM3 model of atmosphere general circulation models (GCM) in LARS-WG model.

Methodology

Study Area

The District of Abidjan is located in the south of Cote d'Ivoire between latitudes 5° 10 and 5° 38 North and longitudes 3° 4 and 5° 21 West (Fig. 15.1). It consists of 13 municipalities since 2001, 10 municipalities in Abidjan and 3 others communes namely Bingerville, Songon and Anyama and covers an area of approximately 2119 km^2. According to the statistic institution of the country records in 2013,

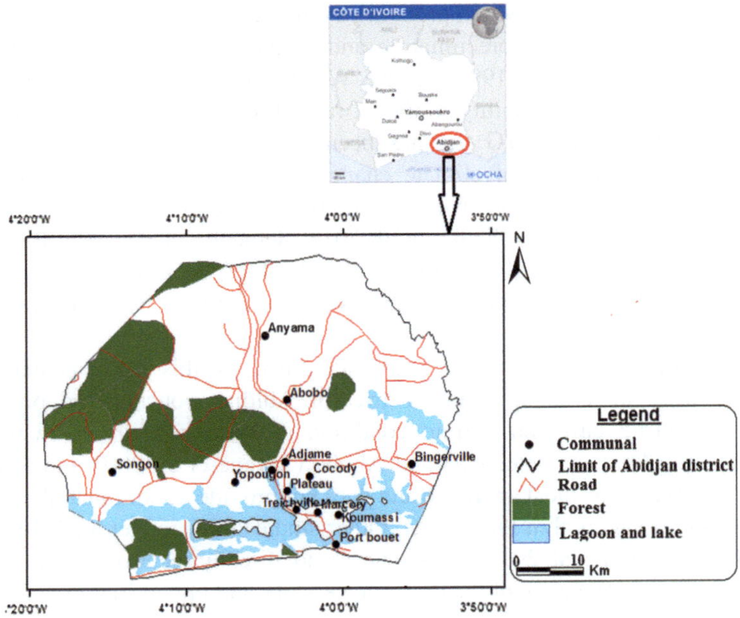

Fig. 15.1 Location of the study area (*Source*: Author)

Abidjan district has an estimated population of 4,739,752 inhabitants in the metropolis, and 4,460,355 inhabitants in the city of about 5 million people or 20.3 % of the national population as at 2013. In addition, this population has a perpetual growth characterized by high industrialization and urbanization but also due to the political crisis through the country. It is limited (Fig. 15.1):

- To the south by the Atlantic Ocean
- In the southwest by the Department of Dabou
- To the west by the Department of Grand Lahou
- To the north by Agboville Department
- South-east by the Department of Grand-Bassam
- To the east by the Department of Alepe

Climatic Variability

The study area has an equatorial climate transition (Climate Attieen), characterized by four seasons: two dry seasons and two rainy seasons within the annual cycle:

- Long dry season from December to April
- Long rainy season from May to July

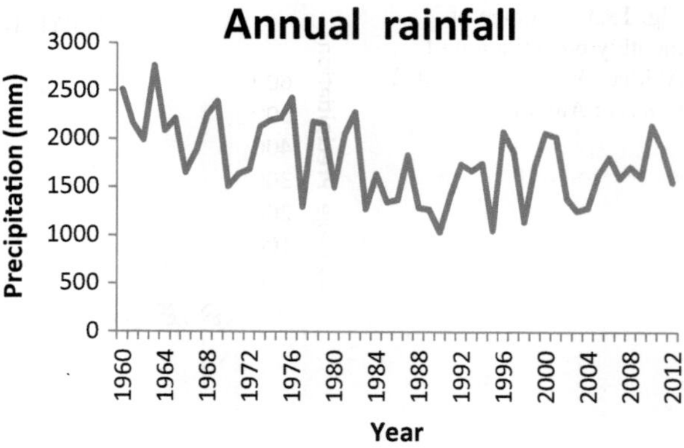

Fig. 15.2 Total annual rainfall variability of Abidjan from 1961 to 2012 (*Source*: Author)

- Short dry season from July to September
- Small rainy season from October to November

Abidjan district has three rainfall stations located at Bingerville, Adiopodoume and Abidjan airport. These stations can be measured on a daily and monthly basis the different climates. So we were able to collect temperature, rainfall and solar radiation data to Abidjan district station.

Precipitation

For the characterization of rainfall fluctuations that have affected the district of Abidjan, we conducted a study of spatio-temporal variation of monthly rainfall and annual rainfall totals (Fig. 15.2).

The high annual rainfall recorded in the district of Abidjan during the period 1960–2012 ranged from 2800 mm in 1963 to 1020 mm in 1990 with an average of 1910 mm. Generally in the 1960s, the annual rainfall ranged between 2000 and 3000 mm. After 1987, there was a drop of this rainfall has oscillated between 1500 and 2200 mm which is a reduction of more than 500 mm of rain compared to the 1960s.

Also, the observation of highest average monthly rainfall shows that June and sometime July are the rainiest month of the District of Abidjan (Fig. 15.3).

Temperature

The annual average temperature of Abidjan is around 26 °C. The temperatures in the district are higher from February to May and November. These temperatures are experiencing a slight decline during the months of June to October.

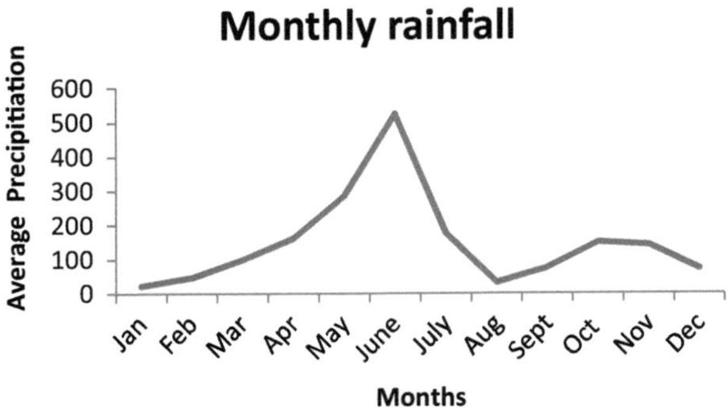

Fig. 15.3 Average of monthly precipitation of Abidjan from 1961 to 2012 (*Source*: Author)

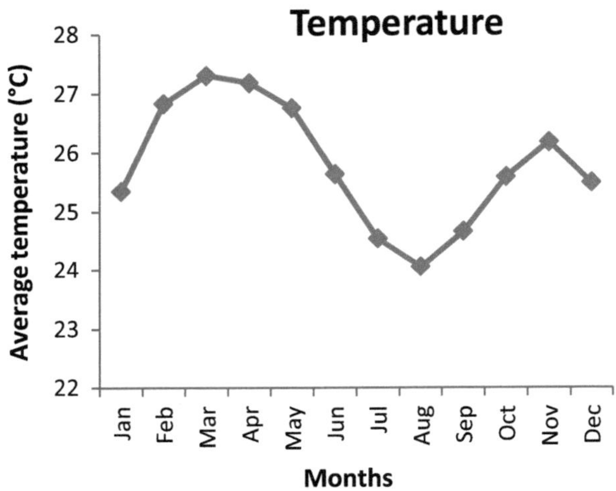

Fig. 15.4 Monthly temperature variability from 1961 to 2012 at Abidjan (*Source*: Author)

The curve show in Fig. 15.4 shows that the average monthly temperatures for the period of 1960–2012 fluctuated between 24 and 28 °C. The months March and April are the hottest months with a monthly average temperature above 27 °C. The months of July to September (rainy season) are the least hot, with a maximum temperature equal to 24.5 °C (Fig. 15.4).

After accomplishment of ombrothermic diagrams which shows comparison between temperature and rainfall charts of Abidjan (Fig. 15.5) over the period 1961–2012, we note that the temperatures are low during the months of heavy rainfall and strong during the months of low rainfall.

Analysis of the Annual Rainfall and the Heavy Rainfall Month

Graphical comparison applies between annual rainfall and heavy rainfall month of the Abidjan District to analyze the trends patterns from 1960 to 2012 bring out that

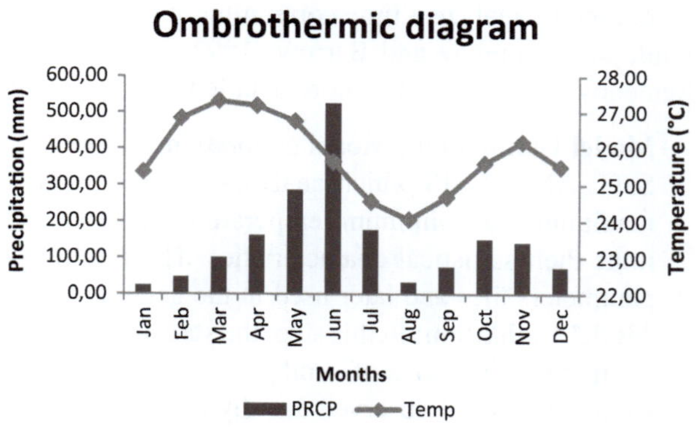

Fig. 15.5 Ombrothermic diagram of Abidjan from 1981 to 2012 (*Source*: Author)

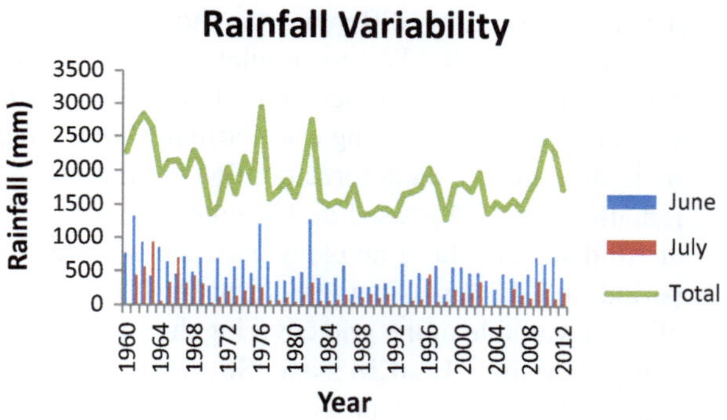

Fig. 15.6 Annual rainfall and heavy rain months of Abidjan from 1960 to 2012 (*Source*: Author)

in general the annual rainfall variability is based on the amount of rainfall in the month June because, it is the most important heavy rainfall month compare to July in the series (Fig. 15.6).

Data and LARS-WG Model Processing

In this study 50 years of observed daily weather data (precipitation, maximum and minimum temperature and solar radiation) have been used with LARS-WG (a stochastic weather generator model) and considered as baseline for the different process. Precipitation is considered as the primary variable in our case for analyzing flood event occurrence in the future and the other three variables come as secondary variables to obtain an overall understanding of future hydroclimatic events.

In LARS-WG, the process of generating synthetic weather data can be divided into three distinct steps: calibration or site analysis, validation or Q-test and Generator or generation of future data based on normal distribution. The seasonal cycle of means and standard deviations was removed from the observed performance, and the residuals approximated by a normal distribution. These residuals were used to analyze a time correlation of each variable. Refer to software manual guide and Semenov and Barrow (1997, 2002), Semenov and Pierre (2010), and Semenov et al. (1998) for more detailed description of the modelling procedure.

- **Model Calibration**: Model calibration is done in LARS-WG by implementing SITE ANALYSIS which analyses observed daily weather data (precipitation, maximum and minimum temperature and solar radiation) of Abidjan to determine their statistical characteristics. The resulting information is stored in two parameters files and later used in the generation process.
- **Model Validation**: In this step, the statistical characteristics of observed data are compared with those of synthetic data generated using the parameters derived from the original observed data by applying the Q-test. Several statistical tests: chi-squared test, Student's T-test and F-test are used in order to determine whether the distributions mean values and standard deviations of the synthetic data are significantly different from those of the original observed data set. Then, the ability of LARS-WG to simulate the data at the chosen site is assessed.
- **Generation of Climate Scenario**: LARS-WG model generates synthetic future weather data by combining a scenario file containing information about changes in hydroclimatic data (precipitation, minimum/maximum temperature, solar radiation and PET) based on baseline weather data (50 years of observed data) at Abidjan site. Baseline parameters are adjusted by the changes for the future period and the emissions predicted by IPCC Global Climate Model for each climatic variable of the grid covering the site. In our study, we used IPCC Global Climate Model: HadCM3 and SRA1B scenario to generate 50 future years predicted weather data for the periods of 2011–2030, 2046–2065 and 2065–2099.

Results

The Table 15.1 shows how the minimum temperature, the maximum temperature, rainfall, solar radiation and the PET will change from the baseline (1961–2010) to three future scenarios 2011–2030 (scenario 1), 2046–2065 (scenario 2) and 2065–2099 (scenario 3). We also observe in Fig. 15.7, the trend variability and

Table 15.1 Minimum/maximum temperature, precipitation, solar radiation and evapotranspiration change prevision at Abidjan

	T_{min}	T_{max}	Rain	RAD	PET
Baseline (1961–2010)	24.18	27.40	3.62	16.91	4.44
Scenario 2011–2030	24.50	27.72	3.75	16.82	4.43
Change	**0.32**	**0.32**	**0.04**	**−0.09**	**−0.01**
Scenario 2046–2065	25.54	28.77	3.82	16.81	4.49
Change	**1.36**	**1.37**	**0.06**	**−0.10**	**0.05**
Scenario 2080–2099	26.72	29.95	3.97	16.76	4.54
Change	**2.54**	**2.55**	**0.10**	**−0.15**	**0.10**

Fig. 15.7 Observed and future weather data variability from 1960 to horizon 2100. (**a**) minimum temperature, (**b**) maximum temperature, (**c**) Rainfall, (**d**) Solar radiation, (**e**) PET (*Source*: Author)

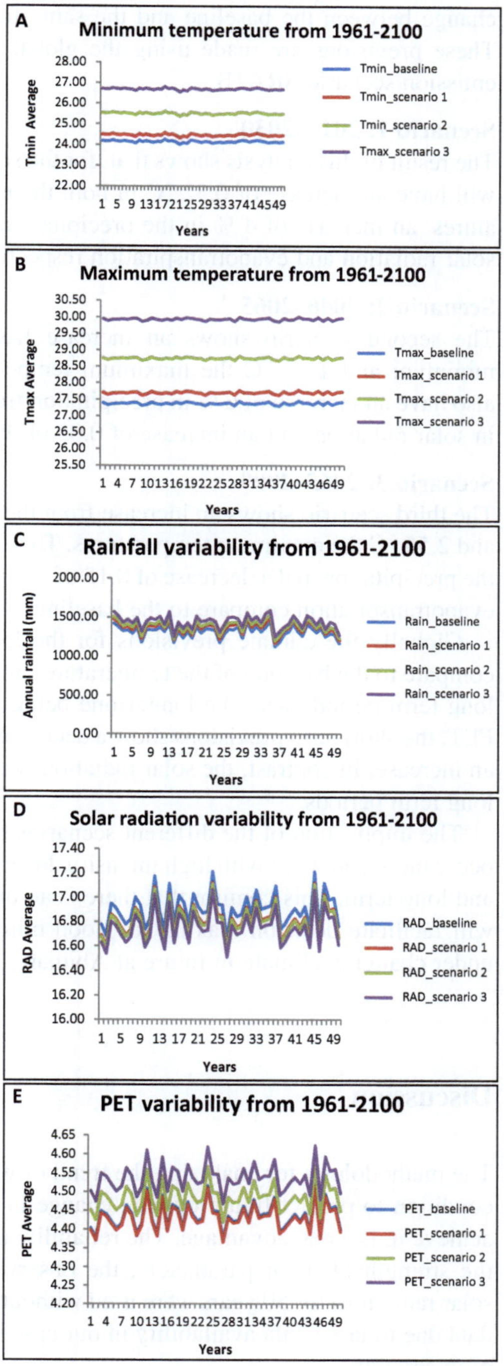

change between the baseline and the same future scenario of each weather data. These previsions are made using the global model HadCM3 and the optimistic emission scenario SRA1B.

Scenario 1: 2011–2030
The result of this analysis shows that, for the scenario 1 compare to the baseline, we will have an increase of 0.32 °C in both the minimum and the maximum temperatures, an increase of 4 % in the precipitation and a decrease of 0.09 and 0.01 in solar radiation and evapotranspiration respectively in the Abidjan district.

Scenario 2: 2046–2065
The second scenario shows an increase from the baseline of 1.36 °C of the minimum and 1.37 °C the maximum temperatures. For this same scenario, we also have an increase of 6 % in precipitation from the baseline and a decrease of 0.1 in solar radiation and an increase of 0.05 of the evapotranspiration.

Scenario 3: 2065–2099
The third scenario shows an increase from the baseline of 2.54 °C of the minimum and 2.55 °C the maximum temperatures. Then, we also have an increase of 10 % of the precipitation and a decrease of 0.15 of solar radiation and an increase 0.10 of the evapotranspiration compare to the baseline.

Globally the climate previsions for the locality of Abidjan show an increase compare to the baseline of the temperatures and the precipitation for short term and long term periods with the longer one being higher than the shorter one. For the PET, the short term previsions show a decrease when the long term prevision shows an increase. In contrast, the solar radiation shows a decrease for the short term and long term periods.

The implication of the different scenarios being modelled is that flood risk will occur more and more with high intensity due to the increase of precipitation in short and long term. This confirm that there is an urgency to put in place a strategy that will facilitate decision making for flood mitigation and better land use planning under changing climate in future at Abidjan.

Discussion

The methodology for analyzing the trend of weather data under present and future condition to predict future climate change in Abidjan using LARS-WG model to achieve it, is a real advantage. The reliability of these future scenarios are linked to the strength of input parameters, the observed daily data (temperature, rainfall, solar radiation) for 50 years were used without the need to generate data as baseline data due to large data availability in our case which reduced the reduced the range of uncertainty.

The HadCM3 model used in this study is from GCMs which is already part of LARS-WG model and subjectivity is also noted in the choice and range of data as input. This subjectivity is reduced by checking the consistency of observed data as baseline by doing calibration and validation of the data before generating the future scenarios which increases our confidence in the models' predictions and reduces the level of uncertainty.

However, GCMs and impact models are undergoing continuous development and improvement because some research have showed some uncertainty and limitations by applying Global Climate model at local scale due methodically model calibration. Thus, these results can be improved by integrating a comparison with others model which are or not statistic model, those from the hydrological modeling for the establishment of extreme events such as flood.

Conclusion

In this research, LARS-WG model has been used to generate specific climate change scenarios (weather data) for prediction of weather events at Abidjan site. HAdCM3 Model and the optimistic emission scenario SRA1B have been used. Integration of weather daily data: precipitation, minimum/maximum temperatures and solar radiation from 1961 to 2010 into the LARS-WG model as baseline data make robust the result. The HadCM3 model was then applied to 50 years observed data to generate 50 years of future daily weather data after calibration and validation for Abidjan site. The Results has shown that for the next 90 years, there will be an increase in short term and long term of minimum/maximum temperature and precipitation and inversely a decrease of solar radiation in Abidjan. In short, the LARS-WG model was proposed for realistic simulation of hydroclimatic data (precipitation, temperature, solar radiation and PET). And it was highly capable to generate future scenario and give better understanding and view concerning weather data variability in future more specifically rainfall in case on flood risk event. However, this study can still be improved for more details by coupling statistical and hydrological model. Finally effective preparedness strategies have to be developed to manage future extreme rainfall or floods occurrence within Abidjan district.

Acknowledgement The authors wish to thank the author of LARS-WG software for the free license received from him to do this work, and the assessors that the relevant comments and criticisms they made which helped improve the quality of this document.

References

Ahoussi KE, Loko S, Koffi YB, Soro G, Oga YM, Soro N (2013) Evolution spatio-Temporelle Des Teneurs en Nitrates Des Eaux Souterraines De La Ville d'Abidjan (CÔTE D'IVOIRE). Int J Pure Appl Biosci 1(3):45–60

Brou YT (1997) Analyse et dynamique de la pluviométrie dans le Sud forestier ivoirien: recherche de corrélations entre les variables climatiques et les variables liées aux activités anthropiques. Thèse 3ème cycle, Univ. Cocody, Côte d'Ivoire, 211 p

Brou YT (2005) Climat, mutations socio—économiques et paysages en Côte d'Ivoire. Rapport d'activités scientifique, pédagogique, administrative et publications. Memoire HDR, 106 p

Goodarzi E, Bavani MA, Dastorani MT, Talebi A (2014) Evaluating effect of downscaling methods; change-factor and LARS-WG on surface runoff (A case study of Azam-Harat River basin, Iran). Desert 19(2):99–109

Goula BTA, Kouassi VJ, Savané I (2006) Impacts du changement climatique sur les ressources en eau en zone tropicale humide: cas du bassin versant du Bandama en Côte d'Ivoire. Agron Afr 18(1):1–11

Hauhouot C (2008) Analysis of the rainfall risks in the precarious districts of Abidjan: Survey of cases in Attécoube. Geo-Eco-Trop 32:75–82

Jourda JP, Kouamé KJ, Saley MB, Kouamé KF, Kouadio BH, Kouamé K (2006) A new cartographic approach to determine the groundwater vulnerability of the Abidjan aquifer. In: Xu Y, Usher B (eds) Groundwater pollution in Africa. Taylor & Francis, London, pp 103–114

Kouamé KJ, Jourda JP, Saley MB, Deh SK, Anani AT, Biémi J (2013) Mapping of groundwater vulnerability zones to pollution in various hydrogeological environments of Côte d'ivoire by drastic method. Int J Sci Eng Res 4(5):9

Kouassi AM, Kouamé KF, Goula BTA, Lasm T, Paturel JE, Biémi J (2008) Influence de la variabilité climatique et de la modification de l'occupation du sol sur la relation pluie-débit a partir d'une modélisation globale du bassin versant du N'zi (Bandama) en Côte d'Ivoire. Rev Ivoir Sci Technol 11:207–229

Molanejad M, Soltani M, Saadat Abadi Ranjbar A (2014) Simulation of extreme temperature and precipitation events using LARS-WG stochastic weather generator. Int J Sch Res Gate 2 (3):121–129

Noori M, Sharifi MB, Zarghami M, Heydari M (2013) Utilization of LARS model for modelling of meteorological parameters in Golestan, province of Iran. J River Eng 1(1):5

Osman Y, Al-Ansari N, Abdellatif M, Aljawad SB, Knutsson S (2014) Expected future precipitation in central Iraq using LARS-WG stochastic weather generator. Engineering 6:948–959. doi:10.4236/eng.2014.613086

Rao CN (2014) Climate change analysis in southern Telangana region, Andhra Pradesh using LARS-WG model. Curr Sci 107(1):54

Saley MB, Kouamé KF, Penven MJ, Biémi J, Kouadio BH (2005) Cartographie des zones à risque d'inondation dans la région semi-montagneuse à l'Ouest de la Côte d'Ivoire: Apports des MNA et de l'imagerie satellitaire. Télédétection 5(1–2–3):53–67

Saley MB, Danumah JH, Sorokoby VM, Kanohin FO, Atcheremi D, Denean SP, Kouame FK, Djagoua VE (2013) Méthodologie pour la cartographie des zones vulnerables a l'inondation par systeme d'information géographique et analyse multicritère: application au bassin versant de la Davo a Sassandra (sud-ouest de la Cote d'Ivoire). Africa Geosci Rev 20(1):21–33

Savané I, Coulibaly KM, Gioan P (2001) Variabilité climatique et ressources en eaux souterraines dans la région semi-montagneuse de Man. Sécheresse 12(4):231–237

Savane I, Coulibaly N, Soro N (2003) Analyse multicritère spatiale pour l'identification de la vulnérabilité aux risques naturels d'Abobo (banlieue de la ville d'Abidjan, Côte d'Ivoire). Rev Ivoir Sci Technol 4:94–107

Semenov M (2014) Delivering CMIP5-based climate scenarios for impact assessments in Europe. In: EGU general assembly conference abstracts, 1

Semenov MA (2007) Simulation of extreme weather events by a stochastic weather generator. Clim Res 35(3):203

Semenov MA, Barrow EM (1997) Use of a stochastic weather generator in the development of climate change scenarios. Clim Change 35:397–414

Semenov MA, Barrow EM (2002) LARS-WG: a stochastic weather generator for use in climate impact studies, 28 p

Semenov MA, Brooks RJ (1999) Spatial interpolation of the LARS-WG stochastic weather generator in Great Britain. Clim Res 11:137–148
Semenov MA, Pierre S (2010) Use of multi-model ensembles from global climate models for assessment of climate change impacts. Clim Res 41:1–14
Semenov MA, Brooks RJ, Barrow EM, Richardson CW (1998) Comparison of the WGEN and LARS-WG stochastic weather generators for diverse climates. Climate Res 10:95–107
Sobhany B, Fateminiya FS (2014) Modelling of climatic parameters in province of Southern Khorasan. Phys Geogr 46(3):9
Trnka M, Rötter RP, Ruiz-Ramos M, Kersebaum KC, Olesen JE, Žalud Z, Semenov MA (2014) Adverse weather conditions for European wheat production will become more frequent with climate change. Nat Clim Chang 4:7. doi:10.1038/NCLIMATE2242
Txomin H, Luis RA, José LG-Y, Jorge AR, Josep EP-P (2013) Multi-level object-based urban mapping from remote sensing and GIS data. In: GIS Ostrava 2013 - Geoinformatics for city transformation, 10 p
Yan G, Qi F, Wei L, Aigang L, Yu W, Jing Y, Aifang C, Yamin W, Yubo S, Li L, Qianqian M (2014) Changes of daily climate extremes in Loess Plateau during 1960–2013. Quat Int 371:5–21. doi:10.1016/j.quaint.2014.08.052

Website

http://www.unocha.org/cotedivoire/maps-graphics/thematic-maps

Chapter 16
Refining NHS Climate Change Adaptation Plans: Central Manchester University Hospitals Foundation Trust (CMFT) Case Study

Oscar Nieto-Cerezo

Abstract In this paper, an innovative climate change adaptation plan for Central Manchester University Hospitals is presented. This adaptation plan is structured in four blocks: (1) stakeholder engagement, (2) information gathering on local climate change impacts, local health inequalities spatial planning and attitudes to more sustainable mode of transport of patients, visitors and staff, (3) decision making and implementation, and (4) monitoring.

The estimate temperature change of the summer warmest day is about +3 °C by 2099 and the central estimate of change in the precipitation on the wettest day is about +18.8 % by 2099 in Manchester. This change in the climate can harshly affect the health of 71 % of the population in Manchester who lives in deprivation. NHS, as the largest public health organisation in the UK, has the responsibility to reduce health inequalities as well as to enhance resilience in the local community by promoting behavioural change. The travel survey carried out at Manchester Royal infirmary (MRI) from travel activities revealed that the main barriers to switching to a more sustainable mode of transport are poor services, tickets affordability, child-care commitments, convenience and shift patterns. The most popular sustainable travel plan initiative is the interest free public transport season ticket loan, with the bike scheme and car-sharing schemes the second and third most popular respectively. Implementing successful a sustainable travel plan at CMFT is contingent on the successful cross-sectoral work with Greater Manchester Public Transport and Manchester Corridor, and a significant investment in onsite facilities such as showers or safe bike shelters. CMFT has also set progress monitoring tools through carbon footprinting from travel related activities. The carbon footprint exercise shows that MRI travel carbon emissions are 5,783,427 kg CO_2e, which represents 1.7 % of national NHS travel total. Staff contributes most significantly to the MRI travel carbon footprint, with 51 % of the total carbon emission, followed by visitors with 26 %, and patients with 23 %. Nurses are the largest contributing staff group with 1,384,590 kg CO_2e but medical staffing is the largest contributor per employee with 919 kg CO_2e.

O. Nieto-Cerezo, B.Sc., M.Sc. (✉)
School of Science & The Environment, Manchester Metropolitan University, Manchester, UK
e-mail: nieto@liverpool.ac.uk

W. Leal Filho (ed.), *Innovation in Climate Change Adaptation*, Climate Change Management, DOI 10.1007/978-3-319-25814-0_16

Keywords Climate change adaptation • Health inequalities • NHS • Travel behaviour change • Travel carbon footprint

NHS Adaptation: Mitigation in a Twin-Track Approach to Addressing Climate Change

On 27th January 2009, the NHS Sustainable Development Unit (NHS SDU) published a NHS Carbon Reduction Strategy for England, "Saving Carbon, Improving Health". The guidance has been developed in response to the global challenge of climate change and in order to promote systematic action by the NHS to meet the Climate Change Act's targets.

The NHS is the largest employer in Europe and is responsible for 25 % of England's public sector emissions (NHS SDU 2009a, b). According to NHS SDU, The carbon footprint of the NHS in England for 2012 is 25 Mt CO_2e. This is composed of energy (17 %), travel (13 %), procurement (61 %), and commissioning (9 %) (NHS SDU 2008). Therefore, there is clearly a huge potential to adopt measures not only for adaptation to the increasing demands placed on health services but for mitigation of such demands through the promotion of healthy behavioural patterns too. The NHS is required to play a leading role in the public sector to encourage UK population to switch to healthier lifestyles and to make population more resilient to the threat of climate change.

The diagram below represents a CMFT carbon reduction and climate change adaptation plan dual approach to climate change impacts. See Fig. 16.1.

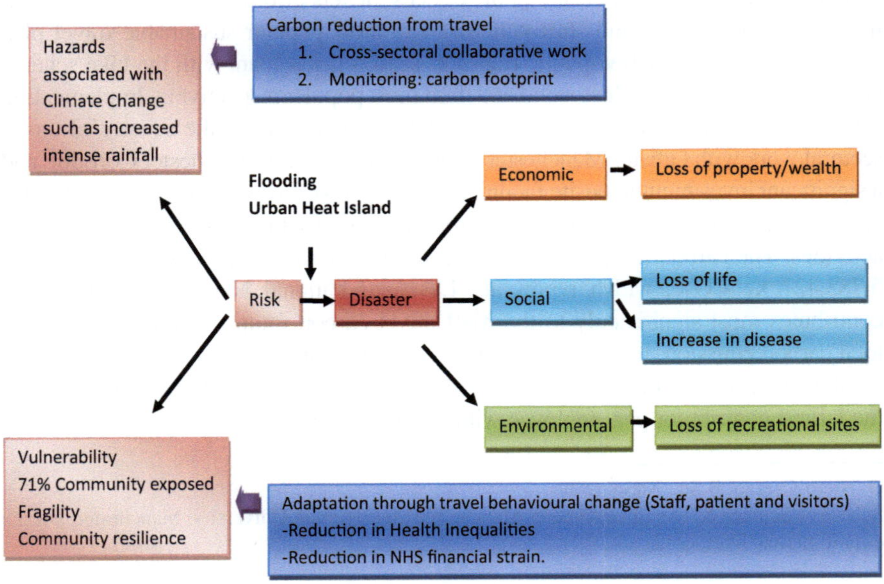

Fig. 16.1 CMFT mitigation and adaptation dual approach to the climate change impacts

Risk is the probability of the hazard times the consequences in monetary terms. The consequences of climate change depends on how vulnerable is the population affected and can be measured by taking into account the fraction of community exposed to the flooding and to the effects of the urban heat island, the socio-economic characteristic of the population and the capacity of the community to recover from the climate change impacts (Confalonieri et al. 2007). When an event strikes, the disaster can be of different nature: economic, social and environmental. Economic consequences such as loss of property can significantly affect the mental health of those people within the community with limited financial resources. Heatwaves and flooding can make people in the community suffering from cardio-vascular and circulation problems more vulnerable to heart attacks. The loss of recreational areas may result in an increase of mental health cases in the community (Bates et al. 2012).

This study considers that promoting travel behaviour change can make population in Manchester more resilient to the impacts of climate change such as flooding and heat urban island because minimising travel time to the hospital premises and incentivising active travel and the use of sustainable mode of transport can reduce the level of deprivation and the health inequalities associated with mental health, circulation problems and cancer.

The innovative component of this climate change adaptation plan is the integration of health inequalities spatial data within the CMFT Clinical Commissioning Groups (CCG) boundaries and the analysis of attitudes to sustainable travel plans by different types of patients, visitors and staff using a bottom-up approach in order to enhance the resilience of the local community to climate change impacts by promoting a travel behaviour change.

Climate Change Adaptation Plan: CMFT Case Study

As shown in Fig. 16.2, CMFT is a large multi-site organisation in Manchester which comprises of the: Royal Manchester Children's Hospital, University Dental Hospital of Manchester, Manchester Royal Eye Hospital, Manchester Royal Infirmary and Saint Mary's Hospital, Trafford Hospitals and Community Health Services (CMFT 2009). CMFT sites are located in seven CCG: Salford, Trafford, Central Manchester, South Manchester, Stockport, Tameside and Glossop and Oldham. CCG are clinically led groups responsible for the delivery of NHS services in a particular geographical area.

SDU launched a guidance called Adaptation to Climate Change for Health and Social Care organisations "Co-ordinated, Resilient, Prepared" on how to draw climate change adaptation plans highlighting the actions to implement to enhance the resilience of NHS organisations across the UK (NHS SDU 2012). This guidance consists of four main blocks; stakeholder engagement, information gathering, decision making and implementation and monitoring.

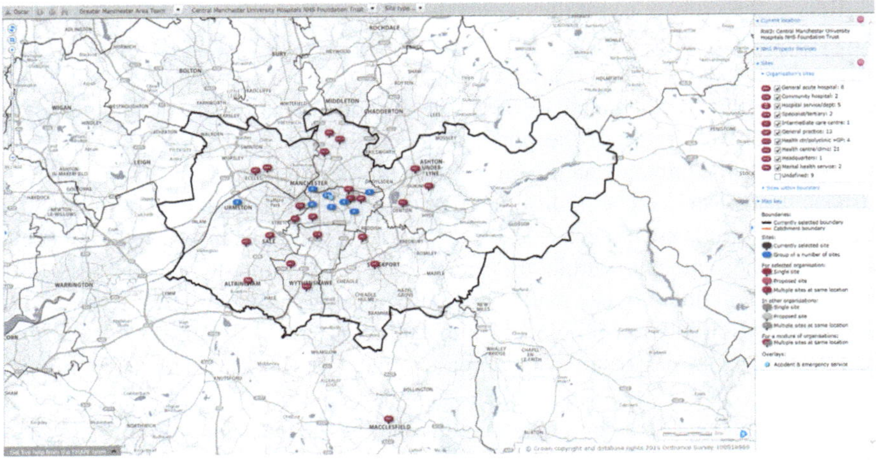

Fig. 16.2 Geographical location of CMFT's sites. ©Crown copyright and database rights 2015 Ordnance Survey 100016969

Stakeholder's Engagement

CMFT's adaptation plan is incorporated into the organisations' Sustainable Development Management Plan (SDMP). It is essential the involvement of senior staff in the design of adaptation plans to ensure longer term implications and to check alignment with the whole organisation. CMFT has created a board approved SDMP including the Chief Nurse/Deputy Chief Executive, the Director of Nursing (adults), the Head of PFI Contract and Commercial Planning and the Chairman & Board Sustainability Lead (CMFT 2013).

The optimal implementation of adaptation plans requires a cross sectoral participation and collaboration of local authorities, public organisations and the community. CMFT alongside Manchester Metropolitan University, The University of Manchester and Bruntwood have created a unique partnership called Corridor Manchester with the objective to become a leading UK location for knowledge-based business, research and development and learning. Some of the themes they are working on are sustainable transport by promoting and implementing sustainable travel plan in partnership with transport for Greater Manchester and environment and infrastructure by transforming the physical environment and infrastructure of the corridor (Corridor Manchester, undated).

Information Gathering

CMFT identified information on climate change impacts in the city of Manchester, health inequalities spatial planning and attitudes to more sustainable mode of transport of patients, visitors and staff essential to design an adaptation and resilience plan.

Potential Climate Change Impacts in Greater Manchester

The following climate prediction graphs have been created using UK Climate Projections 09 (UKCP09). UKCP 09 is based on sophisticated scientific methods provided by the Met Office, with input from over 30 contributing organisations. UKCP 09 can be used to help organisations assess potential impacts of the projected future climate and to explore adaptation options to address those impacts.

One way the UKCP 09 presents climate projections is using cumulative distribution functions (CDF) showing the probability that change will be below some specific values. We have chosen climate projections under high emissions. This scenario is characterized by a rapid economic growth, a global population that reaches nine billion in 2050 and then gradually declines, the quick spread of new and efficient technologies and extensive social and cultural interactions worldwide with an emphasis on fossil-fuels.

Figure 16.3 represents the change in temperature of the warmest day in summer time of the time period between 2010 and 2099 in the city of Manchester under a high emissions scenario. Under High emissions, the central estimate of change in the temperature of the summer warmest day is about +3 °C by 2099. It is very

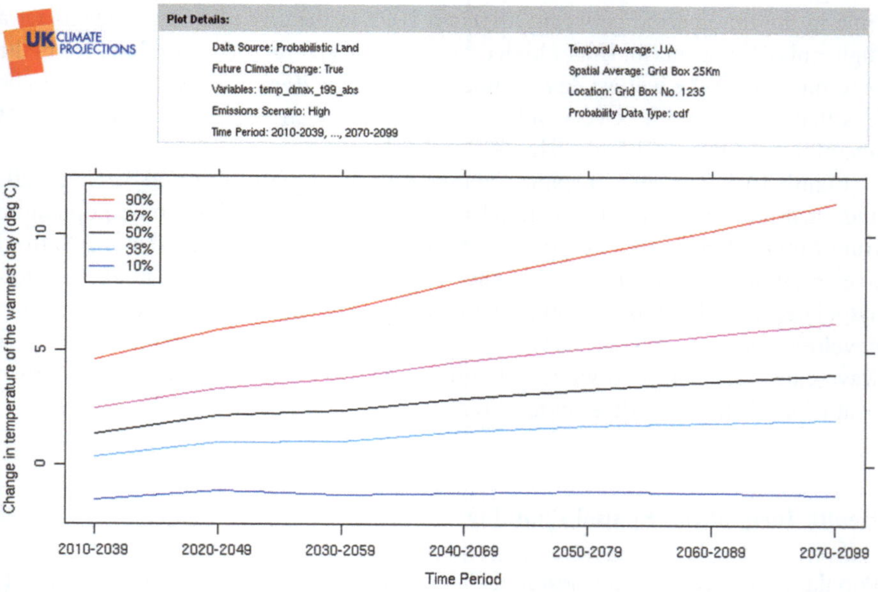

Fig. 16.3 Temperature change of the warmest day in summer time in the time period between 2010 and 2099 using cumulative distribution functions. ©UK Climate Projections (2009)

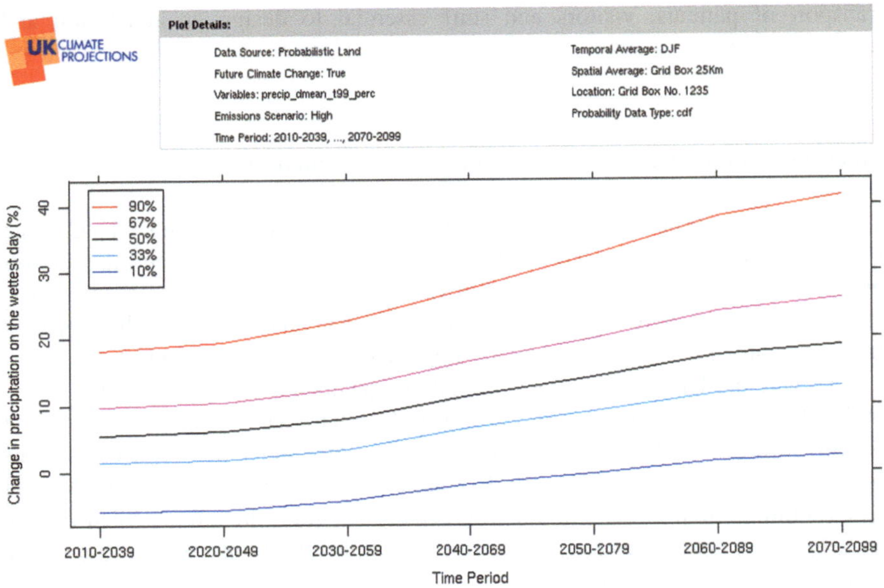

Fig. 16.4 Change in precipitation of the wettest day in winter time in the time period between 2010 and 2099 using cumulative density functions. ©UK Climate Projections (2009)

unlikely to be less −2 °C and it is very unlikely to be more than +11.5 °C. A wider range of uncertainty is from −4.6 °C to +19.9 °C.

Figure 16.4 represents the change in precipitation on the wettest day in winter time in the time period between 2010 and 2099 in the city of Manchester under a high emissions scenario. Under high emissions, the central estimate of change in the precipitation on the wettest day is about +18.8 % by 2099. It is very unlikely to be less than 2.3 % and it is very unlikely to be more than +40 %. A wider range of uncertainty is from −7.6 to +64.0 %.

Figure 16.5 represents a spatial dataset with flood plain split into 50 × 50 cells and each allocated one of four flood risk likelihood categories. Some population from South Manchester, Central Manchester, Trafford and Stockport and Salford live in an area where there is a chance of flooding greater than 1 in 30. Impacts associated with flooding, particularly fluvial flooding, pose a greater risk to housing development areas in Manchester than other hazards. Impacts associated with heat wave cause heat stress to vulnerable groups in areas where exposure to the urban heat island is highest (Bates et al. 2012).

Health Inequalities Spatial Planning

Population in Greater Manchester has higher levels of deprivation and lower levels of health compared to other UK regions making them more likely to be negatively affected by the impacts of Climate Change. In Manchester, 71 % of the population

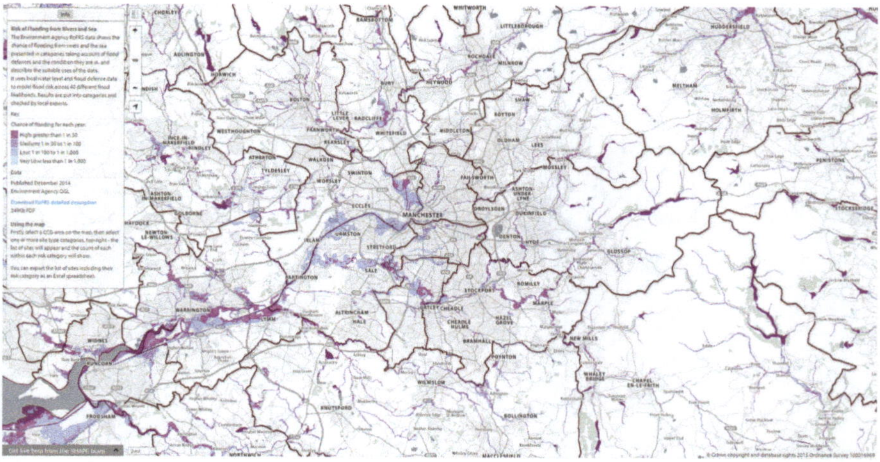

Fig. 16.5 Risk of flooding from Rivers and Sea in Greater Manchester. ©Crown copyright and database rights 2015 Ordnance Survey 100016969

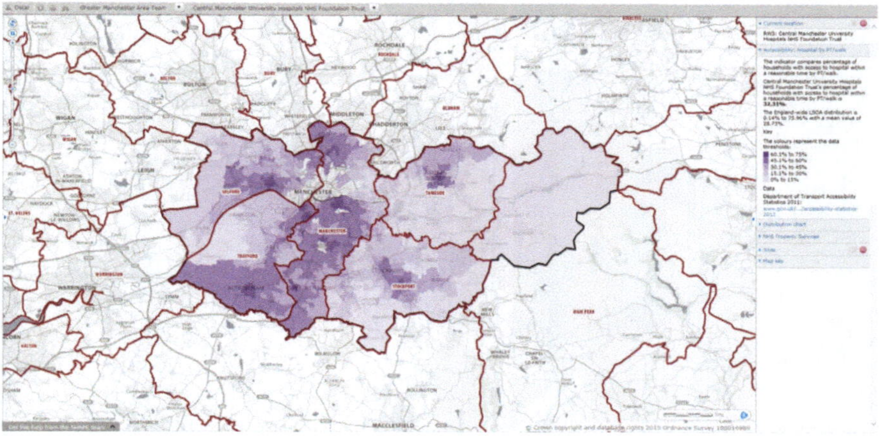

Fig. 16.6 Index of multiple deprivation (IMD). ©Crown copyright and database rights 2015 Ordnance Survey 100016969

lives in the 20 % most deprived areas of England (Bates et al. 2012). Figure 16.6 shows that the Index of Multiple Deprivation (IMD) for CMFT is 29.94 which is higher than the national average of 21.67. A large number of suburbs in Central Manchester and South Manchester, Salford and Tameside score in the top quintiles of the IMD.

Figure 16.7 shows that most of the population in Manchester, Salford and Trafford has better accessibility to CMFT by public transport or walk than residents of Tameside, Stockport or Glossop and Oldham.

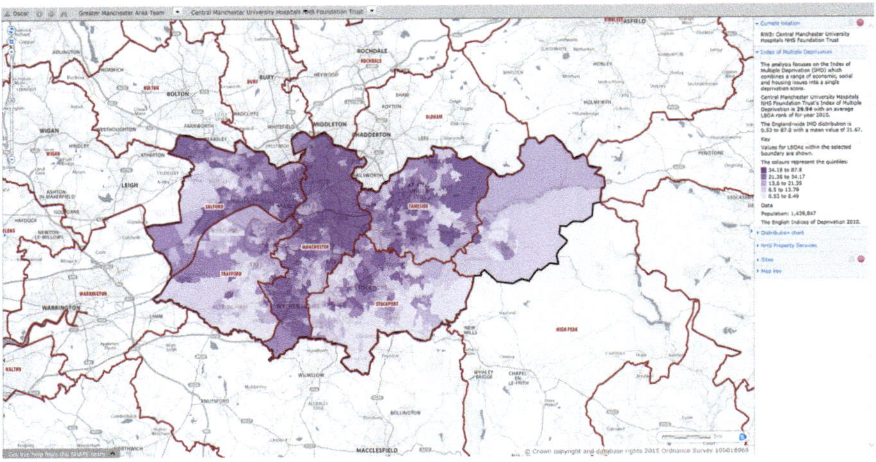

Fig. 16.7 Access to CMFT within a reasonable time by public transport and walk. ©Crown copyright and database rights 2015 Ordinance Survey 100016969

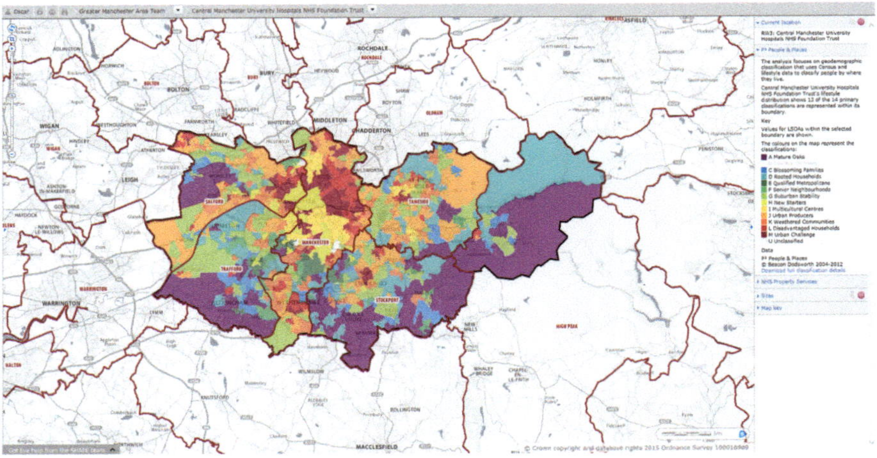

Fig. 16.8 Geodemographic classification of people by where they live. ©Crown copyright and database rights 2015 Ordnance Survey 100016969. Beacon Dodsworth (undated)

Some of the other factors affecting the ability to recover from events are the level of income, housing mobility, general mobility, social networks and well-being of the population and that change from one location to another (Lindley et al. 2011). A description of the classification of Fig. 16.8 is presented in Table 16.1.

In 2009, lancet stated that climate change could be the biggest global health threat of the twenty-first century putting further financial strain on the NHS (Lancet 2009).

Table 16.1 People and place characteristics according to Beacon Dodsworth's classification

	Income	Transport	Health
Mature oaks	Self-employed Management and professional workers	Large proportion work from home	Good standard of living
Blossoming families	Employers Management and professionals occupations	Travel to work by car, driving over 20 km	Little ill health Some older members with long-term illness
Rooted household	Skilled manual workers in the manufacturing industry	Many travel to work by car	Good health and low deprivation levels
Qualified metropolitans	Very well qualified managers and professionals	Most travel to work by train, while others work from home	Health problems are scarce. Little deprivation
Senior neighbourhoods	Most are retired, having previously worked in mining, construction and manufacturing	Many household will have one car Some people work from home	There is little ill health. There is some deprivation and poor living standards
Suburban stability	Most people are not highly qualified Manual workers	They often travel to work in the car	There are no major long term illness problems. There is little deprivation
New starters	Many people do not work although a good proportion of the unemployed are students	Many live near their work or place of study and walk there whilst others go by bus	There are poor living conditions and high level of deprivation in some households
Multicultural centres	The unemployment rate is twice the national average. Semiskilled manual workers	They mainly travel to work by train or bus	Illness levels are above average. Many live in deprived households
Urban producers	The unemployment rate is above the national average. Semiskilled manual workers in manufacturing	Many live near enough to their work to walk there, whilst other travel by bus	Many people smoke and live in deprived households. Poor health
Weathered communities	Most people are retired, having previously worked in construction. Unemployment is high	They get around mainly by bus and some on foot	With older population and some below the standard living conditions, the health is poor
Disadvantaged households	There is high rate of unemployment. Some unskilled workers	Workers commute short distances by bus or on foot	Despite young population, high level of ill health with long-term illness being common
Urban challenge	This group has the highest level of employment with long-term unemployment. Semi-skilled workers in manufacturing	People commute to work by bus or walk	Long-term population illness is common due to high level of deprivation

Tackling social inequalities can be an effective mechanism to make communities more resilient and less vulnerable for the future climate change impacts. These policies should focused on developing healthy and sustainable places and communities by improving active travel across the UK and improving the availability of good quality open green spaces across the social gradient to reduce obesity, improving circulation and mental health (Marmot 2010).

Attitudes to More Sustainable Modes of Transport of Patients, Visitors and Staff

Two hundred and thirty travel surveys were completed by patients, visitors and staff accessing the hospital premises to gather the relevant data for the calculation of carbon footprint and to understand the travel patterns of patients, visitors and staff. Three versions of the questionnaire were created, with one for each of the participants group: patients, visitors and staff. The staff questionnaire comprises eight questions. In order to simplify the data collection process, data was only collected from Manchester Royal Infirmary (MRI). Further research will be needed to obtain travel information of the other CMFT sites and investigate any possible interaction between sites (CMFT 2009).

In this section is only concerned with the seventh question which analyses the reasons for staff, visitors and patients for not using the public transport network or bikes to access CMFT and the eighth question of the survey which investigates staff attitude towards the implementation of sustainable travel plans and the popularity of low carbon initiatives such as car sharing schemes and bike schemes.

From Fig. 16.9 shows that the car is the most popular mean of transportation for staff except for nursing support and ancillary and maintenance workers in which car

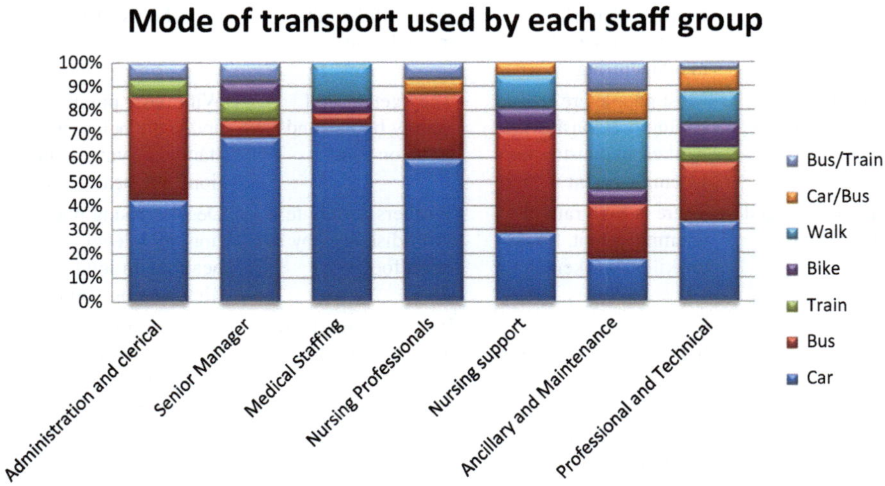

Fig. 16.9 Mode of transport used by each staff group

comes second after bus and walking respectively. Medical staff, senior managers and nursing support are by far the most dependant on private motorised cars whereas clerical administration and nursing support staff show a wide use of public transport network with a significant 43 % using bus as a the primary choice to travel to work. A significant number of staff commutes to work using a combination of bus/taxi or tram which ranges from a 3 % of journeys made by professional and technical to a 12 % made by ancillary and maintenance.

Car dependency is associated with the level of income and distance of the residence. As hospital workers move further away from the hospital premises when financial situation allows this, more staff relies upon private car to commute to work. In general term, women show more dependency on cars than men. This may be due to child care commitments and safety concerns.

In seeking to understand and identify what are the main barriers for private car users to switch to more sustainable modes of transport staff response shows that the main barrier is the lack of integrated public transport in Greater Manchester as well as poor services making it difficult to rely on it at certain times of the day (Fig. 16.10).

People living offside the main routes struggle to make their way to the MRI by using public transport. In spite of the fact that Oxford Road is the busiest route in Europe, people still complain about the lack of reliable public transport in the outskirt of Manchester. The second main response was "not considered" except for administration and clerical, professional and technical staff who showed a good knowledge of the options available in their local area.

Figure 16.11 shows staff popularity to sustainable travel plan initiatives. It was evident that some staff did not have any prior knowledge of the concept and function of green travel plans in NHS organisations. In fact, a bike scheme was rolled out at MRI 2 years ago and it was notable that hardly any of the interviewees

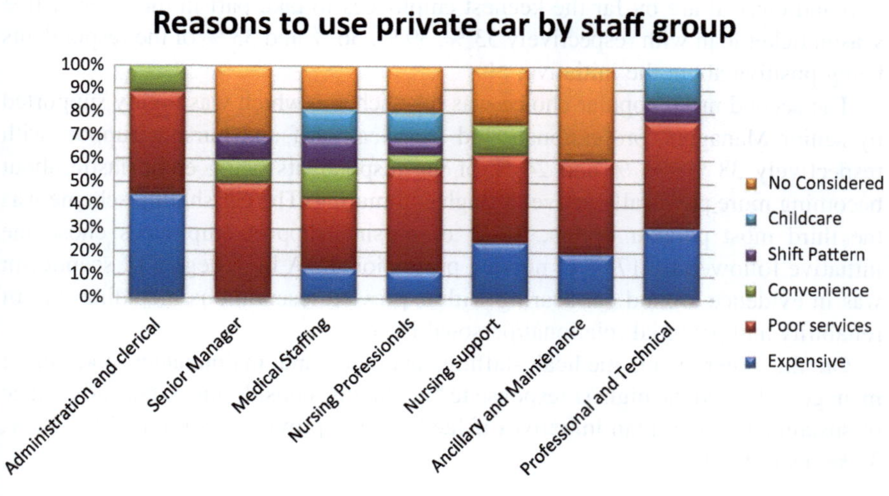

Fig. 16.10 Reasons to use private car by staff group

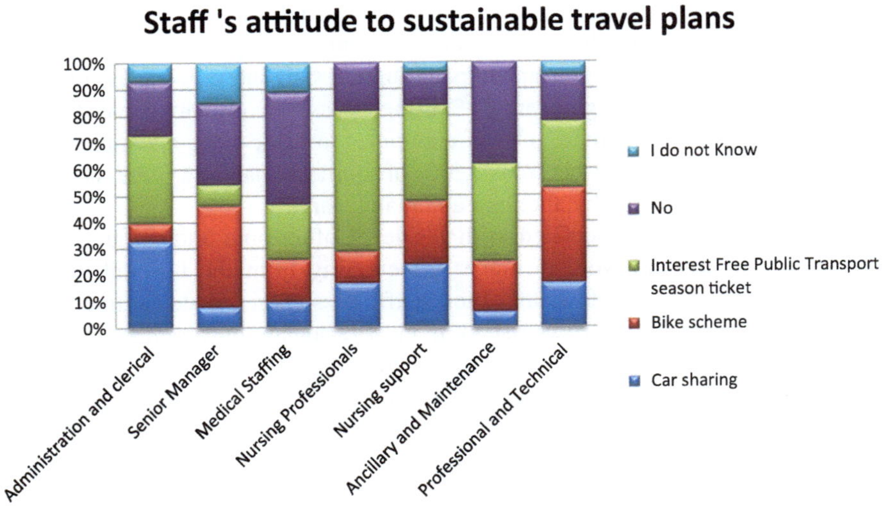

Fig. 16.11 Staff's attitude to sustainable travel plans

were aware of the schemes existence. Sustainable travel plans have the purpose of facilitating a switch to more sustainable means of transportation by providing low cost and low carbon alternatives such car sharing schemes, bike schemes, interest free public transport season ticket loan and so forth. This initiatives enable the commuter through sacrificing a very small amount of salary every month to access tax free bikes and public transport to become more active, greener and healthier as well as reducing considerably their carbon contribution. Overall, the most popular choice is the interest free public transport season ticket loan scheme followed by the bike scheme except for senior managers and professional and technical staff. Nursing professionals, ancillary and maintenance, nursing support and administration and clerical are by far the keenest employees to take part in the interest free season ticket loan with respectively 53 %, 39 %, 36 % and 33 % of the respondents being positive about the initiative.

The second more popular choice was bike scheme which was highly supported by senior Managers, professionals and technical staff and nursing support with respectively 38 %, 36 % and 24 % of the respondents being enthusiastic about becoming more physically active through commuting. The car sharing scheme was the third most popular option, 24 % of nursing support employees chose the initiative followed by 17 % of nursing professionals. A high degree of scepticism was in evidence around car sharing with expressed reservations including that of reliability and potential relational/personal issues.

On the other hand, medical staffing, ancillary and maintenance and senior managers showed the highest response to "not having considering to take advantage of sustainable travel plan initiatives if facilities are provided" with a 42 %, 38 %, 32 % respectively.

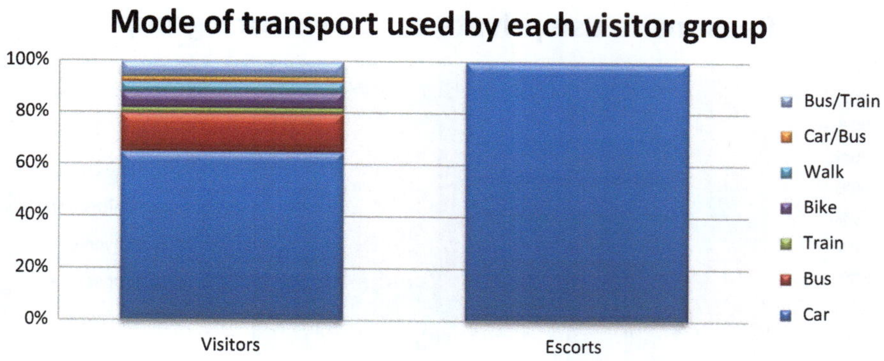

Fig. 16.12 Mode of transport used by each visitor group

From Fig. 16.12 we learn that nearly two out of three visitors use the car to visit relatives and friends and only 15 % travel by bus. Escorts interviewed during the data collection said that they travelled by car and some of them revealed that they were escorting relatives and friends with reduced mobility or elderly patients.

More than 41 % of the respondents who are car users did not consider public transport as an option to travel to the NHS premises. Time constrains and long distances mean that people have a high dependency on cars. Poor services are the main barrier for visitors, especially those people who travel from remote areas where limited facilities are provided. The cost of trains and bus fares are not affordable for some families who are unemployed and travelling from deprived areas. Some families report that it cheaper to share the cost of petrol and car park between the members of the family that buying individual train or bus tickets.

Thirty-three percent of Escorts never considered travelling to MRI by public transport because they have been explicitly asked by the patient to provide them with a journey by car to the hospital due to a health condition. On the other hand, 33 % of the respondents found that the public transport services provided were not reliable and were a cause for concern regarding missing their appointment time. Some escorts stated that the journey to the hospitals involved taking both train and buses, making the public transport experience to expensive. Other people reported that reducing carbon from travel activities is not a priority for them at the moment due to the circumstances (Fig. 16.13).

The three factors which determines car dependency rate are the distance between people's residence and MRI, staff income and clinical or non-clinical staff status. As shown in the previous section, there is a positive correlation between the first two factors which means that more highly paid employees commute longer distances than those on lower pay rate.

More than 60 % of MRI staff lives within the M60 perimeter; 70 % of whom are located in the South Manchester areas of Chorlton, Burnage, Withington, Rusholme, Levenhulme in the South West Manchester area of Heaton Moor and in the South East areas of Stretford or Trafford. The remaining 30 % within the M60

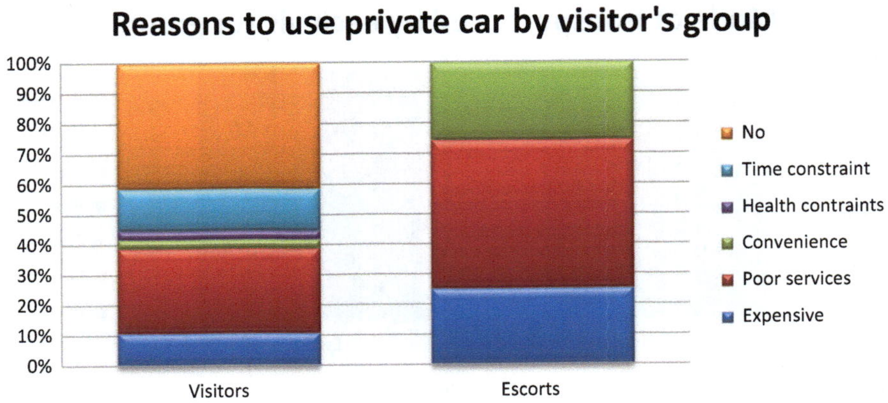

Fig. 16.13 Reasons to use private car by visitor's group

perimeter live predominately in Gorton, Clayton, Miles Platting, Audenshaw, Moston, Middleton, Prestwich, Salford in Northern, Eastern and Western areas of Manchester. In addition, some clinical staff such as junior doctors lives in the hospital hall of residence within MRI premises.

According to the results, staff car users living beyond M60 perimeter reports that the main reasons for driving to work is the cost and lack of provision of frequent public transport services at off peak times. The most concerning finding lies in the fact that a significant percentage of staff who belong to the largest contribution group to the staff travel carbon footprint stated that they would not consider public transport as an alternative to commuting even if facilities were provided.

Decision Making and Implementation

CMFT is implementing a new sustainable travel plan covering the years 2014–2018. This plan includes improving cycling infrastructure, including provision of a new cycle hub and including drying, changing and showering facilities. CMFT is also reviewing the current car sharing database and look to promote more widely or adopt a new system if appropriate.

CMFT is working closely with the Manchester Corridor partners and Transport for Greater Manchester to deliver joint sustainable travel initiatives and attend joint travel forums. Some of these initiatives are focused on encouraging and support cyclists through hosting the Corridor Monthly Cycle to Work events in conjunction with University of Manchester, Manchester Metropolitan University and the Royal Northern College of Music.

CMFT is aiming to improve the current Good Corporate Citizenship Travel score from 42 to 50 % by March 2015, and to achieve a score of 75 % by March 2020. See Fig. 16.14. The Good Corporate Citizenship is a tool to help NHS

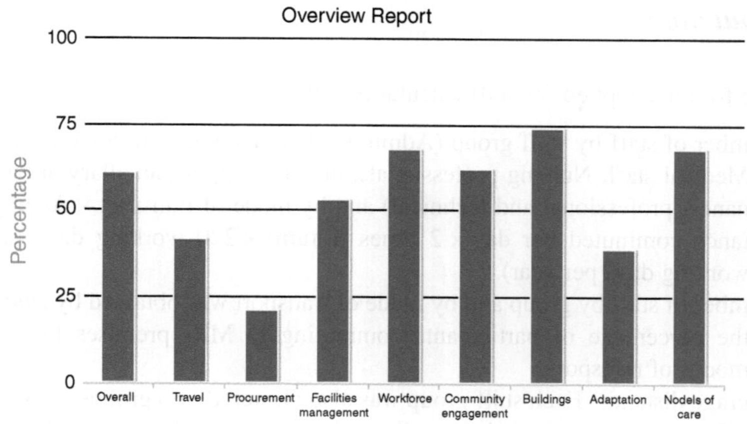

Fig. 16.14 CMFT good corporate citizenship score (Travel 42 %)

organisation to evaluate their sustainability performance against eight criteria; Travel, Procurement, Facilities Management, Workforce, Community Engagement, Buildings, Models of Care and Adaptation.

Monitoring: MRI Carbon Footprint

Carbon footprint can be regarded as a robust benchmarking tool for decision makers to facilitate the efficient implementation of carbon measures as well as monitoring progress of reducing emissions over time.

Calculation of MRI carbon footprint from travel related activities.

Identification of Relevant Data for the Purpose of the Project

- The travel survey collects information on the age and gender of the participant, travel distance, job titles, type of contract, car engine type.
- Manchester Royal Infirmary statistics: number of staff by type, number of patients admitted by type.
- National Travel Survey: Number of trips per person per year to visit friends in NHS premises.
- Office for National Statistics: England and North West population data.
- DEFRA: GHG Conversion Factors (data needed for visitor's estimations).

Formulation

- The formula applied for staff calculations was:

  Number of staff by staff group (Administration and Clerical, Senior Managers, Medical staff, Nursing professionals, nursing support, ancillary and maintenance, professional and technical) and by mode of transport × Average distance commuted per day × 2 times (return) × 220 working days (average working days per year).

  Number of staff by group and by mode of transport was obtained by calculating the percentage of participants commuting to MRI premises by different modes of transport.

  Average distance: Each staff group was also clustered into groups (<10, 10–20, >20) according to their travelling distances in kilometres. The average distance of each cluster was then calculated for each mode of transport.

  220 working days was calculated by subtracting to the 365 days of the year, 104 weekend days, 8 bank holidays and 33 days of annual leave.

- The formula applied for patients calculation was:

  Number of patient groups (outpatients, inpatients and accident and emergency patients) by mode of transport × Average distance commuted per visit × 2 times (return).

  Number of patients by group was obtained by calculating the percentage of patients traveling to MRI premises by different modes of transport.

  Average distance has been obtained for each patient group and for each mode of transport from National Travel Survey.

- The formula applied for visitors calculation was:

  Number of visitors by visitors group (visitors and escorts) and by mode of transport × Average distance commuted per visit × 2 times.

  Number of visitors by group and by mode of transport was obtained through the addition of each district population within the Central Manchester area and calculating the percentage of visitors by mode of transport.

  Average distance has been obtained for each visitor group and for each mode of transport.

  Two times means the estimated number of trips made by UK population to visit relatives or friends to the hospitals according to National Travel Survey.

Conversion of Distance Values into CO_2e

DEFRA has produced a comprehensive methodology to calculate emissions from travel related activities. The DEFRA conversion factors spreadsheet calculates the travel Carbon Footprint through the provision of fuel information or estimated distance travelled using UK average emissions factors for different modes of transport. Bus emissions per passenger km are influenced by the vehicle model, the driving cycle (proportion of urban, rural and highway driving) and by vehicle average occupancy. The previous emission factor of 89.1 g CO_2/pkm was calculated as an average across all UK buses and coaches. This calculation was based on fleet average g CO_2/km for all bus class and journey data from the UK Greenhouse Gas Inventory and an average load factor of 9.2 calculated using total bus vehicle km and passenger km from Transport Statistics Great Britain (TSGB).

DEFRA Green House Gases conversion factors spreadsheet produces information of direct emissions of CO_2, CH_4, and N_2O from the combustion of fuel from owned/controlled transport. It also provide indirect emissions associated with the extraction of transport of primary fuels as well as the refining, distribution storage and retail of finished fuels. A tonne of carbon equivalent (CO_2e) is the most precise unit because it represents the emissions of the all greenhouse emissions involved in that activity based on the Global Warming Potential ("GWP"). GWP represents the effect of a GHG on climate change, relative to the same amount of carbon dioxide (DEFRA 2009a, b).

Results

Manchester Royal Infirmary carbon footprint associated with travel activities is 5,783,427 kg CO_2 and represents 0.17 % of National total for NHS travel activities which is 3,450,000,000 kg CO_2e. The largest contributor participants group is staff with annual emissions of 2,943,487 kg CO_2 representing 51 % of the total workforce carbon footprint.

The nurses represent the largest professional group employed by the trust with more than 1550 people working on a full time basis. Their annual carbon contribution is of 1,384,590 kg CO_2 which means that on average each nurse is responsible for emitting 885 kg CO_2 into the atmosphere. Nurse staff contribution represents up to 47 % of staff total emissions, in spite of representing just 34 % of the total MRI workforce.

The second largest contributor to MRI Carbon Footprint is the nursing support group with 464,518 kg CO_2 which represents 16 % of the total workforce emissions. Nursing support is also the third largest workforce group with 831 people working full time for the trust. Each nursing support staff emits annually 559 kg CO_2 for commuting purposes.

The third largest contributor after nursing staff are administration and clinical which also represent the second largest staff group working for the Trust with 1426 professionals. Administration and clinical staff contribute to the MRI Carbon Footprint by emitting more than 430,000 kg CO_2 every year which represents 14.7 % of the total workforce emissions. Nevertheless, this staff group is one of the lowest contributors per individual with just 303 kg CO_2 per year.

Medical staff contribution is of 384,394 kg CO_2 and represents 13 % of the total. MRI employs 418 doctors and consultants and their contribution by individual is by far the largest of the hospital workforce with 919 kg CO_2 per individual per year. Professional and technical come fifth in terms of carbon emissions contributing just 7 % of the total staff emissions. Professional and technical emissions associated with commuting are up to 212,304 kg CO_2, however, their individual emissions are one of the highest with 718 kg CO_2 for each of the 273 professional and technical staff employed by the trust.

Finally, the lowest two staff contributors to the total staff carbon footprint are senior managers and ancillary and maintenance with 43,666 kg CO_2 and 20,705 kg CO_2 respectively. The senior managers' contribution is of 1.5 % whereas the ancillary and maintenance contribution is of just 0.7 % of the total.

These two staff group contains with the lowest number of employees; 53 senior managers and 67 ancillary staff. On average, each senior manager is responsible for 823 kg CO_2 whereas each ancillary and maintenance member emits only 309 kg CO_2.

Patients are the smallest contributor with annual emissions of 1,309,236 kg CO_2 which represents 22.6 % of the total travel activity carbon footprint.

Accident and emergency patients is the highest group contributor with 49 % of the total patient's emissions which is equivalent to 635,523 kg CO_2. Approximately, 145,000 Accident and Emergency patients are treated every year at MRI. Each A&E patient contributes on average 4.4 kg CO_2 per visit to MRI. The second largest patient group contributor is outpatients. There are approximately 70,457 first outpatient attendances every year and an additional 203,682 outpatients travelling to MRI for follow up appointments. The outpatient's carbon footprint is of 599,607 kg CO_2 which represents 45 % of the total patients' carbon footprint. An outpatient travelling to the hospital contributes approximately 2.2 kg CO_2 per visit.

The lowest patient group contributor is inpatients whose contribution is just 74,376. In 2009–2010, there were 9705 elective patients, 20,007 day case patients and 29,243 none elective patients treated at the MRI. On average, these patients contributed to the total emissions with just 1.26 kg CO_2 per visit.

Visitors contribute to the travel carbon footprint with emissions of around 1,309,236 kg CO_2 per year which represents 26.5 % of the total carbon footprint associated with travel activities. Visitors travel from all over the country to visit relatives and friends at the MRI. For calculation purposes, it has been assumed that the number of people traveling to MRI premises every year is the same as the population of the Central Manchester area. Therefore, an estimated 195,944 people have been travelling last year to the MRI for visiting or escorting purposes. Furthermore, the number of visitors have been multiplied by two on the basis that according to the National Travel Statistics, on average, people in the UK travel to NHS premises for visiting purposes two times every year.

Each visitor is responsible for contributing 3.9 kg CO_2 emissions to the travel carbon footprint. Whilst this seems to be a plausible figure, there is a risk of double counting in this calculation; therefore further research is needed to develop more precise mechanisms to monitor visitors' number to NHS premises.

Conclusion

This case study has shown a transferable route map approach to climate change adaptation underpinning the importance of collecting geographical spatial data and stablishing cross sectorial collaborative work in order to promote travel behaviour change by successfully implementing sustainable travel plans. The implementation of a climate change adaptation plan in the NHS requires the development of bespoke mechanisms for each NHS organisation, due to their varying profiles in terms of geographical location, healthcare services provided and infrastructure.

Nevertheless, a number of limitations have been identified and further research should focus on addressing these for more accurate results.

UKCP09 integrates the Met Office Hadley's simulation of global climate uncertainties in model parameters and the ensemble of other climate models simulating uncertainties in model structure. Some variables of the other climate models include data outside the CCG's boundaries by the corresponding perturbed physics ensemble of the Met Office Hadley Centre Model. Such inconsistencies imply that robust probabilistic projections cannot be provided by UKCP09. In addition, the existence of three separate emissions, low, medium, high is due to the uncertainty of future greenhouse gas and aerosol emissions.

A second set of limitations arise from the data collected through the questionnaire. Perhaps, the most important factor affecting the accuracy of travel distances lies on the fact that data was only collected from one of the sites comprising CMFT. Some specialised sites may receive patients from longer distances seeking for treatment. Attempts have been made to obtain geospatial data by converting patients, visitors and staff's postcode into coordinates in order to analyse and display reference data geographically using the Geographical Information System software (GIS). Nevertheless, in order to understand the demographics in more detail, a larger database is required.

Another limitation is the fact that part-time employees, who represent 7 % of the total workforce, have been classified as full time employees for calculation purposes. This has proved necessary due to a lack of data regarding the working shifts of part-time employees and the divergent nature of their schedules. Some part time employees commute to the MRI on 2 or 3 days each week, working 7 h shifts or, alternatively, commute to work 5 days per week, working half day shifts. In addition, some full time clinical staff only commute 3 days a week to work 12 h shifts.

Perhaps, the most challenging aspect of influencing a behavioural change of healthcare professionals in their travel habits is to send the most suitable messages for people from distinct professional backgrounds. For example, communicating the importance of climate change adaptation plans to hospital board members must be embedded into a business case which should include cost and benefit analysis of the proposed carbon reduction plan. According to the Department of Transport, implementing green travel plans can save money to the organisation; while the annual cost of maintaining a parking space can be £300–£500, the cost of running travel plan was typically £47 a year for each full time employee. NHS organisations can also reap the benefits from reducing health inequalities. The carbon reduction plan should of course highlight the savings associated with health enhancement and a reduction in the demand for parking spaces.

Doctors and other health professionals are best placed to promote healthy transport policy individually with patients and also collectively, as part of a broader aim of promoting healthy public policy. The messages in this case will therefore be tailored to focus on the health benefits of using cycling, walking and other sustainable modes of transport, such as the reduction of risk of CHD, stroke, type II diabetes by up to 50 %, the reduced risk of premature death by 20–30 %, and the potential improvements in mental health and wellbeing for all and in particular for maintaining independence and good health in older people.

A significant percentage of MRI staff and visitors live within 2 miles of the MRI perimeter; therefore, the provision of a network of cycle/pedestrian routes, combined with the reallocation of road space to give priority to buses, cyclists and pedestrians will support people in attempting to change their travel habits.

Comprehensive measures are required to support safe routes throughout the urban area, which should include on-street infrastructure (traffic calming/crossings) and on-site infrastructure (secure cycle parking, lockers, showers), provision of quality bus stops and shelters with raised boarding areas to allow level access, good lighting, and the provision of improved timetable travel information.

Manchester Corridor alongside Councils in Greater Manchester are to provide a reliable, safe and affordable public transport network to facilitate the access to CMFT in areas such as Trafford, Salford, Tameside, Stockport or Glossop and Oldham.

References

Bates G, Chadborn N, Jones L, McVeigh J (2012) The impacts of climate change upon health and health inequalities. North West Public Health Observatory

Beacon Dodsworth (undated) Beacon Dodsworth geodemographic classification. Understanding people by where they live. www.beacon-dodsworth.co.uk. Accessed 4 June

Central Manchester University Hospitals (CMFT) (2009) Annual report and summary financial statements

Central Manchester University Hospitals (CMFT) (2013) Sustainable development management strategy and plan 2014–2020. www.cmft.nhs.uk. Accessed 10 Feb 2015

UK Climate Projections (2009). www.ukclimateprojections.metoffice.gov.uk. Accessed 10 Mar 2015

Confalonieri U, Menne B, Akhtar R, Ebi KL, Hauengue M, Kovats RS, Revich B, Woodward A (2007) Human health. Climate change 2007: impacts, adaptation and vulnerability. In: Parry ML, Canziani OF, Palutikof JP, van der Linden PJ, Hanson CE (eds) Contribution of working group II to the fourth assessment report of the Intergovernmental Panel on Climate Change. Cambridge University Press, Cambridge, pp 391–431

Corridor Manchester (undated) Strategic vision to 2020. www.corridormanchester.com. Accessed 10 Mar 2015

Department of Environment of Environment, Food and Rural Affairs (DEFRA) (2009a) Guidelines to DEFRA/DECC's GHG conversion factors for company reporting; methodology paper for emissions factors. www.defra.gov.uk. Accessed 20 Sept 2010

Department of Environment, Food and Rural Affairs (DEFRA) (2009b) Guidance on how to measure and report your greenhouse gases emissions. HMSO, London. www.defra.gov.uk. Accessed 4 Mar 2010

Lancet (2009) Lancet and University College London Institute for Global Health Comissioning. Manag Health Eff Clim Change 373:1693–1733

Lindley S, O'neill J, Joseph K, Lawson N, Christian R, O'neill M (2011) Climate change, justice and vulnerability. Joseph Rowntree Foundation

Marmot M (2010) Fair society, healthy lives. The Marmot review. Strategic review of health inequalities in England Post-2010

NHS Sustainable Development (SDU) (2009a) NHS England carbon emissions: carbon footprint modelling to 2020. www.sdu.nhs.uk. Accessed 10 Jan 2010

NHS Sustainable Development Unit (SDU) (2008) NHS England carbon emissions: carbon foot printing report. www.sdu.nhs.uk. Accessed 5 Feb 2015

NHS Sustainable Development Unit (SDU) (2009b) Saving carbon, improving health. NHS carbon reduction strategy for England. www.sdu.nhs.uk. Accessed on 7 Sept 2013

NHS Sustainable Development Unit (SDU) (2012) Climate change for health and social care organisations "Co-ordinated, Resilient, Prepared". www.sdu.nhs.uk. Accessed 7 Feb 2015

Chapter 17
Strategies for Climate Change Adaptation in Metropolitan Areas: Initiating, Coordinating and Supporting Local Activities: The Approach of Stuttgart Region, Germany

Thomas Kiwitt and Silvia Weidenbacher

Abstract Metropolitan areas exhibit a concentration of population, economic activities and infrastructure. The different impacts of climate change can jeopardize the different functions these areas have for their population, their hinterland and the economy. Adaptations strategies are therefore highly important for densely populated and prospering regions that are subject to these impacts.

Located in the south-western part of Germany, Stuttgart Region expects a higher risk of flooding and heat stress. Due to the topographic situation, the most significant impacts are predicted within the core area of the region, with a population of approx. 2 m. people and some 100,000 jobs.

This paper presents from a practitioner perspective the adaptation strategy pursued by the Verband Region Stuttgart—the regional planning and development authority of one of Germany's most prospering areas. The implementation of this strategy is based on close cooperation between different administrative tiers, optimized use of all available instruments, the joint development of new instruments and best practices.

It focuses on practical adaptation measures and their co-ordinated implementation as part of urban development, the protection of open spaces, public debate and political decision-making.

Keywords Climate atlas • Comprehensive regional planning • Regional adaptation strategy • Stuttgart Region • Verband Region Stuttgart

T. Kiwitt, Dipl.-Ing. (✉) • S. Weidenbacher, Dipl.-Ing. (FH), M. Eng.
Verband Region Stuttgart, Kronenstr. 25, Stuttgart, Germany
e-mail: Kiwitt@region-stuttgart.org

W. Leal Filho (ed.), *Innovation in Climate Change Adaptation*, Climate Change Management, DOI 10.1007/978-3-319-25814-0_17

Joining Forces Stuttgart Region's Adaptation Strategy

Like many other Metropolitan areas, Stuttgart Region is characterised by a high concentration of population, economic activities and infrastructure: 2.5 m inhabitants, approx. one third of the GDP of Baden-Württemberg and significant elements of the transportation and administrative infrastructure are located within the region. This densely populated and industrialised area is subject to different impacts of climate change, but a higher risk of flooding and additional heat stress are considered the most important issues. Due to the important role the region plays for the economy and the supply of services for a wide hinterland, the possible effects can jeopardize an area far beyond local administrative borders.

However, the responsibility for implementing necessary adaptation measures lies mainly with the local authorities. This paper presents the strategy pursued by the Verband Region Stuttgart as the regional planning and development authority in initiating these measures and coordinating their implementation.

This approach makes use of the broad scope of comprehensive planning on regional level and integrates all instruments available:

- Mandatory (formal) planning instruments to protect relevant open spaces (including public outreach and consultation)
- Studies on microclimate and Provision of GIS-based data with information for city/land use planning on a local level
- Studies on the vulnerability of specific forms of land use (e.g. housing, infrastructure, agriculture)
- Implementation of pilot projects on different scales (e.g. land use planning, city quarters)
- Dissemination of and information on lessons learned from these pilot projects for other municipalities
- Cooperation with relevant research institutes and universities
- Exchange with international partners on different issues of climate change in metropolitan areas on a regular basis (via formal cooperation and EU-funded projects)
- Regularly informing political decision-makers on the progress and outcomes of the strategy

However, these adaptation measures are not intended to replace the necessary—and equally important—reduction of greenhouse gas emissions. They rather form an integral part of an overall strategy, together with a variety of mitigation measures such as the promotion of renewable energies or provision of infrastructure for electro-mobility.

The following report describes how these aforementioned eight elements are used as parts of an integrated regional adaptation strategy. Starting from legally binding instruments (1), the provision of relevant data (2), studies on the vulnerability and the impact of climate change on Stuttgart Region (3), pilot projects for different scopes of local planning and dissemination of the findings (4), cooperation with researchers and exchanges within networks (5, 6), activities for communication and political advisory (7).

1. Mandatory (Formal) Planning Instruments: Protecting Relevant Open Spaces on a Regional Level

The Regional Plan (Regionalplan Region Stuttgart 2009) is a comprehensive document on a supra-municipal level. The regulations are legally binding for all land use plans adopted by the municipalities. Therefore, regional planning regulations are an efficient instrument for coordinating development activities beyond local administrative boundaries. The provisions in this comprehensive plan include (among others) the development of built-up areas, transport corridors, areas reserved for the large-scale use of renewable energies and the protection of open spaces.

The mandatory planning regulations protect many open spaces in the Stuttgart Region: "*Regionale Grünzüge*" (large open spaces) and "*Grünzäsuren*" (green belts or corridors between built-up areas). Both types are strictly protected and cannot be used for development. This distinction between open spaces and urbanized zones has been made for more than 40 years and is well established both in legal and methodological terms. Figure 17.1 shows the application of large and small scale instruments to protect open spaces.

One important adaptation issue in densely populated areas is reducing the effects of urban heat islands. This can be achieved most effectively by protecting areas

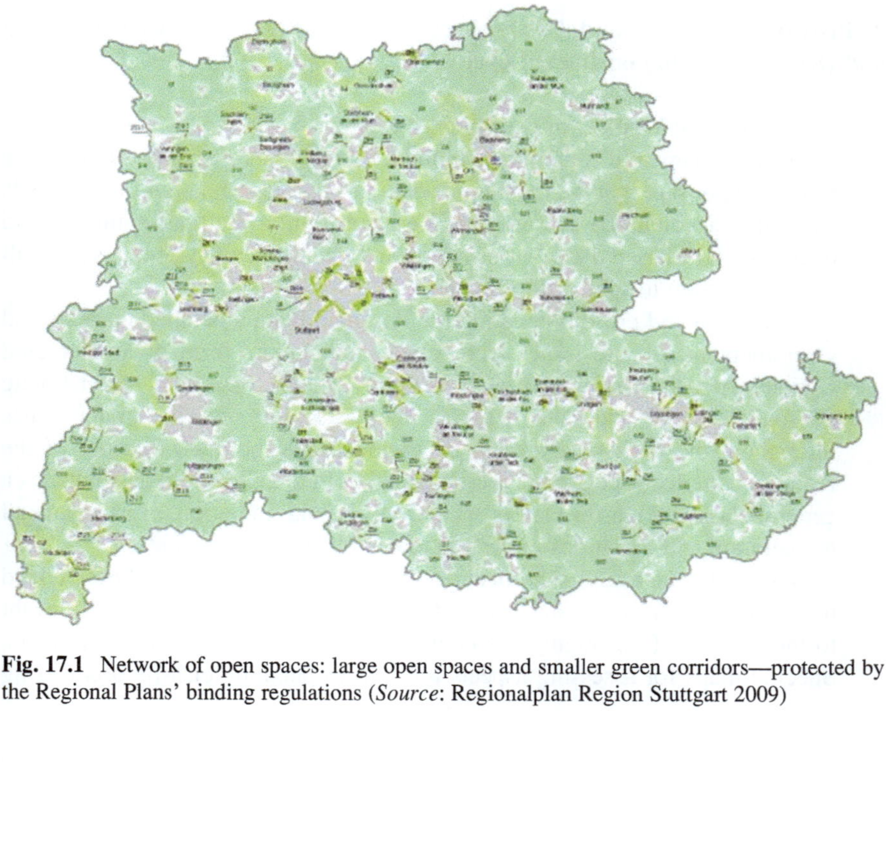

Fig. 17.1 Network of open spaces: large open spaces and smaller green corridors—protected by the Regional Plans' binding regulations (*Source*: Regionalplan Region Stuttgart 2009)

relevant for cooling and the flow of air into the core areas against development. Instead of introducing new planning instruments to prevent these effects, existing instruments for protecting (multi-functional) large and small-scale open spaces have been extended to meet these additional requirements.

A region-wide analysis mapped all relevant areas for cooling and air flow (see the following description of the "Regional Climate Atlas") and used them as one key aspect in defining the areas protected by the aforementioned instruments.

A further challenge is to protect residential areas, industrial zones and important infrastructural facilities. Here again, the Verband Region Stuttgart uses mandatory planning instruments. Besides regulations to direct urban development into areas that are not or less prone to flooding, additional areas have been zoned for flood retention. These are important precautions in dealing with the higher risk of flooding that is expected to accompany climate change.

Like all mandatory instruments, the respective elements of the Regional Plan have been subject to public consultation, political debate among elected officials on different administrative tiers and the final decision making process of the Regional Assembly.

Regional directives which are binding for municipalities and local land use plans—especially when they have a restrictive effect on further urban development—are rarely appreciated by those concerned and limited in their planning options. However, the additional argument of a positive impact on the quality of life and future resilience, which was new at the time and is nowadays proven, has garnered significantly stronger acceptance for these directives.

2. Providing Reliable and Easily Accessible Knowledge for City Planning and Decision-Making on Local Level

(A) Regional Climate Atlas for the Stuttgart Region

The environmental components of the Regional Plan are based on extensive research to ensure that the most up-to-date information on natural functions is taken into account. To this end, the Verband Region Stuttgart compiled and operates a GIS-based digital information and management system covering all aspects relevant to spatial development.

In addition and prior to updating the Regional Plan, a detailed analysis and assessment of climate issues were carried out in the form of a "Regional Climate Atlas" (Baumueller et al. 2008). This is based on an elaborate geospatial model and data network that track and record climate issues throughout the region, regardless of administrative borders. Because of the region's complex topography, the information in the Climate Atlas includes a general classification for factors such as wind, solar radiation, average annual temperature and variations in surface temperature and precipitation.

Designed as an analytical tool and technical support for comprehensive land use planning, it also contains geo-coded information on areas that are relevant to the cooling of air, night-time cool air flows or heat stress, among others. Specific maps for assessing aspects relevant to planning classify areas using different indicators. The characteristic effects of structural uses and possible measures for significantly improving the climate situation have been defined for eight categories, reaching from undeveloped open spaces to intensively built-up areas.

Within the built-up areas, so-called "climatopes" are allocated with certain microclimatic characteristics and land use features, including information on bio-climate and ambient air quality. Any zoning and development decisions in the scope of spatial planning procedures can therefore draw on concrete and sound information with regard to their specific climatic relevance. The evaluation of local wind conditions, characterized by speed and direction, provides knowledge of the possible distribution of air pollutants. This allows a main factor of air hygiene to be taken into account in the further development of built-up areas.

Open spaces that are important for preserving the climate and directly neighbouring highly sensitive residential or mixed-use areas are considered critical and have been protected from further development in the mandatory Regional Plan. Small-scale open spaces (*Grünzäsuren*) seem to be most important for preserving local fresh air corridors—especially across the administrative boundaries of different municipalities. Figure 17.2 shows how the analytical results of the Regional climate atlas have been used to define protected open spaces.

Fig. 17.2 Green corridors and belts in the Regional Plan protect fresh air corridors and areas producing cold air (*Source*: Klimaatlas Region Stuttgart 2008; Regionalplan Region Stuttgart 2009)

The Climate Atlas is based on data from the Baden-Württemberg State Institute for Environment, Metrology and Nature Conservation (LUBW) and the German Meteorological Service. The local wind conditions were evaluated using synthetic wind field calculations. The magnitude and flow of cold air were calculated with the help of a digital elevation model, supported by measuring flights equipped with infrared thermographic cameras for thermal mapping. Stuttgart City's Department of Urban Climatology supervised the overall evaluation and processing of the data and compiled the Climate Atlas.

In general, the different maps in the Climate Atlas provide regional and city planners, politicians and other stakeholders with sound information on options and constraints for further development, taking account of the health and welfare of the population under given conditions. However, global climate change enhances the importance of these considerations. This is why expected changes have also been considered and taken into consideration, in addition to the current climate situation.

Annual precipitation is likely to change, both in terms of quantity and distribution. This is expected to lead to a significantly higher risk of flooding, especially in autumn and winter, but also less rain in summer. The forecasts also examine the change in average annual temperatures and the implications for the bio-climate.

While under current conditions only 5 % of the area is expected to experience more than 30 days with heat stress per annum, this perimeter will increase to 57 %—assuming that the number of days with heat stress will double by 2100. The densely populated core area, with a topography characterized by a narrow valley basin, is predicted to have more than 60 days of heat stress.

A significantly higher percentage of inhabitants will be exposed to severe heat stress in summer. Epidemiological studies show that people sensitive to heat will be less able to adapt under these circumstances—especially those having a predisposition for cardiovascular and respiratory diseases. This may lead to an increasing number of fatalities. The expected development of the number of bio-climate days with heat stress from 2000 to 2070 shown in Fig. 17.3 underlines the importance of adaptation measures and has been subject to many political debates.

Against this backdrop protecting the compensatory functions of open spaces is even more important. In addition, the microclimatic situation in the built-up areas has to be enhanced.

(B) GIS-based data for urban/land use planning

Urban and land use planning on a local level is based on a broad database. Unlike demographic, fiscal, ecological or other spatial aspects, however, information on current or future climate aspects is not always available in a form that can be efficiently processed in planning and (political) decision-making processes. To support the integration of climate-related data on a local level—and hence the awareness of future challenges and requisite adaptation measures—the Verband Region Stuttgart provides a corresponding database.

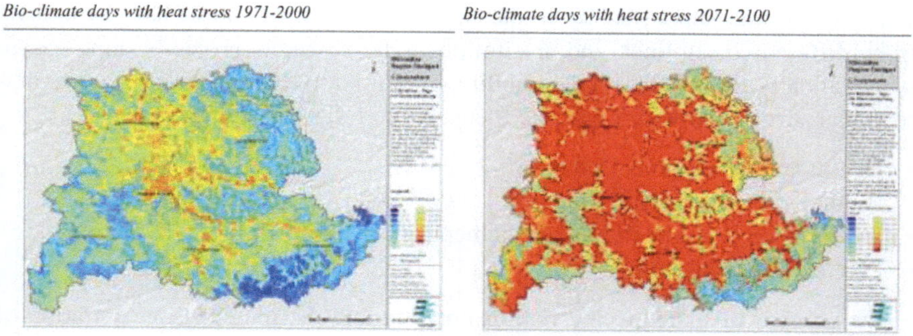

Fig. 17.3 The bioclimate maps compare the actual situation as measured by the German Meteorological Service with the forecasts (*Source*: Klimaatlas Region Stuttgart 2008)

All municipalities in the region have access to a reliable, easy-to-integrate database that meets the requirements of land use planning in terms of content, form and scale and is free of charge. It contains the findings of the Regional Climate Atlas as well as the current results of the vulnerability analysis and its evaluation by the aforementioned network.

The data form an integral part of the GIS-based spatial information system of the Verband Region Stuttgart. This way, climate-relevant information can easily be connected with other spatial or ecological data used for urban planning on a regular basis and in well established procedures, contributing to the integration of adaptation information at an early stage of the planning and decision-making processes.

Despite the platform's user-friendly design and experience of users, several workshops were organized to train local practitioners. The input focused not only on technical issues but also on the methodology and integration in strategic decision-making processes on a local level.

The feedback confirms the success of this customized form of data provision: information on climate change is increasingly integrated in analytical and conceptual land use planning procedures. As a result, there is a growing demand for measures to cope with the challenges of climate change in built-up areas and open spaces.

3. Demonstrating the Impact of Climate Change in Stuttgart Region: Regional/ Local Studies on the Vulnerability of Soil and Land Use

The findings of the Climate Atlas raised the awareness of practitioners and politicians for the expected impact of global warming. Stimulated by this prognosis, a pilot project on the vulnerability of Stuttgart Region in relation to climate change was implemented in 2010.

The Verband Region Stuttgart worked with Stuttgart University's Institute for Regional Planning and Development, the cities of Esslingen and Ludwigsburg and the Regional Economic Development Corporation (WRS) on a project funded by the Baden-Württemberg Ministry of Transport, Building and Urban Development.

The project was supported by a stakeholder network of more than 50 experts from different disciplines and institutions. The research project's aim was to systematically analyze the vulnerability of Stuttgart Region based on a cross-sectional survey and with a specific focus on the following aspects:

- Health and development schemes
- Biodiversity, agriculture, forestry
- Water protection and flood management
- Planned: Transportation

On account of the expected risk of flooding and the massive concentration of residential areas, infrastructure and production facilities along the rivers, a specific in-depth analysis has been conducted on this topic. Based on existing surveys analyzing the risk of flooding, a methodology has been developed to assess the possible impact on public and private property and the functionality of important infrastructure. This cost calculation is to be used as an additional element to raise acceptance and gain sufficient funding for adaptation measures and improved flood protection.

In addition, a systematic evaluation of the possible impact on transportation is planned. Besides a traditional analysis of the road and railway network, this will also include a check on obstacles to traffic flow in the entire transport grid using the region's traffic model for simulating several scenarios.

All analyses were conducted with the support of an extended network of experts from various disciplines, administrative tiers and scientific institutions. Three thematic working groups (biodiversity/agriculture, health, water) consisting of relevant stakeholders were set up to find appropriate methods for assessing the region's vulnerability in those fields. Each group developed and used specific methods, ranging from indicator-based spatial data processing to discursive processes.

The results of the working group focusing on health issues generated the greatest public interest. It became obvious that the quality of life in bigger cities is already lower as a result of the microclimate. According to the experts, the health-related vulnerability of Stuttgart Region is correspondingly high. In addition to the increasing threat of problems such as strokes, exhaustion and collapses caused by heat, the loss of well-being and work output in hot summer periods will also play a bigger role. The findings and subsequent discussion underlined the need for concepts to improve the urban microclimate.

Apart from specific expertise, the broad panel of specialists provided a wide range of different perspectives and ensured sufficient scope to meet the demands of the complex cross-sectional approach, without neglecting local characteristics. This network also supported the development of possible measures to reduce the vulnerability and/or improve the adaptation process. Cooperation between the different stakeholders proved to be crucial for the successful implementation of the adaptation strategy. Besides expertise, the involvement of different stakeholders guarantees that the overall approach to tackling the challenges of climate change raises the awareness of relevant players in the different fields. Thus, the network can also be seen as an instrument to coordinate adaptation activities far beyond the usual formal structures.

Some of the most important measures have to be implemented on a local level. For this reason, the active contribution of city planners and other practitioners from this administrative tier was highly relevant to ensure the practical use of the outcome. Thanks to their support, the content of the vulnerability analysis as well as suggested measures for analytical adaptation not only cover regional scale and scope, but are also focused on the city and municipality level. Special attention is paid to all issues connected with land use planning and urban development. Esslingen and Ludwigsburg, two cities well known for their innovative sustainability policy, are supporting the network. Therefore, these partners' input also makes a significant contribution to enhancing the quality and progress of the network's outcomes. The recommendations produced by the expert group include, for example, a systematic inventory of all greening and unsealing potentials, guidelines for further tree planting, incentives such as subsidies for green projects and specific planning standards for climate change adaptation measures (Weis et al. 2011).

The specific constellation of experts from different levels and fields was the key element of the project. On account of the highly valuable contribution made by the network, regular meetings are envisaged to evaluate the results of the analysis and monitor the impact and implementation of different measures. These future outcomes will also provide an important input to the political debate on the ongoing adaptation process.

4. Pilot Projects for Different Levels of Action: From City to Quarter to Individual Property

The Regional Climate Atlas shows that the current combination of heat island effects in summer with a high population density is already significant. The impacts of climate change will greatly increase this risk—and therefore reduce the quality of life and competitiveness of the region. In addition, the vulnerability analysis outlines a major need for action both on regional and local level.

This is why the Verband Region Stuttgart coordinated several follow-up projects to study how resilience against global warming could be enhanced. While focusing on different levels of actions—from measures on a regional level to small-scale properties—special attention was paid to an integrated approach with a functional interconnection across administrative tiers.

The overall strategy is designed with a view to providing pilot projects at all relevant levels to guarantee seamless vertical integration. As a first step, a more coordinated procedure to close the gap between regional and land use planning in methodology and scale is developed. Starting from the binding regulations in the comprehensive Regional Plan, e. g. the need to protect open spaces relevant to the microclimate and prevent or mitigate future impacts of extreme weather events, possible measures are compiled for concrete planning levels and demonstrated in pilot projects. To implement this objective from regional to local level through land use planning and zoning regulations, a model approach is put in place with the financial support of the Federal Ministry for Environment (*Bundesministerium für*

Umwelt, Naturschutz, Bau und Reaktorsicherheit) and scientific coordination of the Stuttgart University of Applied Sciences (*Institut für Angewandte Forschung der Hochschule für Technik Stuttgart*).

Within the KARS project (*Klimaanpassung in der Region Stuttgart*/Climate Adaptation in the Stuttgart Region), a special thematic layer is developed to be integrated into the local land use plan. Covering the whole local jurisdiction—built-up areas and open spaces alike—specific requirements to improve adaptation and/or reduce vulnerability are implemented in all relevant decisions on urban development and open-space protection at a very early stage. As part of public outreach activities, information on this layer is discussed in a public debate on future urban development.

It became clear that the complexity of urban planning and development increases when the predicted microclimate changes in cities are taken into consideration. Competing aims such as the planning premise of brown-field site development versus the preservation of climate-active open spaces within a city require profound substantiation. However, the consequences of climate change and its impact on the quality of life and potential results for urban development found their way into political debate and public discussion.

A closer focus is put here on an urban development project in Ludwigsburg. A city quarter is especially designed to meet the specific demands of climate change: buildings, public areas and open spaces demonstrate how adaptation requirements defined at higher levels can be transformed in adequate zoning and building regulations.

On a very concrete scale, a pilot project called "TURAS" (Transitioning towards Urban Resilience and Sustainability) is carried out under the 7th European Research Framework Programme. The project aims at bringing together urban communities and businesses with local authorities and researchers to collaborate on developing practical new solutions for more sustainable and resilient European cities (TURAS website, TURAS Website. http://www.turas-cities.eu/, www.turas-cities.eu). One of its research focuses is the adaption to climate change in urban areas. Being part of the project adds to Stuttgart Region's climate strategy by providing innovative, holistic and creative design solutions for green urban infrastructure. A local demonstration site serves as an example of a multifunctional green space which can accommodate climate change adaptation as well as recreation and biodiversity needs (Refer to Fig. 17.4). The Stuttgart University's Institute for Landscape Planning and Ecology (*Universität Stuttgart, Institut für Landschaftsplanung und Ökologie*) is responsible for the scientific steering of the project.

As mentioned above, Stuttgart Region is not only predicted to suffer strongly from heat stress but already experiences heat island effects in summer. Besides the development of a concept for so-called urban climate comfort zones, the main objective is to implement real-life mitigation measures that ideally serve multiple functions (ecosystem services + social services) within a green infrastructure concept.

Fig. 17.4 The multifunctional green living room in Ludwigsburg

The pilot project, located in the City of Ludwigsburg, was originally intended to be a "Green Wall" but was transformed for many reasons into a free-standing three-dimensional structure that is now labelled the "Green Living Room".

This green structure is situated on top of an underground car park in the city centre. It consists of a modular construction made of steel baskets, filled with special substrate and planted with mature plants. It uses connation and also includes plane trees. Planted as small and young trees, they will merge rapidly to form an artificial collective organism (plant addition). They will soon provide the same filtering and cooling functions as a big tree without needing the same space (and time) to grow.

The new and innovative aspect of this demonstration site is the fact that it uses engineering with living plants as a form of urban greenery.

Further pilot projects are envisaged under the regional co-funding scheme for the development of "green infrastructure". This programme aims at developing open spaces mainly for recreation purposes. Besides this, a strong focus is also put on accompanying the effects of climate change. Most important in this field are the redesign of river banks and the preparation of multi-functional open spaces for flood protection and retention.

5. Regional Dissemination: Information and Best Practice for All Municipalities

The Verband Region Stuttgart works as a multiplier to disseminate findings from pilot projects, ongoing research and network activities. All information activities are focused primarily on the municipalities in the region. The use of well-established communication structures, e.g. regular planner meetings, workshops and information published by the Verband Region Stuttgart, allows to reach out to many different stakeholders. Even smaller municipalities (with less than 10,000 inhabitants) have easy access to the results.

Beyond regional borders, outreach activities are carried out together with partner organizations on metropolitan, federal state and national level.

6. Continuous Innovation: Cooperation with Research Institutes and Exchange with Partners

Beyond the monitoring of ongoing activities within and outside the region, research activities are supported to deepen and improve the knowledge of possible spatial responses to climate change. In this context, the Verband Region Stuttgart cooperates with several universities and research institutions in Germany.

In addition, we have an intensive dialogue with planning and development institutions. At regular meetings with partner organizations from other German regions or within METREX—a network of more than 50 metropolitan areas and regions in Europe—the issues of climate change and climate protection are addressed.

A continuous and structured exchange with partners from different European areas is an essential part of the TURAS project. The "Transatlantic Urban Climate Dialogue" has allowed to extend our geographical focus. Together with the Northern Virginia Regional Commission—a long-standing partner of the Verband Region Stuttgart in the USA—and partners from Ontario/Canada and the German Rhine-Ruhr area, the Verband Region Stuttgart organized several workshops with specific views from politicians, practitioners and researchers from either side of the Atlantic, supported by the German Federal Ministry of Economics and Technology *(Bundesministerium für Wirtschaft und Technologie)* and the Environmental Policy Research Centre *(Forschungszentrum für Umweltpolitik)* of Freie Universität Berlin.

With the "Adaptcity" project co-funded from the EU Life + programme, an adaptation strategy for all major cities in Poland is currently being developed, using several elements from Stuttgart Region as examples. In return, new findings from the approach in Poland should then be used as the basis for a structured evaluation and updating of the model used in Stuttgart.

7. Policy Counts: Regular Information About the Progress/Outcomes of the Strategy

The implementation and further development of the climate adaptation strategy had strong political support among the members of the Regional Assembly. After two elections and for the third parliamentary term, the approach is still among the important issues on the political agenda and provided with the necessary financial and human resources.

The relevant commissions of the Regional Assembly are informed on a regular basis about ongoing activities and the overall progress of the adaptation strategy. Moreover, individual measures to increase adaptation capacities are especially highlighted whenever related to other activities of developing built-up areas or open spaces. Therefore, the implementation of the respective strategy is becoming everyday business, in the sense of being accepted as routine work that does not require special justification.

To underline the political dimension of adaptation as a strategic aim has been crucial for this success. Mapping the potential changes and defining the possible consequences of climate change for the population, nature and infrastructure have been important for raising awareness. However, concepts of how to react to these challenges and practical examples of feasibility and impact have fostered political interest—as has the direct exchange with decision-makers from other regions. Continuous networking activities will therefore be kept as a relevant instrument for monitoring and improving the overall approach.

Relevant measures have to be supported—if not implemented at local level. Information and motivation of municipalities is a key issue for successful adaptation. Judging from practical experience, a plausible database, concrete, (economically) feasible measures and high added value with regard to other objectives of sustainable urban development (e.g. quality of life, improved ecological quality, flood protection) are crucial for local adaptation activities.

Regional, supra-municipal planning institutions can provide and update data and know-how, describe possible cross-section effects and ensure long-term spatial coordination.

Conclusion

Stuttgart Region's adaptation strategy is not fully implemented, yet. However, different elements have been successfully prepared and several measures to cope with the challenges have already been implemented. Nevertheless, further development, outreach and implementation activities are required and foreseen.

First results show, that supra-municipal, regional planning can develop and implement mandatory guidelines for local land use planning to improve adaptation to the impact of climate change. The provision of reliable and easy accessible information on the expected impact of climate change supports the acceptance among political decision-makers and professionals. Demonstration and pilot projects with concrete solutions for the typical impacts of climate change are most helpful for supporting local activities. This is especially true if relevant measures also support other objectives of city planning e.g. flood protection, quality of urban/green spaces, design of public spaces.

First results from monitoring the implementation of Stuttgart Regions adaptation strategy underline the long-term dimension of the approach. Therefore a permanent political support is required as well as the provision of sufficient funding and human resources. Cooperation between municipalities and different administrative tiers can foster political perception and allow a more cost-efficient implementation if their approach and the related activities are coordinated.

The climate change adaptation of a densely populated urban area like Stuttgart Region requires also an adaptation of existing structures and development plans. To change these existing elements, holistic concepts and strategies for the implementation on different planning levels are necessary. Owing to the diversity of climate impacts and to different responsibilities and capabilities of the relevant governmental and non-governmental stakeholders, a vertically integrated approach is necessary to guarantee the interconnection between the different tiers.

References

Baumueller J, Reuter U, Hoffmann U, Esswein H (2008) Klimaatlas Region Stuttgart, Schriftenreihe Verband Region Stuttgart, Nr. 26. Ed. Verband Region Stuttgart, Stuttgart (Verband Region Stuttgart, Regionalplan Region Stuttgart, 2009 Ed. Verband Region Stuttgart, Stuttgart)

TURAS Website. http://www.turas-cities.eu/, www.turas-cities.eu. Accessed 12 Feb 2015

Stuttgart VR (ed) (2009) Verband Region Stuttgart, Regionalplan Region Stuttgart. Verband Region Stuttgart, Stuttgart

Weis M, Siedentop S, Minnich L, Baumüller J, Flaig H, Haakh F, Kaule G, Reuter U (2011), Vulnerabilitätsbericht der Region Stuttgart, 2011 Ed. Verband Region Stuttgart. Online http://www.klimamoro.de/fileadmin/Dateien/Ver%C3%B6ffentlichungen/Publikatione_aus_den_Modellregionen/Stuttgart_Vulnerabilit%C3%A4tsbericht_web.pdf. Accessed 16 Feb 2015

Chapter 18
Influence of Climate Change on Cocoyam Production in Aba Agricultural Zone of Abia State, Nigeria

C.C. Ifeanyi-obi, A.O. Togun, and R. Lamboll

Abstract The paper examined the influence of climate change on cocoyam production in Aba agricultural zone of Abia State. Data for the study was collected using a participatory pair-wise ranking technique from Key cocoyam farmers, village chiefs and Agricultural extension agents in a Focused group discussion and in-depth interview. Findings revealed that the major occupation of the people in the area is farming while the major crops grown are yam, cassava and plantain. Cassava ranked first as the major source of both income and food. On the average, size of farm is 0.3 ha while land acquisition is majorly by inheritance and leasing. The major source of labour is family members and hired labored. The cocoyam cultivars grown in the area are Edeocha, Ede Uhie, okpanambe, and Ede ofe. Farmer's previous harvest and neighbor are the major source of planting material. Major cropping pattern done in the area is mixed cropping while crops planted with cocoyam include Maize, groundnut and vegetables. Farmers aim of producing cocoyam is for consumption and sometimes sales. All farmers agreed that there is a change in the climate of their zone. The major climate variables that is changing and as well affecting cocoyam production in the zone according to the farmers are rainfall, heat (atmospheric temperature) and sunshine (solar radiation) while the major influence of climate change on cocoyam production include decline in yield of cocoyam, reduction of soil fertility, uncertainty in planting and harvesting date, stunted growth of cocoyam, increase in decay of planted corms/cormels and increase loss during storage in the barns. The study then recommends that access to and cost of fertilizer should be enhanced and subsidized by the Government, this will help farmers to have access to fertilizer thereby overcoming the problem of soil

C.C. Ifeanyi-obi (✉)
University of Port Harcourt, Port Harcourt, Nigeria
e-mail: clara.ifeanyi-obi@uniport.edu.ng; obiclara@yahoo.co.uk

A.O. Togun
University of Ibadan, Ibadan, Nigeria

R. Lamboll
University of Greenwich, London, UK

W. Leal Filho (ed.), *Innovation in Climate Change Adaptation*, Climate Change Management, DOI 10.1007/978-3-319-25814-0_18

fertility reduction. It further recommends that improved storage facility should be put in place to reduce the huge loss encountered during storage.

Keywords Cocoyam • Influence • Climate change • Production

Introduction

Cocoyam belongs to the monocotyledonous family Araceae known as the Aroids and is ranked fifth most consumed root and tuber crop in the world after potato, cassava, sweet potato and yam (FAOSTAT 2005). According to FAO (2007), Nigeria is the largest producer of cocoyam in the world with annual production of 5.49 million metric tonnes in 2006 equivalent to 45.5 % of world production. And it is ranked third as the most valuable root crop in the country (Echebiri 2004). The two major cultivated and consumed specie of cocoyam in Nigeria are the Xanthosoma and Colocasia (Chukwu et al. 2010) (Figs. 18.1 and 18.2).

Though Cocoyam is a neglected and underutilized species (NUS), it has tremendous potentials to immensely contribute to poverty alleviation and boost food security in Nigeria. In Southern Nigeria, it is attractive to many households, particularly the low income households. This could be because of its cheaper price hence more affordable to low income rural families than yam and cassava. Nutritionally cocoyam is superior to cassava and yam in the possession of higher protein, mineral and vitamin contents in addition to having a more digestible starch (Parkinson 1984; Mwenye et al. 2011). In addition, it possesses a more edible starch than Yam and Cassava. Also, the vegetable parts of cocoyam are used as food in one form or the other in different parts of the country. Okoye et al. (2008) noted that cocoyam occupies a very important place in rural household food consumption in Southern Nigeria. But in recent times, the

Fig. 18.1 Colocasia esculenta. *Source*: Chukwu et al. 2010

Fig. 18.2 Xanthosoma sagittifolium *Source*: Chukwu et. al. 2010

production of cocoyam is facing myriads of militating factors. Amusa, Enete and Okon (2011) listed some of these militating factors to include high cost of farm input, inadequate extension contacts, inadequate access to inputs, high labour cost, poor soil fertility, poor storage facilities, inadequate or lack of access to mechanized services, poor technical know-how of most farmers, poor road network in the area land tenure problem, relatively old age of the farmers, inadequate finance and lack of access to fund to secure farm inputs. A more critical look at these factors shows that some of them are by product of climate change. For instance, poor soil fertility could result from flooding and other events associated with extreme weather events. These militating factors are trying to extinct cocoyam from the agricultural production system in the country (Onyeka and Nwokocha 2010; Chukwu et al. 2010; Ifeanyi-obi et al. 2013). The crop has become neglected and grossly underutilized. This is evident in reduced availability of cocoyam. Rural households in Southern Nigeria that substitute yam and cassava with cocoyam are gradually being exposed to hunger and malnutrition.

Cocoyam is a tropical and sub-tropical crop grown during the summer in temperate regions with optimal minimum temperature requirement of 21 °C and maximiun of 28 °C, rainfall requirement is optimal minimum of 1800 and maximum of 2700 (FAO ECOCROP). In the face of global climate change, cocoyam production faces adverse weather conditions that affect its production, storage and stability over time. According to Schmidhuber and Tubiello (2007), Ziervogel and Ericksen (2010) and Ingram (2011) climate change is likely to affect all four of the recognized components of food security namely: availability, access, utilization, and stability over time. That could imply that every area of cocoyam farming could be affected by the change in climate. The Nigerian climate review (NIMET 2010) published by the Nigeria meteorological agency noted that the review of Nigerian climate has shown extreme weather events like warmer than normal temperatures, rainfall amounts exceeding long term means. These extreme events led to dislocation of families, destruction of farmlands, infrastructure and loss of lives. According to the report, some of these extreme weather events were quite significant in some part of the country and occurred at critical stages of crop development. It noted that if the present trend of Climate variability continues, it is likely that the frequency and intensity of weather related disasters may increase in the years ahead. Therefore, there is need for adequate understanding of past, present and future climate trends especially as it concerns different crops and livestock produced in the different regions of the country, climate risk information derived from analyses of Nigerian climate data need to be integrated into our national planning and decision-making. This will enable policy makers manage the changing climate and mainstream climate information into our national and global development plans.

In addition, Vermeulen et al. (2012) stated that low-income producers and consumers of food will be more vulnerable to climate change owing to their comparatively limited ability to invest in adaptive institutions and technologies under increasing climatic risks. This reveals more the high vulnerability of the rural populace who are the major producers of cocoyam in Nigeria. Their production activities are mainly weather-dependent and this has resulted in huge losses among the farmers in event of changes in climate variables. Furthermore, very little research activity has been carried out on cocoyam compared to other tuber crops like yam and cassava (Onyeka and Nwokocha 2010). This may be due to little importance attached to the crop in the contemporary Nigerian situation. Bearing in mind the many benefits of cocoyam especially to rural household, it has become very necessary for research work especially in the area of climate change to be carried out to enable farmers (particularly poor rural resource farmers) understand better the vagaries of climate change on their cocoyam farming. It is against this background that this research was aimed to examine the influence of climate change on cocoyam production in Aba agricultural zone of Abia state. The broad objective of this study is to identify the influence of climate change on cocoyam production in Aba agricultural zone of Abia State. Specifically the study examined the socio-economic characteristics of cocoyam farmers in the zone as well as the trend in cocoyam production in the zone.

Methodology

This study was conducted in Aba agricultural zone of Abia State. Abia state is one of the five states that made up the Southeast Nigeria. It is located within latitudes 4°47″35‴N and 7°7″44‴N, and longitudes 7°54″26″E and 8°27″10″E in the tropical rain forest zone of Nigeria, with mean maximum temperature of 27 °C, and total annual rainfall exceeding 2500 mm (Ezemonye and Emeribe 2012). The zone is mainly agrarian and inhabitants depend more on land resources, due to its dense population averaged to about 1000 people/km^2. Mbakwe et al. (2004) on the other hand stated that the climate of Southeast Nigeria is influenced by the three major air masses namely; the equatorial Maritime, the Equatorial Estuaries and the Tropical continental air masses. They further explained that the equatorial Maritime originates from the southwest and is warm and very moist, the tropical continental has its origin in the northwest and is warm and very dry while the equatorial estuaries is a cool dry upper air mass which blows from east to west but is occasionally deflated downwards. According to them rainfall is the key climatic variable while soil type is predominantly loam; the main cropping systems in the zone is mixed inter cropping system with the number of crops that are mixed-croppedranging from two to over five. Major crops produced in the area include yam, cassava, maize, cocoyam and vegetables. Diseases, weeds, pests and climate change constituted the major constraint to increased production. Most farmers in the area still depend on locally manufactured tools for their farming activities.

Nine out of thirteen agricultural blocks in Aba agricultural zone were used for the study. They were selected based on rurality (dominance of rural communities in the block). Data was collected for this study using participatory pair-wise ranking technique. This was done in three different focused group discussions (FGD). Each of the focused group discussion comprised of cocoyam farmers from three blocks in the zone. The major limitation of the study was the inability of the respondents to give precise timing of events mentioned, this was observed to be as a result of poor record keeping on the part of the farmers. Also, some of the farmers saw the FGD as a government rapid rural appraisal of their needs and as such were not forth coming with honest responses. It took much explanation to convince the farmers that the activity was purely for research purposes before they could give honest answers. Furthermore, poor state of the roads made the journey a very hectic one.

Results and Discussion

Trends in Cocoyam Production in the Study Area

The area of study is a forested rural area though deforestation has significantly reduced the forested area in the area. Monogamy is the dominant marriage system with average household size of between 6 and 7 persons. The major occupation of the people is farming though presently many of them do other things outside farming to help complement their income unlike in the olden days when they depend solely on farming. Many of the farmers owned farms of between 0.2 and 0.33 ha in size while the major method of land acquisition is inheritance and leasing. Their major source of labour is family members and hired labour where the farm is large or the farmer has grown weak. Other livelihood activities of the people include Weaving, trading and tailoring. Major Crops grown in the zone are yam, cassava, maize, vegetables, cocoyam, plantain and pineapples. These crops were further ranked for their importance as source of food and income. Table 18.1

Table 18.1 Pairwise ranking of the major crops grown (as source of income)

S/N	Crops	1	2	3	4	5	6	7	Scoring	Ranking
1	Yam	XXXX	2	1	1	1	6	1	4	3
2	Cassava		XXXX	2	2	2	2	2	6	1
3	Maize			XXXX	4	3	6	3	2	5
4	Vegetables				XXXX	4	6	4	3	4
5	Cocoyam					XXXX	6	7	0	7
6	Plantain						XXXX	6	5	2
7	Pineapples							XXXX	1	6

Source: FGD (2015)

show that Cassava ranked first as the major source of income followed by plantain and yam. According to the farmers, cocoyam does not have much market value as the demand for it is very low. As regards source of food, cassava ranked first followed by yam and vegetables. Cocoyam still was not recognized as a major source of food. The farmers further explained that in the 1960s and early 1970s, cocoyam was a major food for the people. It was difficult to find a household that does not depend on cocoyam for their daily food intake. But recently, it is no longer same. This is as a result of the introduction of other food like rice which was not commonly found before, also the yield of cocoyam gradually declined on its own, unlike in the past where little effort produces bumper harvested. Recently, even with much effort and manure addition the yield is not encouraging. People gradually began to attach more importance to other crops like cassava, yam, maize and plantain. Cocoyam is now seen as a poor man's food (the last resort, when the other foods are not available) (Table 18.2).

Both men and women are equally involved in cocoyam production. This is surprising as cocoyam is known to be 'a woman's crop' in many parts of Southeast zone of Nigeria. The different types of cocoyam cultivars grown in the area include edeocha, ede uhie, okpanambe, and ede ofe. From the description of the cocoyam cultivars, the five varieties planted by these farmers are from *Colocasia* and *Xanthosoma* family. According to the farmers, planting materials are acquired mainly from the farmers previous year's harvest. Where it is not sufficient, they source from fellow farmers at no cost, pick from the bush and sometimes buy from the market when it is needed in large quantity (which rarely occurs). The recycling of planting materials year by year by cocoyam farmers may be one of the contributing factors to the decline in cocoyam yield. According to Mbanaso (2007) recycling of planting materials results in accumulation of pathogens in them and translates to decline with time. He therefore recommended generation of 'clean' planting material through meristem tip culture and multiplication of these will not only stem this process but increase yield as well.

The major cropping pattern in the study area is mixed cropping. Cocoyam is inter cropped with maize, groundnut and vegetables. In rare occasions, it is planted with cassava or plantain. Similarly, Ugbajah (2013) identified Cassava, maize, yam and plantain/banana to be the major crops planted with cocoyam Anambra state. The major aim of producing cocoyam according to the farmers is consumption and on rare occasion for sales. As regards time of planting cocoyam in the zone, the farmers explained that cocoyam is basically planted in April and May and harvested December and January. April used to be the major time of planting in the 1970s but due to fluctuations in weather events, its planting time seems to be shifting down to May, even to June.

As regards cocoyam yield in the study area, the farmers explained that the yield has not been steady. In some years, it increases while in some it declines. But comparing the yield with what was obtained in the 1960s and 1970s, there is a decline in yield. Their reasons for the decline includes: no interest in cocoyam as a crop among the people, changing weather which affects the production, loss of cocoyam during storage and even in the farm as a result of too much heat, no high

Table 18.2 Pairwise ranking of the major crops grown (as source of food)

S/N	Crops	1	2	3	4	5	6	7	Scoring	Ranking
1	Yam	XXXX							4	2
2	Cassava	2	XXXX						6	1
3	Maize	1	2	XXXX					2	4
4	Vegetables	1	2	4	XXXX				3	3
5	Cocoyam	1	2	3	4	XXXX			1	5
6	Plantain	6	2	6	6	6	XXXX		2	4
7	Pineapples	1	2	3	4	5	6	XXXX	0	6

Source: FGD (2015)

yielding or disease resistance specie (it's the same specie they have been cultivating), Little market demand for cocoyam, inability to find other uses of cocoyam that can make the market to move, feel cocoyam is not a useful crop. From the discussion with the farmers, it is evident that farmers in the study area do not attach much importance to the production of cocoyam. According to them, it's not a useful crop both as source of food and income. They were surprise to hear that cocoyam has good nutritive starch value when compared to cassava and yam that is their cherished crop. Similarly, Chukwu, Mbanaso, Okoye, Onwubiko, and Nwosu identified neglect of cocoyam, declining soil fertility, poor value addition, high deterioration in storage, low genetic base, pests and disease problem as the major challenges of cocoyam farming in Southeast zone of Nigeria. Also, Onwubuya and Ajani (2012) and Amusa, Enete and Okon (2011) in their different identified low soil fertility as one of the major constraints to effective production of cocoyam in Nigeria.

The problem of apathy for the crop and poor value addition to the crop are very evident in the study area. Very low importance is attached to the crop in the study area. Also, cocoyam is mostly consumed through boiling and eaten with palm oil. Only very few process it through drying (flour). No other means of processing was known by the farmers in the study area.

The major disease of cocoyam in this area 'onwu ede', which from the description is leaf blight disease while the common pest is termites.

Cocoyam Farmer's Awareness of Climate Change

The farmers all agreed that they have noticed change in the climate of their zone. According to them, the change gradually started in the 1990s and became very noticeable from the year 2000 upwards. They described the change in climate as a general unpredictable weather manifested in fluctuations in rainfall pattern, unusual heavy rainfall leading to flooding, long duration of drought, very hot sunshine, high atmospheric temperature and high humidity when it eventually comes. The causes of this change in climate according to the farmers include wrath of God on mankind as a result of our sins, changing lifestyle of human being (introduction of all sorts of chemical to our environment), deforestation and use of chemicals for farming. According to the farmers, the major climate variables that are changing and as well influencing cocoyam farming in the area are rainfall, heat (atmospheric temperature) and sunshine (solar radiation).

Influence of Climate Change on Cocoyam Production

The farmers listed the influence of climate change on cocoyam production to include uncertainty in onset and cessation of rainfall affects the planting and harvesting of cocoyam, fluctuation in rainfall affects the growth of cocoyam, the fluctuation in rainfall induces pests and diseases outbreak and weed growth, the heavy rainfall causes flooding and even erosion, reduction of soil fertility, loss of farmlands to flooding events, decay of planted corms as a result of long period of drought between rainy season, decay of planted corms as a result of increased intensity of heat (atmospheric temperature), prolong dry season also increases decay in planted material, decline in yield, increased loss of corms during storage in the barns as a result of increased intensity of heat, also prolong dry season (drought) causes more decay of corms in the barn, increased loss during drying as a result of unusual rainfall in dry season, faster drying as a result of high intensity of the sun. The pairwise ranking of the factors showed that decline in yield of cocoyam as a result of changing climate variables is the major influence of climate change in the study area. This is followed by reduction in soil fertility, uncertainty in planting and harvesting dates, stunted growth of planted cocoyam and increase in decay of planted cocoyam. This finding is similar to the findings of Ukonze (2012) which identified decline in yield, soil fertility reduction and increase incidences of flooding and erosion as impacts of climate change on cocoyam production. The findings of this study are in agreement with Ozor (2009) which stated that the increasing incidence of flooding, erosion, bush burning will definitely lead to low yield, which will change the supply and demand pattern, greater post harvest losses, loss of arable land and increased growth of weeds among others (Table 18.3).

Conclusion and Recommendations

The study examined the trend of cocoyam production and the influence of climate change on cocoyam production in the study area. The findings of this study revealed that cocoyam production in the area is declining as a result of many factors inclusive of climate change. Changing consumption pattern and very low importance attached to the cocoyam can be seen to be a major factor influencing the production of cocoyam. Farmers seem to be ignorant of the nutritive value of cocoyam compared to their major food source (cassava and yam). Inability of farmer's to generate reasonable income from cocoyam as result of low market demand seem to be among the major players of cocoyam production decline. The major influence of climate change on cocoyam farming is a general decline in yield as a result of climate change followed by soil fertility reduction.

Based on the result of this study, it was recommended that there is a need to develop value-added products from cocoyam in order to extend the shelf life, meet consumer acceptability and improve its usefulness thereby enhancing demand of the product. This could be done through enhancing the processing of cocoyam into more appealing by-products as well as developing other uses of cocoyam by-products. This will stimulate the demand for cocoyam products thereby acting as incentive to increased production of cocoyam by the farmers. In addition, introducing cocoyam snacks and dishes that are appealing to children and youths

Table 18.3 Pairwise ranking of the influence of climate change on cocoyam

S/N	Statements	1	2	3	4	5	6	7	8	9	10	11	12	Scoring	Ranking
1	Uncertainty in rainfall affects planting date and harvesting date	XX												7	3
2	Stunted growth of cocoyam is experienced	1	XX											7	3
3	Increased pests and diseases	1	2	XX										5	4
4	Increase weed	1	2	3	XX									4	5
5	Flooding and erosion events are increased	1	2	3	4	XX								3	6
6	Soil fertility is reduced	6	6	6	6	6	XX							10	2
7	Reduced farmlands	1	2	3	4	5	6	XX						2	7
8	Increase in decay of planted corms	8	8	8	8	8	6	8	XX					9	3
9	Decline in yield due to change in climate variables	9	9	9	9	9	9	9	9	XX				11	1
10	Increased loss during storage in the barns	10	2	10	10	10	6	10	8	9	XX			7	3
11	Increased loss during drying due to uncertainty in rainfall	1	2	3	4	5	6	7	8	9	10	XX		1	8
12	Fast drying due to increased solar radiation	1	2	3	4	5	6	7	8	9	10	11	XX	0	9

Source: FGD (2015)

may be another good way of stimulating more production of cocoyam and making it a valued food source as farmers tends to focus more on crops that are either acceptable to family members or generate income for them.

Also, the notion that cocoyam is a poor man's food need to be dispelled and the enlightenment of the people on the nutritive value of cocoyam need to be done more extensively. This could be achieved through massive awareness programme on cocoyam using the existing extension agents.

Furthermore, it was shown that soil fertility reduction is a major influence of climate change on cocoyam farming. This could be tackled through making the procurement of inorganic fertilizers more accessible to the farmers. Other problems that anchor on fluctuations in climate variable could be tackled by educating the farmers more on climate change issues making them to make their farming activities more flexible to accommodate changes. Also, viable research findings and technology development that have the potential to enhance the farmer's cocoyam activity should be disseminated to the farmers to help them overcome the negative influences of climate change on their cocoyam farming. For instance the farmers in the study area are yet to know of the Gocing cocoyam storage system which is an improved storage system developed by the cocoyam unit of the research institute coordinating in the study area (National root crop research institute, Umudike).

Acknowledgements This research is supported by funding from the Department for International Development (DfID) under the Climate Impact Research Capacity and Leadership Enhancement (CIRCLE) programme.

References

Amusa TA, Enete AA, Okon UE (2011) Socio-economic determinants of cocoyam production among small holder farmers in Ekiti state, Nigeria. Int J Agric Econ Rural Dev 4(2):97–109

Chukwu GO, Nwosu KI, Mbanaso ENA, Enyinnaya AM (2010) Field production of cocoyams. In: Ukpabi UJ, Nwosu KI (eds) Yam, cocoyam and sweet potato. Production and post-harvest management. National Root Crops Research Institute (NRCRI), Umudike

Echebiri RN (2004) Socio-economic factors and resource allocation in cocoyam production in Abia State, Nigeria: a case study. J Sustain Trop Agric Res 9:69–73

Ezemonye MN, Emeribe CN (2012) Rainfall erosivity in Southeastern Nigeria. Ethiop J Environ Stud Manag (EJESM) 5(2):112–122

FAOSTAT (2005) Statistical database. Agricultural production of primary crops. Food and Agriculture Organisations of the United Nations (FAOSTAT), Washington, DC

FAO Statistics (2007) Food and Agriculture Organisation database results. FAO, Rome

Ifeanyi-obi CC, Adesope OM, Issa FO, Ugwuja VC (2013) Organic agriculture as a resuscitating option for cocoyam production in Nigeria. Int J Appl Res Technol 2(6)

Ingram J (2011) A food systems approach to researching food security and its interactions with global environmental change. Food Secur 3:417–431

Mbakwe R, Ukachukwu SN, Muoneke CO, Ekeleme F (2004) Problem identification and development of research base. In: Umunna RPA, Onwudike OC, Uwaegbute AC, Edeoga HO, Nwosu AC (eds) Farming systems research and development in Nigeria—principles and practice in Humid and derived Savanna southeast zone. Michael Okpara University of Agriculture, Umudike

Mbanaso ENA (2007) Cocoyam program. www.nrcri.gov.ng/page/cocoyam.html

Mwenye OJ, Labuschangne MT, Herselman L, Benesi IRM (2011) Mineral composition of Malawian cocoyam (Colocassia esculenta and Xanthosoma sagittifolium genotypes). J Biol Sci 11(4):331–335

NIMET (2010) Nigeria climate review bulletin 2010. Nigeria Meteorological Agency (NIMET). http://www.nimetng.org/uploads/publication/2010%20Climate%20Review.pdf. Retrieved 11 Mar 2015

Okoye BC, Onyenweaku CE, Ukoha OO, Asumugha GN, Aniedu OC (2008) Determinants of labour productivity on small-holder cocoyam farms in Anambra State, Nigeria. J Sci Res Essay 3:559–561

Onwubuya EA, Ajani EN (2012) Strategies to improve the production and processing of cocoyam among women farmers in Anambra state, Nigeria. Univers J Educ Gen Stud 1(6):169–173

Onyeka TJ, Nwokocha HN (2010) Important diseases and pests of cocoyam and their management. In: Ukpabi UJ, Nwosu KI (eds) Yam, cocoyam and sweet potato. Production and post-harvest management. National Root Crops Research Institute (NRCRI), Umudike

Ozor N (2009) Understanding climate change: implication for Nigerian agriculture. Policy and education. In: Paper presented at the national conference on climate change and the Nigerian environment, organized by the department of geography. University of Nigeria, Nsukka, 29 July–2 Aug 2009

Parkinson S (1984) The contribution of aroids in the nutrition of people in the south pacific. In: Chandra S (ed) Edible aroids. Clarendon, Oxford, UK, pp 215–224

Schmidhuber J, Tubiello FN (2007) Global food security under climate change. Proc Natl Acad Sci USA 104:19703–19708

Ugbajah M (2013) Enhancing income through cocoyam production, processing, and consumption patterns, in Dunukofia local government area of Ambra state, Nigeria. Green J Soc Sci 3 (6):334–339

Ukonze JA (2012) Impact of climate change on cocoyam production in South Eastern Nigeria. Int J Educ Sci Public Policy Afr 2(1):161–168

Vermeulen SJ, Campbell BM, Ingram JSI (2012) Climate change and food systems. Annu Rev Environ Resour 37:195–222

Ziervogel G, Ericksen PJ (2010) Adapting to climate change to sustain food security. WIREs Clim Change 1:525–540

Chapter 19
Between Intention and Action: Psychosocial Factors Influencing Action on Climate Change in Organisations

Nadine Andrews, Stuart Walker, and Kathryn Fahy

Abstract Whilst global awareness of the importance and urgency of acting to mitigate climate change and its impacts is generally high, actual behaviour has matched neither the scale nor the complex nature of the challenge. Understanding why despite good intentions appropriate action is not forthcoming is critical if we wish to avoid catastrophic consequences for social justice and the wellbeing of humans and other species. Research gaining insight into underlying psychosocial processes has an important contribution to make in this regard, yet it tends to be overlooked.

This paper draws on an empirical interdisciplinary study enquiring into the experience of individuals acting to influence the organisation with regard to environmental decision-making. The study investigated psychosocial factors that may influence motivation, resilience and effectiveness, specifically psychological threat coping strategies, innate psychological needs, identity salience and ways of conceptualising experience.

Our study illuminates the complex nonlinear dynamics between these psychosocial forces, and reveals tensions in satisfying needs, and in the effectiveness of coping strategies such as suppressing 'deep green' identity, suppressing negative emotion about climate change, and in going into nature places.

The findings contribute nuanced insight to the body of knowledge about the dynamics of underlying psychosocial forces that influence approaches to climate change and other pro-environmental behaviours.

Keywords Psychosocial • Coping • Emotion • Organisation • Nature relatedness

---

N. Andrews (✉)
HighWire Doctoral Training Centre, Lancaster University, Bailrigg, Lancaster LA1 4YW, UK
e-mail: n.andrews@lancaster.ac.uk

S. Walker
Imagination Lancaster, Lancaster University, Bailrigg, Lancaster LA1 4YW, UK

K. Fahy
Lancaster University Management School, Lancaster University, Bailrigg, Lancaster LA1 4YW, UK

W. Leal Filho (ed.), *Innovation in Climate Change Adaptation*, Climate Change Management, DOI 10.1007/978-3-319-25814-0_19

Introduction

Climate change is understood in this paper as one of several types of environmental change processes caused by human actions that threaten life on our planet (Rockström 2009). One might think that knowledge of the facts would lead to change in human behaviour to bring the planetary systems back within safe operating boundaries but despite numerous climate change panels and treaties substantial change has not occurred, leading to doubts that staying within the crucial 2° threshold is realistic (Klein 2014).

Understanding why despite good intentions appropriate action is not necessarily forthcoming is critical if we wish to avoid catastrophic consequences for social justice and the wellbeing of humans and other species. Psychology research has an important contribution to make in this regard, in developing our understanding of the psychological dimensions of global climate change (APA 2009).

One such contribution is research on climate change behaviour that finds evidence of a phenomenon of rebound, which is when reduction in energy use in one area is cancelled out by an increase elsewhere (Jenkins et al. 2011; EST 2011). An example of rebound is switching to an energy efficient boiler but then leaving the heating on for longer. The rebound effect has obvious implications for climate change mitigation.

Other studies find that whilst values can be reliable predictors of behaviour, they are not the sole determinants (Maio et al. 2001). Factors that may influence the relationship between values and behaviour have been studied in the context of environmental behaviour (Schultz and Tabanico 2007; Wright et al. 2012) and ethical behaviour more generally (Palazzo et al. 2012) but it is a topic that remains relatively under-researched.

Given this gap in research, and because businesses have a significantly higher ecological impact than households (DECC 2013), we conducted a study exploring psychosocial factors influencing pro-environmental decision-making by individuals in their organisation. Pro-environmental decision-making includes actions to reduce emissions of greenhouse gases such as fossil fuel energy efficiency and use of renewable energy, as well as reducing waste and loss of natural habitats.

Taking an individual-level perspective, we enquired into the experience of six individuals with roles in their organisation concerned with environmental policy, strategy and practice, aiming to illuminate psychosocial processes influencing their motivation, resilience and effectiveness. Of particular interest were psychological threat coping strategies, innate psychological needs and ways of conceptualising experience, building upon previous cognitive linguistics and psychology studies relating to pro-environmental behaviour (Darnton and Kirk 2011; Crompton and Kasser 2009; Lakoff 2010).

This paper focuses on discussing findings relating to psychological threat coping strategies and the implications of these responses. These findings contribute to knowledge about climate change adaptation from a psychosocial perspective that we believe tends to get overlooked. As we dig below the surface layer of behaviour to underlying fears and mental models, we see how the social and the psychological are fundamentally entangled.

We begin with a brief overview of the theoretical background and methodology, and follow with a discussion of key findings illustrated with diagrams and quotes from participants using fictitious names.

Theoretical Background

We drew upon various conceptual resources in our study; of most relevance to the findings discussed in this paper are the following:

Psychological Threat Coping Strategies

Macy (1993) says we are caught between a sense of impending apocalypse and an inability to acknowledge it. The enormity of the environmental challenge facing humanity poses a profound psychological threat including existential threat, threat to the integrity and stability of self-identity, and threat to self-esteem (Crompton and Kasser 2009; Weintrobe 2013; APA 2009).

Psychological threat is stressful and unpleasant; the tendency is to attempt to alleviate stress and associated emotions through defence mechanisms and coping strategies, which in terms of addressing environmental issues may be adaptive or maladaptive. There is often a fear that allowing oneself to feel despair, for example, will break us apart, that we will get stuck in a dysfunctional state. Despair, Macy says, is "tenaciously resisted because it represents a loss of control, an admission of powerlessness. Our culture dodges that by demanding instant solutions when problems are raised" (1993, p. 18).

Regulating strong negative emotion through suppression however takes emotional and physical effort and impairs our ability to think as it diverts cognitive resources away from other tasks. Emotions that aren't accepted steer us unconsciously. Dissonance between felt and expressed emotions may induce stress, and has been shown to have detrimental effects on physiological and cognitive functioning (Rogelberg 2006).

Adaptive strategies for coping with psychological threat involve acceptance of facts and accompanying emotions, and regulating emotions. These strategies promote psychological adjustment and stimulate actions appropriate to the new reality (Crompton and Kasser 2009; Macy and Brown 2014).

Psychological Needs

Self Determination Theory (Deci and Ryan 2000) proposes that we have three main innate psychological needs that require satisfying for psychological wellbeing: need for competency, for relatedness and for autonomy. The extent and the manner in which these needs are satisfied have consequences for vitality, motivation, resilience and effectiveness.

Values and Identity

Numerous studies have found that self-transcendent values and intrinsic goals are associated with pro-environmental behaviour (Schwartz 1992; Schultz et al. 2005; Grouzet et al. 2005; Kasser and Kanner 2004). It seems that values need to be linked to the self in order to be influential in choices that are made (van der Werff et al. 2013). Seeing nature as part of one's in-group, having a sense of self as part of nature, and feeling connected with nature are identities associated with pro-environmental behaviour (Crompton and Kasser 2009; Schultz et al. 2005). However, our identities are multiple and constantly shifting. Identities are arranged hierarchically in the mind: the salience of a particular identity depends on context and commitment to that identity (Clayton and Opotow 2003).

Organisational Context

Wright et al. (2012) highlight the contradictory space within which sustainability specialists are situated due to conflict between climate change and economic growth discourses, and the need for managing identities to overcome this conflict in order to politically influence the organisation. Studies into unethical behaviour in organisations show how context than be stronger than values, reason and good intentions (Palazzo et al. 2012).

Ways of Conceptualising Experience

We draw on the idea of cognitive frames and conceptual metaphors to analyse ways in which research participants make sense of their experience (e.g. see Lakoff and Johnson 1980; Blackmore and Holmes 2013) Studies show that conceptual metaphors and cognitive frames influence perception and behaviour (e.g. Thibodeau and Boroditsky 2011).

Methodology

The study focuses on six individuals with pro-environmental values and formal roles in their organisation in relation to environmental policy, strategy and practice. The organisations were in local and regional government, social housing, credit union and hospital sectors in the UK and Canada.

Using semi-structured interviews, we enquired into their experience as they act to influence their organisations. The 2-hour interviews were audio recorded and transcribed verbatim. Data was analysed using the framework of Interpretative Phenomenological Analysis (IPA), which is concerned with gaining insight into how someone experiences and makes sense of a given phenomenon. IPA takes a critical hermeneutic approach that allows for the development of alternative narratives, informed by extant theory, whilst also leaving space for both intuitive and intersubjective approaches to data analysis (Smith et al. 2009). IPA fits within a critical realist/contextual constructivist epistemology. The interview transcripts were analysed for cognitive frames and conceptual metaphors, coping strategies, emotions and physical sensations, conflicts and ambivalence. Exploratory, emerging and higher order themes were recorded on a case-by-case and cross-case basis.

This kind of in-depth qualitative research with a small number of case studies focuses on nuance of subjective experience. We do not make any claims therefore about the generalisability of our results for a wider population. Furthermore, the participants were from public and third sector organisations in the UK and Canada; none worked within the private sector or in non-Western societal contexts.

Results

Our analysis reveals interesting and complex relationships between various psychosocial processes influencing motivation, vitality, resilience and effectiveness, as outlined in the diagram below, which shows the nonlinear nature of the phenomenon under investigation. We provide an explanation of the diagram, illustrated with quotes from participants, using fictitious names (Fig. 19.1).

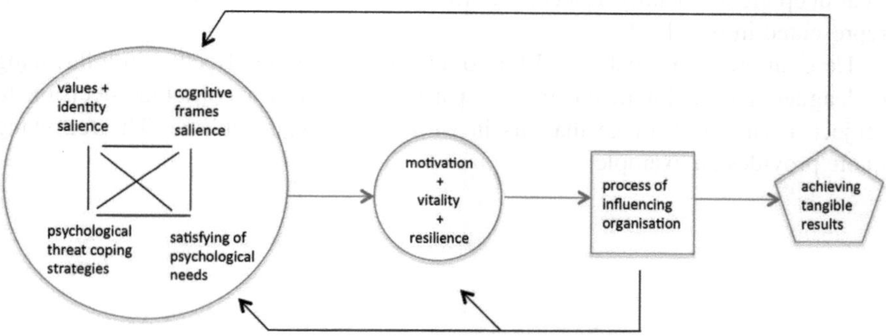

Fig. 19.1 Overview of interrelating processes in participant experience

Working backwards from right to left along the blue arrows, we find first of all the primary concern of the participants is to make a difference, to get tangible results, whatever that meant to them in the context of their work at the time of interview. With financial pressures as extreme as some organisations are under, proving your worth to the organisation by delivering results is critical if you want to keep your job.

> Ash: these aren't the times when you can afford, just someone going around being the green conscience of the organisation you know, got to deliver results to justify your own existence really

Secondly we find that they sought to achieve these results through the process of influencing thinking and decision-making. Thirdly, motivation, vitality and resilience are psychological inputs that they bring and which affect their experience of influencing and interacting with the organisation. Fourthly, interrelating factors of salience of pro-environmental values and identities, salience of particular ways of conceptualising experience (cognitive frames), strategies adopted to cope with psychological threat, and the extent to which innate psychological needs were satisfied or thwarted, have implications for participants in terms of their motivation, vitality and resilience in doing their work effectively.

Now to the black arrows: achieving tangible results influences the circle of factors, for example by satisfying competency needs. The participant experience of interacting with the organisation affects their motivation, vitality and resilience, and influences the circle of factors, creating feedback loops as shown. The following quote is an example of how values, competency, motivation and vitality are linked:

> Rosemary: being in the system and trying to alter it is energetically exhausting... because I care about that I want to do a really good job so the pressure is immense

Dynamics of Psychosocial Forces

To illuminate the relationships in more depth and detail, we modelled some of the key dynamics that appear to be in play in the situations described by the participants. We will now focus on one model that shows how suppression or expression of a 'deep green' identity as a coping strategy may interrelate with other factors, as represented in Fig. 19.2.

Here, in order to satisfy need for competency by being effective in influencing colleagues and senior managers, the participant makes a conscious decision to project a particular image that fits in with the dominant culture. The following quote provides an example:

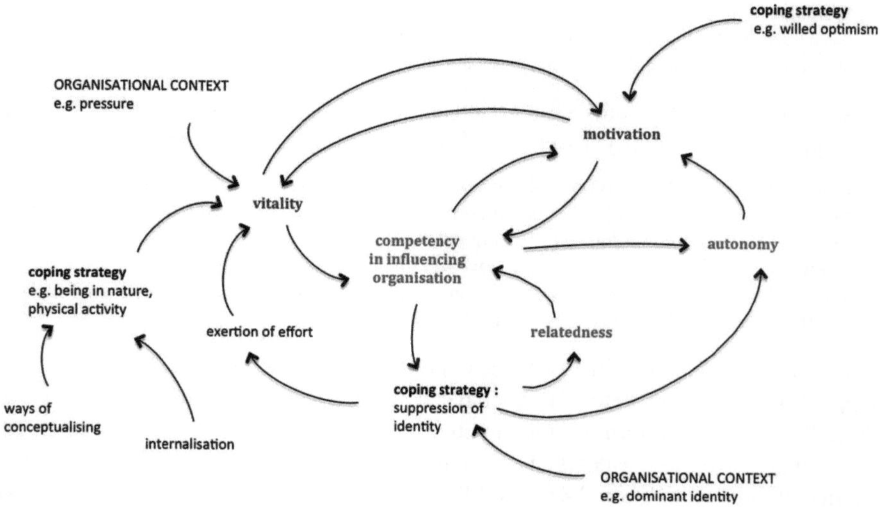

Fig. 19.2 Diagram of factors influencing competency

> Ash: I think the idea of a kind of deep green where you get into an almost spiritual sense of it, of every species has its right to life and should be respected and so on um I have some sympathy with that view... I've made a priority in my role getting the ear of (senior executives) which requires get- being credible... so early days fresh out of university very clear I'm not going to wear sandals bring lentil sandwich to work and have a beard... So being seen as being credible and professional

The participant's concern to maintain good working relationships is conveyed in this quote:

> Ash: But being seen as a kind of agent of change in a kind of positive credible way rather than as- a stirrer, or you know trouble maker

So perceived threat to competency need satisfaction triggers a coping strategy that involves suppressing an identity that might jeopardise credibility. If the strategy works, it can create a self-reinforcing feedback loop where the participant keeps suppressing that identity in order to keep fulfilling need for competency.

The strategy supports relatedness need satisfaction, but perhaps only to some extent:

> Ash: I work with lots of people across the organisation but their priority is saving a million quid for the Property budget or you know they'll do the carbon stuff but we're not peas in a pod or birds of a feather or anything you know so they're not really in the same place as me

With suppression, inner coherence is likely to be undermined, as the following quote suggests:

> Ash: Ah occasionally I wonder I think about it less now but occasionally sense of whether I did the right thing. Harking back to a sense, I was leaving university and I remember reading stuff by Jonathan Porritt ... whether you can be part of the system and change it from within... And another view saying no it's like everything else that starts outside the

system as a force for change all of these things they just get absorbed and you know you'd have been out there on the barricades and instead they've got you in a nice professional job sitting in an office maintaining the status quo broadly

Deficiencies in autonomy satisfaction are likely to have an impact on motivation. Suppression of identity also takes effort, and effort involves energy.

Ash: I could, I could call myself Ash Corporate, you know and I live at number 1 Corporate Street Corporate Town and I love doing corporate and just talk about- and he still wouldn't believe I was anything other than Friends of the Earth in residence and he's just got that despite my best efforts not to you know wear a green shirt and uh open toed sandals or try and speak with too much passion or emotion about why we need to act, which is why I find it difficult-because I so consciously shut that out

As previously noted, in diverting cognitive resources away from other tasks, suppression of strong emotion can have consequences for cognitive functioning (Rogelberg 2006), undermining competency.

There are two more aspects of the diagram we will now discuss: motivation and vitality. Firstly motivation. We found that in order to create personal meaning and greater inner coherence, which contributes to satisfying autonomy needs, participants tell themselves a story that justifies their experience and provides motivation to keep doing what they are doing. Common to all participants, one story we found can be interpreted as a form of willed optimism (Foster 2015) where the logic goes like this: it can't be too late to make a difference because if it is too late then what has all this effort and suffering been for? The stories have narratives such as hero/saviour, doing good, doing your bit, and fulfilling your destiny. But if the story collapses, what then? As one participant said:

Rosemary: my worth is wrapped up in how successful I am about protecting the environment of the region. Risky! (laughs)

Secondly vitality. To maintain levels of personal energy required for action, participants used a variety of coping strategies. Two that we shall focus on in this paper are going into natural places and doing physical activity. Both of these strategies have the potential to create space in the mind to retreat from the stressful and often frustrating experience of working to protect the natural environment in a largely unsupportive organisation, to re-energise, and in the case of being in nature, to re-new motivation. However, we found the effectiveness of these strategies with our participants is influenced by two factors in particular:

- The extent to which their motivation for doing the strategy was internalised
- How they conceptualised the experience

The following quote shows how physical exercise is not fully internalised, so it is not an autonomous behaviour. The participant's preferred coping strategy appears to reduce vitality, not increase it, and evidently he has sufficient self-awareness to recognise the problems of his strategy:

Jay: I'll go home and get drunk (laughs) to get out the frustration... so I'm quite self destructive with it (laughs). Unfortunately. Whereas I need to get to a balance of start-shake that off and start trying to be constructive with that. And go down the gym, take the tension out in the gym Jay because that'll make you stronger. Rather than drinking which will make you weaker

With regards to being in nature, all participants described experiences involving close observation and multisensory interaction with the natural world, and recognised its intrinsic value. With these kinds of intimate embodied experiences, such appreciation is not just an intellectual abstract idea it is subjectively felt. These practices are located externally in a physical place and they stimulate internally a sense of mental spaciousness and emotional stability. It is a fully embodied interactive sensory experience of being in the world.

Heather: when I feel most at peace, calmest and happiest is sort of being outside with some sort of natural environment... I think when I'm outside particularly if I'm sort of in the Lakes on a mountain I feel completely and utterly at one, you can almost feel like the earth beats I know that sounds a bit sort of em a bit odd but I when you can hear everything and you can sort of smell outdoor smells and you can sort of touch the grass I just feel completely at one with sort of the rest of the planet really, which is quite nice. Em it re-energises me, em gives me sense of peace, it just feels fantastic so that's why I try and get outside when I can

It can also influence motivation as well as vitality:

Rosemary: Tremendous sense of calm actually and starts to... give I don't know help me reflect... So its em yeah just going outside just being outside reminding me that's why I come and do this stuff because that's quite hard sometimes to you know well battling within a public organisation that's going through massive budget cuts and lots of pressure to do this quickly and we've got to do that, to just remember ok that's why I'm doing it out there um and so that's why

But these experiences of intimate relatedness are not consistent. There are instances when participants feel closer to or more part of nature and instances when they feel more distant or detached. It appears it is not enough to simply be with the body in a natural place. If the mind is distracted elsewhere then the connection, and the benefits, are diminished.

And if the body is not situated in a natural place then this too seems to make it harder to feel a sense of connectedness with nature:

Ash: I think being in any city makes you feel apart from it (nature)... I think it's I think its easier to forget about it here (in the office in town)... there's almost a sort of once there's an out of sight out of mind isn't there and the risk I think with some of this is if its just not around you at all you'd think you miss it but you don't wander around going 'where are the bees'.

Robin: I'm probably thinking about it more when I'm in it (nature) that when I'm not. Yeah so it's probably a natural consequence of actual location, the physicality of the environment.

The implication is that for connection—and hence the restorative benefits—to be felt most strongly, mind and body and natural place need to be present all together. Which is clearly problematic with continuing destruction of natural habitats.

In the above examples, nature is conceptualised as a place, as an object: 'it'. Objects have boundaries that separate them from other objects. This creates the possibility for nature being somewhere where we are not. In other words it sets up the possibility for separation and alienation, which is precisely what the participants are seeking to overcome by going 'into' nature. This is one way of framing experience that we propose may influence not just the subjective experience but also as suggested in the quotes, how much or little the participants are actually thinking about the natural world. This raises the question, if they are not thinking as much about the natural world when indoors at work, how is this affecting their decision-making?

Emotional Responses to Climate Change and the Ecological Situation

There is one further finding we would like to discuss: the coping strategy of suppression of emotion.

When thinking about the natural world and what humans are doing to it, participants described emotions such as *sadness, frustration, angry, worrying, overwhelmingly distressing, depressing, upset, gloomy, melancholy* and *deeply disturbing.*

As people working with the express aim of environmental sustainability and protection of natural habitats and species, the participants seemed to accept the reality of the ecological situation, and the resulting threat it poses to human existence, modern way of life or standards of living, and to loss of wildlife and natural spaces. In light of this, these emotions are an appropriate response. However, whilst the facts as provided by science appear to be accepted, we find that the emotions accompanying these facts are not fully accepted or engaged with.

Consider the following quotes from participants:

> Robin: I think the initial emotion is anger or frustration. But I try and not retain those feelings really because they're quite self-destructive... I try to remain positive... I try to find reasons to be optimistic

> Ash: Emotions I suppose when I think about it which I try and avoid think- it's rare that I- it makes me feel sad. Uh... I think I just feel a sense of melancholy about it really... Um... I tend not to explore them or I think I've got them in a box in my head and I carry on um being a parent and being a sustainability manager and trying to do good stuff. So and I kind of I you know in a box sitting in an attic in my mind really so I don't think its particularly helpful to explore because its kind of disabling in a way really and disheartening. So I think I put it in a box in the attic

> Heather: I think the key thing is recognising it so I'm getting better at recognising when I feel that sense of tension or worry, so if I can recognise it I'll try do something quickly about it so like I say if I can go for a run I'll go for a run, or I'll do an exercise or I'll do something positive so I'll try and not stay with it too long really otherwise I get a headache or just sort of feel depressed or unhappy so. I think negative emotions is sort of it's healthy to feel a reaction to it but em I think it's trying to sort of then turn it into either a positive action or a way of dissipating that negative sensation in your body so you're not carrying carrying it too much, which I think makes you less effective because the more you worry and get tense em and feel unwell you cant do positive change can you, you cant make a difference so

The quotes above show a reluctance to engage with or stay with these emotions out of fear doing so will lead to dysfunction and disablement—a clear threat to competency need fulfilment, which as previously established is of paramount concern. That suppression of emotion requires effort is indicated by use of the word *try*. *Disheartening* is an interesting term, invoking not just a physical feeling in the body but also a weakening of resolve, possibly affecting vitality and motivation.

The following example shows a link between suppression of emotion and lack of autonomy:

> Jay: I cant watch distressing programmes, habitat destruction or... because I'm powerless to do anything about them and I put too much I've invested too much emotional energy in them in the past.

Although suppression may alleviate stress for the individual, as noted earlier it can also have detrimental effects (Rogelberg 2006) and it may also be maladaptive in terms of addressing environmental issues (Macy and Brown 2014).

But perhaps the participants are wise to suppress these difficult emotions, in the absence of a safe container for holding and working through them. Consider this:

> Ash: (silence) how do I feel about it (in quiet voice) as local government officers its all bashed out of us in our day job because what we feel about things is completely irrelevant its about what the business case is, and you know pragmatic

Who would feel safe to open up in a context that you experience as violently hostile to emotion?

Conclusion

We discovered through our process of inquiry a rich and complex set of nonlinear interrelating factors that appear to be influencing participant effectiveness in particular contexts. Driven by concern for the natural world and human wellbeing, as well as innate needs and perceived threats, participant experience when observed closely is far from coherent and consistent. Tensions were found in all areas of experience that we studied.

In this paper we have shown how some of the tensions in participant experience come from using coping strategies that have potential to be both adaptive and maladaptive on an individual and ecological level. Suppressing negative emotion about climate change out of fear that engagement will affect competency may be adaptive if the organisation does not present a safe container for holding and working through emotions, but it is maladaptive in that it is effortful and diverts cognitive resources away from other tasks, and the dissonance may induce stress thereby reducing competency. Similarly with suppressing a 'deep green' identity: with the advantages come disadvantages. Some strategies may simultaneously support and thwart need satisfaction or deplete inner resources, and seeking relief by going into natural spaces can be problematic if nature is conceptualised as an object, as objects have boundaries that separate them from other objects i.e. us.

Discussion

As we stated earlier, understanding why despite good intentions appropriate action is not necessarily forthcoming is critical if we wish to avoid catastrophic consequences of climate change for social justice and the wellbeing of humans and other species.

Our findings contribute nuanced insight to the body of knowledge on climate change behaviour and factors influencing congruent enactment of pro-environmental values, specifically the dynamics of underlying psychosocial forces.

In bringing the dynamics that may be in play to conscious awareness we suggest this allows for the possibility of intervening in order to increase inner coherence and more congruent enacting of values through behaviour. A vicious circle may be transformed into virtuous circle, a self-reinforcing feedback loop into a self-regulating loop.

Without awareness, it is not possible to self-regulate (Shapiro and Schwartz 1999). With awareness, people who wish to act effectively to mitigate climate change and restore ecological balance may then notice when their innate needs are not being satisfied, when coping strategies are veering towards maladaptive or when unhelpful ways of conceptualising experience become salient in their mind. They can then choose to intervene and respond differently, in ways that enhance rather than undermine their effectiveness in achieving their goals.

References

American Psychological Association (2009) Psychology and global climate change: addressing a multifaceted phenomenon and set of challenges (report). American Psychological Association, Washington, DC

Blackmore E, Holmes T (2013) Common cause for nature: values and frames in conservation (report). Public Interest Research Centre, Powys

Clayton S, Opotow S (2003) Identity and the natural environment: the psychological significance of nature. MIT Press, Cambridge

Crompton T, Kasser T (2009) Meeting environmental challenges: the role of human identity. WWF-UK, Godalming

Darnton A, Kirk M (2011) Finding frames: new ways to engage the UK public in global poverty. Bond, London

DECC (2013) 2012 UK greenhouse gas emissions. Department of Energy and Climate Change, UK

Deci EL, Ryan RM (2000) The "what" and "why" of goal pursuits: human needs and the self-determination of behavior. Psychol Inq 11:227–268

Energy Saving Trust (2011) The elephant in the living room: how our appliances and gadgets are trampling the green dream. Energy Saving Trust, London
Foster J (2015) After sustainability. Routledge, London
Grouzet FM et al (2005) The structure of goal contents across 15 cultures. J Pers Soc Psychol 89 (5):800–816
Jenkins J et al (2011) Energy emergence: rebound and backfire as emergent phenomena. Breakthrough Institute, Oakland
Kasser T, Kanner AD (eds) (2004) Psychology and consumer culture: the struggle for a good life in a materialistic world. American Psychological Association, Washington, DC
Klein N (2014) This changes everything: capitalism vs. the climate. Simon & Schuster, New York
Lakoff G (2010) Why it matters how we frame the environment. Environ Commun 4(1):70–81
Lakoff G, Johnson M (1980) Metaphors we live by. University of Chicago Press, Chicago
Macy J (1993) World as lover, world as self. Parallax Press, Berkeley
Macy J, Brown M (2014) Coming back to life. New Society Publishers, Gabriola Island
Maio G et al (2001) Addressing discrepancies between values and behaviour. J Exp Soc Psychol 37:104–117
Palazzo G, Krings F, Hoffrage U (2012) Ethical blindness. J Bus Ethics 109(3):323–338
Rockstrom J (2009) A safe operating space for humanity. Nature 461:472–475
Rogelberg SG (ed) (2006) Encyclopedia of industrial and organizational psychology. Sage, London
Schultz PW, Tabanico J (2007) Self, identity and the natural environment. J App Soc Psychol 37 (6):1219–1247
Schultz PW et al (2005) Values and their relationship to environmental concern and conservation behaviour. J Cross-Cult Psychol 36(4):457–475
Schwartz SH (1992) Universals in the content and structure of values. Adv Exp Soc Psychol 25:1–65
Shapiro SL, Schwartz GER (1999) Intentional systemic mindfulness: an integrative model for self-regulation and health. Adv Mind-Body Med 15:128–134
Smith J et al (2009) Interpretative phenomenological analysis: theory, method and research. Sage, London
Thibodeau PH, Boroditsky L (2011) Metaphors we think with. PLoS One 6(2):e16782
van der Werff E et al (2013) The value of environmental self-identity. J Environ Psychol 34:55–63
Weintrobe S (ed) (2013) Engaging with climate change: psychoanalytic and interdisciplinary perspective. Routledge, London
Wright C et al (2012) Hippies on the third floor. Org Stud 33(11):1451–1475

Chapter 20
The Impacts of Climate Change on the Livelihood of Arable Crop Farmers in Southwest, Nigeria

B.G. Abiona, E.O. Fakoya, and J. Esun

Abstract Agriculture places heavy burden on the environment in the process of providing humanity with food and fiber, while climate is the primary determinant of agricultural productivity. Given the fundamental roles of agriculture in human welfare, concern has been expressed by federal agencies and others, regarding the potential effects of climate change on agricultural productivity. The study examined the perceived impact of climate change on the livelihood of arable crop farmers in Southwest, Nigeria. Various strategies adapted to reduce the effect of climate change on their crop and livelihood includes: crop rotation ($\overline{x} = 2.68$), planting of leguminous crop ($\overline{x} = 2.05$), application of organic fertilizers ($\overline{x} = 2.28$), mulching ($\overline{x} = 2.38$) and by planting drought resistance crops ($\overline{x} = 2.30$). Reported among the effects of climate change on crop and farmers' livelihood were: discoloration of crop leave ($\overline{x} = 2.49$), increase infestation of pests and diseases ($\overline{x} = 2.50$) and reduction of crop yield ($\overline{x} = 2.49$). Experience gathered during the course of research will be useful at both local and international level as a way of reducing climate change threat.

Keywords Climate change impact • Livelihood • Southwest arable crops

Introduction: Some Facts About Southwest, Nigeria

Southwest Nigeria is made up of Lagos, Ogun, Oyo, Osun, Ondo and Ekiti States. It is located in the coastal region of the country and is characterised by humid to sub humid eco-climate. The vegetation cover ranges from forest to savanna woodland or forest-savanna transition zones. Average annual rainfall is 1200–1500 mm. Majority of inhabitants are farmers and practice farming enterprises ranging from crop production to livestock breeding, forestry practices, fisheries and aquaculture

B.G. Abiona (✉) • E.O. Fakoya • J. Esun
Department of Agricultural Administration, Department of Agricultural Extension and Rural Development, Federal University of Agriculture Abeokuta, 2240 Ogun State, Abeokuta, Ogun State 1110001, Nigeria
e-mail: dolace6ng@gmail.com

W. Leal Filho (ed.), *Innovation in Climate Change Adaptation*, Climate Change Management, DOI 10.1007/978-3-319-25814-0_20

as well as agricultural processing. A field survey was carried out in Southwest Nigeria to confirm the perspectives of farmers, extension officers and researchers on climate change issues and the efficacy of some emerging and indigenous technologies that farmers adopt and adapt to cope with the phenomenon of climate change. Southwest Nigeria is bounded in the north and east by the Kwara and Kogi States of Nigeria; in the West by the Republic of Benin and in the south by the Atlantic Ocean.

Climate Change and Its Impact in Southwest, Nigeria

Agriculture places heavy burden on the environment in the process of providing humanity with food and fibre, while climate is the primary determinant of agricultural productivity. Given the fundamental roles of agriculture in human welfare, concern has been expressed by federal agencies and others, regarding the potential effects of climate change on agricultural productivity. Interest in this issue has motivated a substantial body of research on climate change and agriculture over the past decade (Wondwossen 2008). Climate change is expected to influence crop production and other components of agricultural systems. However, the nature of the biophysical effects and the human responses to them are complex and uncertain.

It is evidenced that climate change will have a strong effects on Nigeria, particularly in the areas of agriculture; land use, energy, biodiversity, health and water resources. Nigeria, like all other countries of Sub-Saharan Africa, is highly vulnerable to the effects of climate change (IPCC 2007a, b). It was also noted that, Nigeria specifically ought to be concerned by climate change because of the country's high vulnerability due to its long (800 km) coastline that is prone to sea-level rise and the risk of fierce storms.

Moreover, rain-fed agriculture practiced and fishing activities from which 2/3 of the Nigeria population depend primarily on foods and livelihoods are also under serious threat besides the high population pressures of 140 million people surviving on the physical environment through various activities within an area of 923,000 km^2 (IPCC 2007a).

Food crop farmers in South-West Nigeria provides the bulk of arable crops that are consumed locally, also, major food crops supplies to other regions in the country. The local farmers are experiencing climate change even though they have not considered its deeper implications. This is evidenced in the late arrival of rain, the drying-up of streams and small rivers that usually flows year-round, the seasonal shifting of the "Mango rains" and the fruiting period in the Western part of Ogun State (Ipokia), and the gradual disappearance of flood-recession cropping in riverine areas of Ondo State are among the effects of climate disturbances in some communities of South-Western Nigeria. The study specifically addressed the following objectives

Specific Objectives

(a) To describe the socio-economic characteristics of arable crop farmers in the study area.
(b) Determine perceived the impact of climate changes.
(c) To identify the coping strategies, if any, towards climate change in the study area.

Hypotheses of the Study
Ho1: There is no significant relationship between personal characteristics of the farmers and perceived impact of climate change.

Climate Change and Its Impact in Southwest, Nigeria

The major direct impact of climate change on agricultural production in Southwest, Nigeria will be through changes in temperature, rain, growing season, and timing of extreme or critical threshold events. Suitable weather condition is essential at every stage of arable crop production. Sensitivity of most perishable crops production to hours of sunshine, rainfall, soil conditions and temperature therefore makes it vulnerable to climate change. Changing climate can also alter development of pests and diseases and change the host's resistance. Unfavorable climate promotes pest infestation and disease outbreak on arable—crop farms. Because farming contributes significantly to farm income and rural employment, adaptation is necessary. It is a clear associate of impact mitigation. It seeks for workable adjustment alternatives to the expected impact of climate change. When changes occur rapidly and because of poverty, farmers cannot respond properly to the impacts of climate change. Effective adaptation needs to make vulnerable people resilient, and able to return to normal status quickly, even after a major hurt. Therefore, we have to identify some other causes of vulnerability such as low income, low assets, illiteracy, resource depletion, poor governance, economic instability, disease, socio economic factors and poor risk management (UNEP 2006).

Methodology

The study was carried out in Southwest, Nigeria which has most of its inhabitants as arable crop farmers who embark on different activities using traditional methods of farming. Data were collected from 80 respondents with the aid of well-structured interview guide. Crop cultivated was measured with yes = 2 and No = 1. Coping strategies was measured with three point rating scale of often used = 3, rarely used = 2 and not used = 1. Data were further analyzed with descriptive and inferential statistical analysis.

Results and Discussion

Personal Characteristics of Respondents

The mean age of the respondents was 43.8 years meaning that the farmers are within economically active age (Abiona et al. 2011). More than half (63.8 %) of the respondents' were male while 36.2 % were female. Also, 55.0 % of the farmers were married. This finding is in line with the finding of (Bello 2010), who explained in his work that the highest percentage of farmers in the rural area was married. This result was further supported the findings of Akinbile et al. (2006), that marriage is an important factors in the livelihood of individual in our society because it is perceived to confer responsibility on individual. The mean household size was five persons living under the same roof. This result is line with the report of Banmeke (2003) which explains that household size is an important factor in agriculture as it serves as source of labour in the rural areas. It was discovered that less than half (36.2 %) of the respondents had elementary education (Table 20.1).

Table 20.1 Socio-economic characteristics of the respondents (n = 80)

Categories	Frequency	Percentage (%)	Mean
Age (years)			
≤30	2	25.0	
31–40	16	20.0	
41–50	24	30.0	43.8 years
51–60	23	28.8	
61 above	15	18.8	
Sex			
Male	51	63.8	
Female	29	36.2	
Marital status			
Single	4	5.0	
Married	44	55.0	
Divorced	14	17.5	
Widowed	18	22.5	
Household size			
1–5	18	22.5	
6–10	40	50.0	
11–15	18	22.5	5.3
16 above	4	5.0	
Educational qualification			
Primary	17	21.3	
Post primary	29	36.2	
Post-secondary	12	15.0	
None	22	27.5	

Table 20.2 Types of crop grown by the respondents (n = 80)

Variables	Frequency	Percentage (%)
Types of crop grown		
Maize	4	3.8
Tomato	4	3.8
Vegetable	7	8.8
Yam	8	10.0
Pepper	2	1.3
Maize and cassava	13	15.0
Maize and yam	8	10.0
Cassava and tomato	7	6.3
Yam and pepper	5	5.0
Maize, cassava and tomato	6	7.5
Maize, cassava and yam	9	10.0
Maize and others	3	2.5
Others	4	5.0

Type of Crop Grown by the Respondents

The cultivation of arable crops and its demand was high in the study area. Based on this, major crop grown by farmers were: maize, yam and cassava. This result is expected because farmer in the rural area depend mainly on intercropping system (Table 20.2).

Coping Strategies Towards Climate Change

Various strategies adapted to reduce the effect of climate change on their crop and livelihood include: crop rotation (53.8 %), planting of leguminous crop (35.0 %), application of organic fertilizers (45.0 %), mulching (56.3 %) and by planting drought resistance crops (46.5 %).

Coping strategies is an adjustment or self-insurance pursued by households to ensure future income generating capacity rather than simply to maintain current level of food consumption (Corbett 1991). Coping strategies are strategies taken to ensure minimal effects of climate change on crop production, to ensure continued production and food security to the increasing population of the study area (Table 20.3).

Table 20.3 Coping strategies used by respondents (n = 80)

Statements	Never used	Rarely used	Often used	Mean
By planting drought resistant crop	13(16.3)	30(37.5)	37(46.5)	2.30
By irrigation when there is no rain water	30(37.5)	33(41.3)	17(21.3)	1.87
By constructing living fences to control sun intensity	44(55.0)	20(25.0)	16(16.0)	1.65
By planting crop that are resistance to flooding	18(22.5)	44(55.0)	18(22.5)	2.00
By mulching	15(18.8)	20(25.0)	45(56.3)	2.38
By practicing crop rotation	15(18.8)	22(27.5)	43(53.8)	2.34
By planting leguminous crop	24(30.0)	28(35.0)	28(35.0)	2.05
By planting cover crops	8(10.0)	10(12.5)	62(77.5)	2.68
By practicing mixed farming	22(27.5)	40(50.0)	18(22.5)	1.95
By practicing bush fallowing	22(27.5)	40(50.0)	18(22.5)	1.95
By application of fertilizer e.g. organic or inorganic	14(17.5)	30(37.5)	36(45.0)	2.28
By fumigating with pesticide and insecticide	8(10.0)	13(16.3)	59(73.8)	2.64
By having access to weather forecasts	44(55.0)	18(22.5)	18(22.5)	1.68

Effect of Climate Change on Crop Production

Effect of climate change on crop production was examined during the course of study. Based on this, the effect was: Reduction in farmers' income in agricultural production ($\bar{x} = 2.60$), increase infestation of pests and diseases ($\bar{x} = 2.50$), leads to discoloration of crop leaves ($\bar{x} = 2.49$), encourage mixing farming ($\bar{x} = 2.44$), reduce crop growth ($\bar{x} = 2.35$) and ease harvesting of agricultural products ($\bar{x} = 2.32$) (Table 20.4).

Hypotheses Testing

The result of Chi-square analysis in Table 20.5 shows that there is significant association between the impact of climate change and respondents' age ($\chi^2 = 51.57$, $p < 0.05$), religion ($\chi^2 = 36.6$, $p < 0.05$), educational qualification ($\chi^2 = 94.98$, $p < 0.05$) and sex ($\chi^2 = 22.78$, $p < 0.05$). The implication of educational qualification and impact of climate change is that arable crop farmers are more educated and hence they take advantage of acquiring more knowledge on climate change to improve their livelihood. Also, the significance of age with impact of climate change was that arable crop farmers 'had farming experience that can help them to cope with any environmental challenges (Table 20.5).

Table 20.4 Perceived effect of climate change on crop production (n = 80)

Statement	High extent	Little extent	None	Mean
Reduce crop growth	38(47.5)	32(40.0)	10(12.5)	2.35
Leads to discoloration of crop leaves	51(63.8)	17(21.3)	12(15.0)	2.49
Reduce grain quality	12(15.0)	46(57.5)	22(27.5)	1.85
Reduce grain quantity	15(18.8)	45(56.3)	20(25.0)	1.93
Increase infestation of pests and diseases	47(58.8)	26(32.5)	7(8.8)	2.50
Increase in death rate of rural farmers	11(13.8)	24(28.8)	45(57.5)	1.57
Reduction in farmers' income in agricultural production	50(62.5)	28(35.0)	2(2.5)	2.60
Reduce infestation of diseases on farm	12(15.0)	32(40.0)	36(45.0)	1.70
Reduce crop yield	48(60.0)	23(28.8)	9(11.3)	2.49
Encourage mixing farming	45(56.2)	25(31.3)	10(12.5)	2.44
Ease harvesting of agricultural products	42(52.5)	22(27.5)	16(20.0)	2.32
Ease processing of agricultural products	17(21.2)	31(38.8)	32(40.0)	1.82

Table 20.5 Chi-square analysis of respondents' personal characteristics and impact of climate change

Variables	χ^2	Df	P	Decisions
Age	51.57	2	0.02	Significant
Religion	36.63	2	0.04	Significant
Educational qualification	94.98	3	0.01	Significant
Marital status	1.62	2	0.07	Not significant
Sex	22.78	2	0.02	Significant

Conclusion and Recommendations

Climate change affects Southwestern agricultural crops in special ways. The extent climate change seen in the region calls for immediate efforts and investment. In order to address the problem of climate change, it is paramount that machineries are put in place so as to allow various agricultural regions to be in a better position to handle the many constraints climate change poses to them. Also, adequate education through training to facilitate farmer's adaptation to the issues of climate change is highly paramount. This can be achieved with the involvements weather forecasts and other useful information for impact mitigation will be a way out. It was concluded from the study that climate change is an impinging factor that seriously affect arable crop production and hence farmers' livelihood despite coping strategies to minimize its effect. It was however recommended that Agricultural policies and practices that could minimize or eliminate its effect should be seriously enacted to boost production and increase farmers' livelihood.

References

Abiona BG, Fakoya EO, Alegbeleye WO, Fapojuwo EO, Adeogun SO, Idowu AA, Arommolaran AK (2011) Constraints to integrated and non-integrated fish farming activities in Ogun State, Nigeria. J Agric Sci 3(4):233–240, Published by Canadian Center of Science and Education, Canada

Akinbile LA, Oladoja MA, Awoniyi FM, Adisa BO (2006) Effect of community participation on perception of sustainability of rural water project in Oyun L.G.A of Kwara State, Nigeria. J Food Agric Environ Sci Technol. www.world-food.net, 4(3&4), July–Oct 2006. WFL Publisher, Helsinki, pp 257–260

Banmeke TOA (2003) Accessibility and utilization of agricultural information in the economic empowerment of women farmers in South Western Nigeria. Unpublished Ph.D. Thesis submitted to department of agricultural Extension and Rural Development, University of Ibadan, Ibadan

Bello K (2010) Perception of farmers towards climate change. Unpublished Agri. Thesis submitted to Department of Agricultural Extension and Rural Development, Federal University of Agriculture, Abeokuta

Corbett D (1991) Famine and household coping strategies. World Dev 16:1099–1112

IPCC (2007a) Impacts, adaptation and vulnerability working group II contribution to the intergovernmental panel on climate change 4th assessment report summary for policy makers, Brussels

IPCC (2007b) Climate change 2007 – impacts, adaptation and vulnerability: the working group II contribution to the intergovernmental panel on climate change 4th assessment report. Cambridge University Press, Cambridge

UNEP (2006) Adaptation and vulnerability to climate change: the role of the finance sector. CEO Briefing, New York

Wondwossen S (2008) Climate change: global and national response. In: Green Forum (ed) Climate change-a burning issue for Ethiopia: proceedings of the 2nd green forum conference held in Addis Ababa, 1st Oct–2nd Nov 2007. Green Forum, Addis Ababa, pp 37–69

Chapter 21
Grassroots Technologies and Community Trust in Climate Change Adaptation: Learning from Coastal Settlements of Bangladesh

Momtaj Bintay Khalil, Brent C. Jacobs, and Natasha Kuruppu

Abstract This paper reports doctoral research that explores grassroots technologies as an asset for poor coastal communities of Bangladesh, how local knowledge contributes to the creation of such technologies, and how they can be useful to build a community's trust in its own adaptive capacity. Bangladesh is one of the most disaster vulnerable countries in the world due to its deltaic morphology and frequent climate-induced hazards (storm surge, annual flooding, salinity intrusion, frequent cyclones, etc.). Southwestern coastal settlements are especially vulnerable because people considered among the poorest in the world inhabit them. To cope with climate extremes under severe resource limitations, grassroots technologies evolve over generations from autonomous decision-making processes and creative experimentation. However, communities often fail to recognize the value of these technologies and may have little trust in their innate capacity for climate change adaptation.

A conceptual framework will be presented that identifies the interactions among grassroots technology, local knowledge, community trust and climate change adaptation. The framework will be validated in case studies of specific grassroots technologies identified through field observations, and explored through qualitative methods to understand the importance of indigenous knowledge to the development of community-based climate coping strategies.

M.B. Khalil (✉) • B.C. Jacobs
Institute for Sustainable Futures, University of Technology Sydney, Building 10, Level 11, 235 Jones Street Ultimo, PO Box 123, Broadway, NSW 2007, Australia
e-mail: Momtaj.B.Khalil@student.uts.edu.au

N. Kuruppu
Institute for Sustainable Futures, University of Technology Sydney, Building 10, Level 11, 235 Jones Street Ultimo, PO Box 123, Broadway, NSW 2007, Australia

Institute for Global Health, United Nations University, IIGH Building, UKM Medical Centre, Jalan Yaacob Latif, Bandar Tun Razak, 56000 Cheras, Federal Territory of Kuala Lumpur, Malaysia

W. Leal Filho (ed.), *Innovation in Climate Change Adaptation*, Climate Change Management, DOI 10.1007/978-3-319-25814-0_21

Keywords Grassroots technologies • Local/indigenous knowledge • Community trust • Climate change adaptation

Introduction

Bangladesh is one of the most vulnerable countries to sea level rise[1] due to its very low elevation, deltaic morphology (Agrawala et al. 2003). The long coastline comprising 32 % of the country's total area[2] (Karim and Mimura 2008) has exposed a large population to climate variability (e.g. frequent cyclones, annual floods, sudden storm surges and salinity intrusion) exacerbated by the impact of global warming. Nearly seven million people in Bangladesh's coastal area face the impacts of extreme climate events such as super cyclone 'Sidr' (in 2007), cyclone 'Bijli' (in 2008) and more recently cyclone 'Aila' (in 2009). Scientific evidence suggests that with one meter of sea level rise[3] the vast majority of coastline will be inundated, which would affect mostly the ultra-poor living at the margin[4] (Rashid 2014). Due to the inherent vulnerability arising from the anticipated climate change impacts, a risky geographic setting and comparatively low economic strengths investigations are overdue of how these coastal communities adapt, and what facilitates or hinders their adaptive responses.

Although the coastal communities possess rich local knowledge that largely facilitates their responses to climate change and utilization of natural resources, over the years outside actors (i.e. various government and non-government, local and international organizations) have also influenced their adaptation strategies. Well documented, government-led disaster preparedness and response initiatives have largely been top-down and fail to recognize the role of community in formulating adaptive responses (Parvin and Johnson 2015). The significant role of other actors (i.e., international organizations, NGOs) remains under-represented and under-researched. In addition research is limited on coastal communities' climate knowledge-base and responses, the effectiveness of the incentives from outsiders to influence grassroots responses and the overall trustworthiness of community-based processes to generate holistic bottom-up approaches to climate change adaptation.

This paper reports doctoral research that explores grassroots technologies as assets for poor coastal communities of Bangladesh, the interaction of such technology with available livelihoods capitals (e.g. Ellis 2000), how local knowledge

[1] Bangladesh is the third most vulnerable country to sea level rise in the world and in next two or three decades the average rising temperature will be expected 2 °C and will worsening intense cyclones recognized by World Bank, 2013.

[2] The long coastline covers an area about 47,201 km^2 (Karim and Mimura 2008).

[3] 0.3 m Sea Level rise would increase 22.7 % depth of flooding and will flooded 20 km from the coastline and 5690 km^2 is a High Risk Zone (HRZ) (Karim and Mimura 2008).

[4] Coastal zone of Bangladesh comprises distinctive problems and prospects but relatively income-poor in comparison with rest of the country (Rashid 2014).

contributes to the creation of such technologies, and how they can be useful to build a community's trust in itself and outside actors to enhance adaptive capacity. A conceptual relationship will be established that identifies the interactions among livelihood resources, local knowledge, and community trust in the process of climate change adaptation. A process to validate this framework, through co-learning with communities in the Gabura region of coastal Bangladesh, will be described under which preliminary field observations of specific grassroots technologies are already in progress.

Research Needs

Adaptation to climate change is the adjustment with the environment or surroundings, in the ability, competency or capacity of a system to adapt with climatic stimuli (Burton 1996). Adaptation encompasses adjustments in ecological, social and economic systems in response to the impacts of climate change (Briggs 2005; IPCC 2007). The concepts of adaptation, adaptive capacity, vulnerability and resilience are interrelated and have a wide application to global change. In a developing context adaptation occurs in different levels in society and may be autonomous or planned change that is anticipatory, reactive, passive or spontaneous, through individual activity or group behavior in order to reduce society's vulnerability to climate change stimuli and impacts (Fankhauser et al. 1999; Smit et al. 1999, 2000; Burton et al. 2005). Community adaptation to climate change in developing countries generally happens via autonomous processes facilitated by social capital and access to natural resources (Adger et al. 2003).

As the nature of autonomous adaptation is a dynamic social process with cultural and place-based specificities, sensitivity is needed to understand how coastal communities behave within changing climate regimes and weather extremes to adopt radically different adaptation strategies. Adger (2001), in the context of coastal settlements of South Asia and the Caribbean, argues for adaptive responses that focus on social capital and collective action. However, in the context of Asia's mega-deltas collective actions and their outcomes may remain unachievable. Conway (2015) argues that inequitable and hidden social policies undermine (autonomous) adaptation, and politics leads to unequal and sometimes contradictory outcomes for intended beneficiaries. Such contradictory and unintended outcomes further lead to increased mistrust among the beneficiaries and other stakeholders dealing with disasters. According to Roy et al. (2012), together the impoverished geography, lack of a workable socio-political platform, ineffective support from public institutions, aid and dependency on NGOs, and the limits to the communities' own agency and structures reduce the capacity for autonomous responses to changed climate circumstances.

Recent post-disaster experiences from the coastal settlements have unveiled gaps in perceptions about disaster preparedness and recovery among different stakeholders. Parvin and Johnson (2015) argue that the political economy of climate change initiatives in Bangladesh leaves little scope to address vulnerability of grassroots communities, and existing policy responses are not able to address the structural dynamics prevalent in disaster relief, response and mitigation pathways. This happens because of the government agencies' surface level understanding of vulnerability and the top-down nature of response design, which fails to acknowledge the contribution of grassroots stakeholders. Mahmud and Prowse (2012) claim that cyclone preparedness and relief interventions are subject to corrupt practices that may lead to significant lack of trust. However, research in the pro-poor context of Bangladesh has created the hope that lessons can be drawn from grassroots experiences of coping with extreme weather for reducing vulnerability to climate change (Jabeen et al. 2010). Post-disaster experiences from the specific context of Gabura have established that grassroots responses are mostly effective as adaptive measures during typical hazards and the responses require little intervention from the outside actors except in extreme hazards (Alam et al. 2015).

Woolcock and Narayan (2000) identify the importance and usefulness of social capital, especially the bonding (of strong ties or intra-community) and bridging (of weak ties or extra-community) aspects respectively among family members, friends, neighbors and community people in the recovery phases. They also recognize the importance of linking relationships (both intra-local and extra-local) in long-term recovery with outside actors who have greater political or economic power, for example, the national and international NGOs, local government, and community-based organizations. Recently, Islam and Walkerden (2014), in the specific context of coastal Bangladesh, identified the importance of social capital, especially the bonding, bridging and linking relationships among family members, neighbors and community people. Accordingly, a framework developed to direct research in these coastal areas should triangulate locally-based knowledge, livelihood resources and community trust that results in varying grassroots responses within a broader framing of autonomous climate change adaptation.

Conceptual Framework

Indigenous Knowledge Overlapping Local Knowledge

Indigenous and local knowledge are concepts that overlap, have much in common in their definitions and are often used interchangeably. Both are place-based, contextual, and experiential, most often transmitted orally and through social learning and engagement across generations, and evolve over time and observation (Agrawal 1995; Ellen et al. 2000; Sillitoe 2006, 2007; Cleveland and Soleri 2007 in Orlove et al. 2010). They may also be considered as the 'personal' assets of local communities (Smith 2001; Briggs and Sharp 2004). Bicker et al. (2003) views local knowledge as embedded in indigenous knowledge that embodies grassroots' traditional lifestyles. Orlove et al. (2010) considers indigenous knowledge to be deeply rooted in local culture and generally associated with long-settled communities that have strong ties to the natural environment. Perhaps it is the type of community rather than the characteristics of the knowledge that is the fundamental difference between these two concepts. Local knowledge could be a product of any place-based community's relationship with its local environment, whereas indigenous knowledge implies a community's long-term cultural ties (Agrawal 1995) or traditional ownership of a place. In that sense, the term indigenous knowledge appears more appropriate in association with the Gabura community in Bangladesh and will be used hereafter.

An appreciation of the importance of indigenous knowledge appears critical to support the formulation of bottom-up approaches to climate adaptation. Recognition of indigenous knowledge is associated with community empowerment (Briggs and Sharp 2004). It plays a central role in traditional approaches to management of natural resources (Berkes et al. 2000) and in local-level decision-making in agriculture, health care, food preparation, education, and community sustainability (Warren 1991; Warren and Cashman 1988). Orlove et al. (2010) describe four major components of local experience that can help to understand climate change impacts as: the awareness of weather signals; the observation of meteorological events; cultural and regional information on climate change; and, opportunities for co-learning with local people (e.g. farmers) to understand seasonal impacts on local agricultural production. Mercer et al. (2010) suggest that the integration of indigenous knowledge and scientific knowledge can help to reduce vulnerability and disaster risk.

Adaptive Capacity and Grassroots Response

Grassroots responses are embedded in the community's store of indigenous knowledge from past experience of climate variability (annual flooding, storm surge, cyclone, consistent rainfall) that stimulates autonomous responses to cope with extreme events (Alam et al. 2015). Such responses rarely result from planned or designed adaptation or policy-making. Rather they evolve through short-term and long-term dynamic processes of local experimentation utilizing intimate indigenous knowledge of natural systems (Alam et al. 2015; Mercer et al. 2010; Warren and Cashman 1988). Orlove et al. (2010), for example, noted the significant role of grassroots responses emerging from indigenous knowledge of climate variability and weather conditions in crop production.

Specific grassroots responses are therefore most likely the product of a community's adaptive capacity, which depends on the availability, transformation and substitution of individual capitals from the range of livelihood resources (Nelson et al. 2007). Livelihood resources are generally considered in terms of a suite of capitals that may include human, social, cultural, financial, physical, political and natural capital (Scoones 1998; DfID 1999). Adaptive capacity of the least developed countries (LDC), such as Bangladesh is considered low and vulnerability is high (McCarthy 2001). For successful adaptation strategies, limitations in one type of capital, such as financial capital in coastal communities of Bangladesh, must be overcome by finding novel ways to combine the resources at hand. This process is likely to rely heavily on trust in the exchange of knowledge through local social capital and networks (Woolcock and Narayan 2000) to develop grass roots innovations in the use of local natural capital (Adger et al. 2003), which effectively embody a community's collective wealth (Scoones 1998).

Community Trust

The conceptualization of trust by different authors shows its specificity with the scale of interaction (i.e., personal, family, neighborhood, community levels, etc.). Narayan-Parker (1999) referred to 'trust' as social capital by using the term 'cross cutting ties' (as bonding and bridging networks) between the formal and informal groups among family, relatives, neighbors and outside actors for collective actions to establish the identity, trust, values and access to power and economic welfare. Woolcock (1998) and Woolcock and Narayan (2000) outlined community trust as operating in two ways, horizontally or the 'strong ties' (through bonding and bridging at the micro level) and vertically or the 'weak ties' (network relationships among local governments, international organizations or NGOs and others at the macro level). Furthermore, Putnam (2001) observed the importance of three elements, namely trust/social value, norms and networks, in upholding good social bondage and community trustworthiness.

In the context of developing countries, Nooteboom (2006) noted that development of trust has an intrinsic value and relies on a combination of personalized trust and local trustworthiness that requires strong support from institutions for its development. However, most often trustworthiness in indigenous communities rests on their past experience, intuition and historical experimentation of coping with environmental conditions using local materials (Orlove et al. 2010). The building of trust between government and communities is generally accepted as creating productive relationships that lead to socially acceptable planning and positive management outcomes (Smith et al. 2013a). Community trust can be established if there is sufficient transparency among community stakeholders and among the community and outside actors.

In the context of Bangladesh, NGOs' such as Action Aid among others are focusing on social and economic development, trying to secure natural resource management (NRM), ecosystems, mangrove regeneration (Sarker 2010) and livelihood resources and empowerment through the introduction of saline-tolerant crops[5] to enhance adaptive capacity (Iftekhar and Islam 2004). International NGOs, may enter positions of trust with the community in relation to adaptation through provision of transitional shelter, initiatives to support the education of cyclone affected children, coordination with government agencies[6] to provide safe water and environmental sanitation facilities or provision of potable water through the installation of tanks for rainwater harvesting.

Smith et al. (2013b) suggest that for some aspects of an individual's level of trust, high levels may be negatively related to public involvement in resource-related activities. Mahmud and Prowse (2012) claimed that ultra-poor households' disaster interventions are significantly affected by corruption, particularly in relation to public works and non-governmental interventions. Therefore, political capital emerges as an important factor in grassroots' disaster responses. Issues such as political nepotism and corruption can erode trustworthiness both horizontally (among community members) and vertically (among the coastal community and outsiders). So in the context of grassroots' climate change responses in coastal Bangladesh, the 'community trust' built over the relationships of bonding-bridging-linking networks need attention to understand how local knowledge can be further mobilized to ensure sustainable management of livelihood resources.

Conceptualizing Autonomous Climate Change Adaptation

Figure 21.1 shows a conceptual framework that links changing climate as a driver of autonomous adaptation, mediated through indigenous knowledge, livelihood resources and community trust that results in the development of grassroots technology at a range of temporal scales and social hierarchies.

The conceptual framework conceives livelihood resources as central to any autonomous response, particularly in a developing-country context, such as Bangladesh, where state-sponsored, planned adaptation plays only a minor role in supporting its citizens' adaptive responses. In this context, empirical evidence suggests that the ultra-poor, who are often landless with no significant financial or physical capital, depend heavily on natural resources and social and cultural supports from their peers for survival. Local knowledge and community trust, although generally considered part of the stock of human and social capitals respectively, are therefore shown as separate components of the model because of their importance in the adaptive capacity of indigenous communities. Together

[5] Saline-tolerant crops like coconut, guava, plum, corn meal, tamarind, cabbage, celery, basil, bindweed, spinach and paddies like bina-9, bina-10 are suitable as salinity intrusion in Gabura after cyclone Aila.

[6] The Government agencies like UNICEF, DPHE (Department of Public Health Engineering) providing temporary learning center and drinking water, local NGOs like NGF, JCF, CREL, CNRS are advocating the local people for the empowerment and helping to secure the livelihood resources and natural resource management at a minimum scale in Bangladesh costal community.

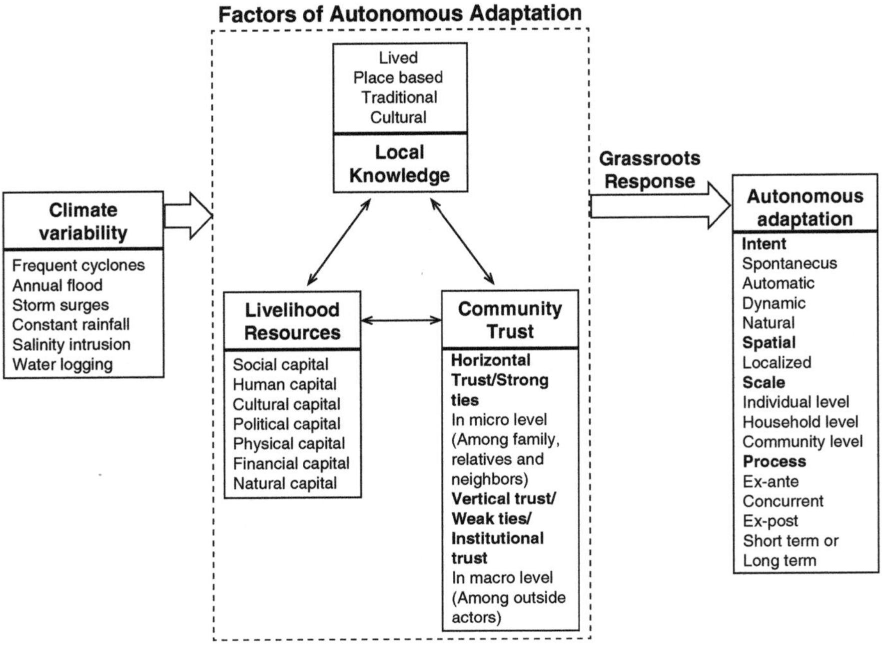

Fig. 21.1 Conceptual frameworks linking to local knowledge, livelihood resources, community trust and grassroots technologies. *Source*: authors

their triangulations determine the level of grassroots responses and form the nature of autonomous adaptation to climate variability.

Preliminary Evidence of Grassroots Technology

Gabura, is an island of the coastal 'Union' under Shyamnagor Upozila at Satkhira district of Khulna division bounded by the Kholpetua and Kopotakhsa Rivers (Fig. 21.2). Once part of the world's largest mangrove forest, now it is the fringe of the Sundarbans. In 2009, the region was hit by super cyclone Aila that burst protective the embankment allowing saline water inundation of the island up to 6 m deep (Choudhury 2009). Significant human and livestock casualties occurred, local communities were displaced, and local ecosystems and natural resource livelihoods were disrupted. The cyclone's impacts continue to disrupt the island's culture, as the historically practiced cultivation systems were rendered useless for up to 4 years.

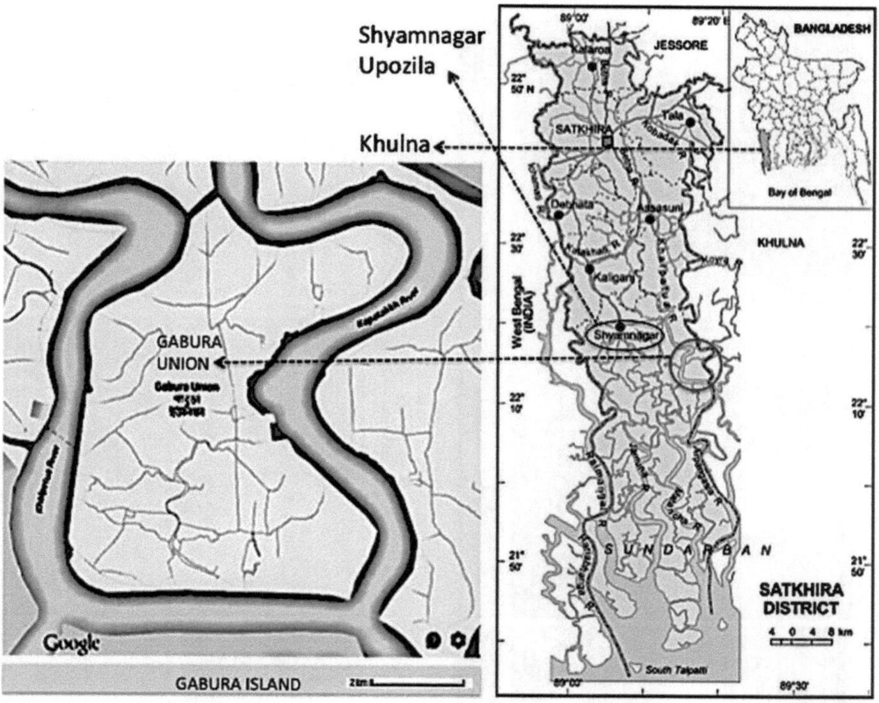

Fig. 21.2 Location of Gabura Island in Shyamnagar Upozila of Khulna Division, which is the site of field studies in Bangladesh. *Source*: Google Map

International aid agencies have offered local assistance. However, preliminary field observations suggest that their interventions are transforming indigenous lifestyles and may be influenced by complex political relationships among multiple actors with positive and negative outcomes for the local community.

The livelihoods of coastal people in Bangladesh are natural resource-dependent and they continually struggle to increase adaptive capacity to cope with climate variability (Adger et al. 2003). However, grassroots technologies (Fig. 21.3), in response to both climatic stimuli and non-climatic drivers, have appeared spontaneously despite chronic limitations in access to resources (Alam et al. 2015). Table 21.1, summaries the field observation of selected grassroots responses by coastal people in Gabura Island that emerged after the cyclone Aila at a range of scales, and the role of external assistance from aid agencies and without outside interventions in the development of the response.

Fig. 21.3 Examples of grassroots technologies observed in Gabura. (**a**) Tying down boats to secure from storm surge; (**b**) Raised plinth height of the houses; (**c**) Tethering to secure house roofs; (**d**) Protection of mud walls from rainfall; (**e**) Plumbing modified to allow upstairs toilet during flooding; and, (**f**) Tanks for rainwater harvesting as drinking purpose

Conclusion

The literature review and preliminary field observations from the coastal community of Gabura, Bangladesh have demonstrated the importance of local knowledge, livelihood resources and community trust in the emergence of grassroots technology for autonomous adaptation to climate change in developing contexts. Grassroots responses are spontaneous and dynamic. This paper identifies some of the grassroots innovations of coastal communities and identifies the contribution of local knowledge that can be an asset to the creation of such technologies. It also

Table 21.1 Grassroots responses during the post-disaster situation in Gabura, Bangladesh

Grassroots response		With aid and outside interventions	Without outside intervention
Livelihoods		Elevated land is used for mangrove plantations and regeneration. Low lying land is used for shrimp and crab farming and fishing and fish processing	At household level people are engaging in fishing and boat puller. The boats are tying down to secure from gusty winds
Food/health	Food storage	Having experience of cyclone Aila, underground storage used to preserve the dry food and others utensils in community levels	The dry food, matchboxes, mud chulli (cooker), homestead fuel are stored above flood level on the side shelf, on the bed top or on the wooden false ceiling underneath the roof
	Water	UNICEF and DPHE are providing water tank to the community people for rainwater harvesting	The villagers are Harvesting rainwater by using plastic sheet on roof and sealed water clay pots and preserve sweet water pond for cooking and potable purpose
	Plantation	Plantation can protect the embankment from gusty wind and storm surge and the soil erosion from excessive rainfall; few local NGOs are helping to secure the natural resources	Plantation in household level as a source of housing construction like sidewall, source of food and vegetable or firewood or physical boundary of housing
	Poultry/ farming	In community level a common space is preserved for homestead animal	The household women are making a small mud/ brick house for poultry farming in two levels in a temporal scale
	Toilet	A shared tin-shade toilet is allocated in community level for 4/5 families	In household level sometimes plumbing modified to allow upstairs toilet by using a pan with a sewerage pipe is adjoined the water flow to outside of the house
	Fuel	The community people are collecting straw and wood from the Sundarbans with controlled access by introducing a 'pass' system of the forest department	In household level the villagers are making some fuel ball with mud and straw for fuel

(continued)

Table 21.1 (continued)

Grassroots response		With aid and outside interventions	Without outside intervention
Shelter/ housing	Walls	Tin partition wall is provided to reduce climate hazards and the lower portion is colored to protect from salinity and rust The houses are relocated on high lands and on stilt	At household level the boundary wall is made by mud and provide shades on the top of the wall as protection from rainfall Battered and inclined walls are used. To strengthen the sidewalls straw, sticks and mud are combined together
	Roof	Roofing materials are changed to CI sheet or asbestos rather than goalpata (a leaf that is easy to collect from the Sundarbans used as thatched roof and it is highly vulnerable) and bamboo bracings are used	The roof is tied by rope to big trees or the embankment as to secure roofs from gusty wind or storm surge and the inclination is kept less than 40°
	Plinth	The plinth of the houses are raised up to 6/7 f. by cement plaster or brick soling experiencing from sea level rise in cyclone Aila as a protection from riverbank erosion	The plinth of the houses are raised up 4/5 f. without plaster and brick soling experiencing the sea level rise in cyclone Aila and a wooden ladder is provided to get inside the houses
	Yard	The gray yards in Gabura Island are now becoming green with lot of saline-tolerant crops and vegetables and mangroves regeneration prescribed by NGOs and AID agencies	In household level the people are using the snail and seashells and covering the yard as to reduce from soil erosion collected from the Sundarbans
Embankment		The embankment is elevated up to 8/10 f. from the sea levels and CNRS has taken initiatives for plantation (Babla tree) to protect from saline water flooding	The other sides of the embankment the people are engaging to regenerate the mangroves and cropping on the low land

Source: Field observation, 2015

explores the idea that technologies are the outcomes of novel uses of livelihood resources. However, it remains to find ways to foster a community's trust in locally developed technology to support the process of climate change adaptation. The role of livelihood resources, in particular indigenous knowledge, is established in the literature, but requires study for a deeper understanding of its place-based significance in Gabura.

While the role of social capital in adaptation process has been articulated at large in climate change literature (Narayan-Parker 1999; Woolcock and Narayan 2000; Woolcock 1998), trust is an important dimension of social capital (bonding, bridging and linking networks) has been less well researched and requires further study in relation to grassroots technology and climate change adaptation.

Real development for communities such as the people of Gabura will be achieved only when local people get involved in the decision making processes (Mercer et al. 2007) and community trust is enhanced and grassroots technologies are acknowledged formally as legitimate attempts to formulate adaptive responses. This can be achieved by taking a holistic approach to adaptation policy development.

References

Adger WN (2001) Social capital and climate change. Tyndall Centre Working Paper

Adger WN, Huq S, Brown K, Conway D, Hulme M (2003) Adaptation to climate change in the developing world. Prog Dev Stud 3(3):179–195

Agrawal A (1995) Indigenous and scientific knowledge: some critical comments. Dev Chang 26 (3):413–439

Agrawala S, Ota T, Ahmed AU, Smith J, van Aalst M (2003) Development and climate change in Bangladesh: focus on coastal flooding and the Sundarbans. OECOD, Paris

Alam AFMA, Asad R, Parvin A (2015) Climate change adaptation through grassroots responses: learning from the "Aila" affected coastal settlement of Gabura, Bangladesh. In: Filho WL (ed) Handbook of climate change adaptation. Springer, Berlin, pp 2011–2034

Berkes F, Colding J, Folke C (2000) Rediscovery of traditional ecological knowledge as adaptive management. Ecol Appl 10(5):1251–1262

Bicker A, Ellen R, Parkes P (2003) Indigenous environmental knowledge and its transformations: critical anthropological perspectives. Routledge, London

Briggs J (2005) The use of indigenous knowledge in development: problems and challenges. Prog Dev Stud 5(2):99–114

Briggs J, Sharp J (2004) Indigenous knowledges and development: a postcolonial caution. Third World Q 25(4):661–676

Burton I (1996) The growth of adaptation capacity: practice and policy. In: Adapting to climate change. Springer, pp 55–67

Burton I, Malone E, Huq S (2005) Adaptation policy frameworks for climate change: developing strategies, policies and measures. In: Lim B (ed) Cambridge University Press, Cambridge

Choudhury DAM (2009) Protecting Bangladesh from natural disasters. Academic Press and Publishers, Dhaka. ISBN 984080252

Conway D (2015) Policy: hard choices and soft outcomes? Nat Clim Chang 5(2):105–106

DfID U (1999) Sustainable livelihoods guidance sheets. UK DFID Department for International Development, London. Available at: www.Livelihoods.Org/info/info_guidancesheets.Html. Accessed 05 Apr 2007

Ellen RF, Parkes P, Bicker A (2000) Indigenous environmental knowledge and its transformations: critical anthropological perspectives, vol 5. Psychology Press, New York

Ellis F (2000) The determinants of rural livelihood diversification in developing countries. J Agric Econ 51(2):289–302

Fankhauser S, Smith JB, Tol RS (1999) Weathering climate change: some simple rules to guide adaptation decisions. Ecol Econ 30:67–78

Iftekhar M, Islam M (2004) Managing mangroves in Bangladesh: a strategy analysis. J Coast Conserv 10(1):139–146

IPCC (2007) Climate change 2007: the 4th IPCC assessment report. Intergovernmental Panel on Climate Change, Valencia

Islam R, Walkerden G (2014) How bonding and bridging networks contribute to disaster resilience and recovery on the Bangladeshi coast. Int J Disast Risk Reduct 10:281–291
Jabeen H, Johnson C, Allen A (2010) Built-in resilience: learning from grassroots coping strategies for climate variability. Environ Urban 22:415–431
Karim MF, Mimura N (2008) Impacts of climate change and sea-level rise on cyclonic storm surge floods in Bangladesh. Glob Environ Chang 18(3):490–500
Mahmud T, Prowse M (2012) Corruption in cyclone preparedness and relief efforts in coastal Bangladesh: lessons for climate adaptation? Glob Environ Chang 22(4):933–943
McCarthy JJ (2001) Climate change 2001: impacts, adaptation, and vulnerability: contribution of working group II to the 3rd assessment report of the intergovernmental panel on climate change. Cambridge University Press, Cambridge
Mercer J, Dominey-Howes D, Kelman I, Lloyd K (2007) The potential for combining indigenous and western knowledge in reducing vulnerability to environmental hazards in small island developing states. Environ Hazard 7(4):245–256
Mercer J, Kelman I, Taranis L, Suchet-Pearson S (2010) Framework for integrating indigenous and scientific knowledge for disaster risk reduction. Disasters 34(1):214–239
Narayan-Parker D (1999) Bonds and bridges: social capital and poverty, vol 2167. World Bank, Washington, DC
Nelson DR, Adger WN, Brown K (2007) Adaptation to environmental change: contributions of a resilience framework. Annu Rev Environ Resour 32:395
Nooteboom B (2006) Social capital, institutions and trust. Institutions and Trust, April 2006
Orlove B, Roncoli C, Kabugo M, Majugu A (2010) Indigenous climate knowledge in southern Uganda: the multiple components of a dynamic regional system. Clim Chang 100(2):243–265
Parvin A, Johnson C (2015) Disaster vulnerability in the policy context of Bangladesh: a critical review. In: Filho WL (ed) Handbook of climate change adaptation. Springer, Berlin, pp 877–899
Putnam RD (2001) Bowling alone: the collapse and revival of American community. Simon and Schuster, New York
Rashid MM (2014) Political commitments and aspirations of grassroots coastal communities: a micro-level study in Bangladesh. Am J Rural Dev 2(2):24–33
Roy M, Jahan F, Hulme D (2012) Community and institutional responses to the challenges facing poor urban people in Khulna, Bangladesh in an era of climate change. Working Paper 163, University of Manchester, Brooks World Poverty Institute, p 63
Sarker S (2010) Mangrove: a livelihood option for the coastal community of Bangladesh. Available at SSRN 1578818
Scoones I (1998) Sustainable rural livelihoods: a framework for analysis. IDS working paper 72, Brighton
Sillitoe P (2006) Introduction: indigenous knowledge in development. Anthropol Action 13(3): 1–12
Sillitoe P (2007) Local science vs global science: an overview approaches to indigenous knowledge in international development. Berghahn, New York, pp 1–22
Smit B, Burton I, Klein RJ, Street R (1999) The science of adaptation: a framework for assessment. Mitig Adapt Strateg Glob Chang 4(3–4):199–213
Smith EA (2001) The role of tacit and explicit knowledge in the workplace. J Knowl Manag 5(4): 311–321
Smith B, Burton I, Klein RJ, Wandel J (2000) An anatomy of adaptation to climate change and variability. Clim Chang 45(1):223–251
Smith JW, Leahy JE, Anderson DH, Davenport MA (2013a) Community/agency trust and public involvement in resource planning. Soc Nat Resour 26:452–471
Smith JW, Leahy JE, Anderson DH, Davenport MA (2013b) Community/agency trust: a measurement instrument. Soc Nat Resour 26:472–477
Warren DM (1991) Using indigenous knowledge in agricultural development. World Bank, Washington, DC

Warren DM, Cashman K (1988) Indigenous knowledge for sustainable agriculture and rural development. International Institute for Environment and Development, Sustainable Agriculture Programme

Woolcock M (1998) Social capital and economic development: toward a theoretical synthesis and policy framework. Theory Soc 27(2):151–208

Woolcock M, Narayan D (2000) Social capital: implications for development theory, research, and policy. World Bank Res Obs 15(2):225–249

Chapter 22
Mainstreaming Resilience into Development Programming: A Practitioner's Perspective

Jacobo Ocharan, Ghislaine Guiran, and Alison Wright

Abstract This paper starts with the acknowledgment that resilience is still a confusing concept, difficult to demonstrate, but a necessity when considering community's development programming. It then presents the innovative approach developed by the INGO Plan International while developing a toolbox addressing both programming and accountability requirements regarding resilience, i.e. combining process-oriented and performance-oriented dimensions. At first it presents the strategic and programmatic vision of Plan for resilience—child-centered approach, based on the characterization of resilient communities and on a two-levels review process (the "what" and the "how") of programming activities. As a second part, specific processes and tools—under construction—are described, that are designed to support programming teams in: (i) screening existing programmes/projects from the resilience perspective, (ii) design projects so that they contribute to resilience building. As a conclusion, the opportunity brought upon by the resilience debate to rethink the development paradigm from a community and human perspective is highlighted. Resilience building must be understood as interfering with causal pathways of change and not only as a linear programming process towards pre-defined outcomes. To make such a shift happen on the ground, tools are needed—translating academic knowledge on resilience into practitioners' toolbox; this paper aims at contributing to connect the dots, providing a framework for a reality check on the field of development programming.

Keywords Resilient communities • Resilience enablers • Development projects • Programming cycle

Introduction: Fishing a Fish Called Resilience

During the last 2 or 3 years, development and humanitarian academics and practitioners have been trying *to fish a fish called resilience*. In fact, most of them are not even using a fishing line. They are trying to capture it with their hands. Nobody has

J. Ocharan (✉) • G. Guiran • A. Wright
Plan International, Dukes Court, Block A, Duke Street, Woking, Surrey GU21 5BH, UK
e-mail: jacobo.ocharan@plan-international.org

W. Leal Filho (ed.), *Innovation in Climate Change Adaptation*, Climate Change Management, DOI 10.1007/978-3-319-25814-0_22

been able to catch it yet but as they are getting close to it, they are starting to realize that, probably, the interest is not in the fish itself but rather on the how it swims (Ocharan 2014).

Hundreds of papers have been produced in the last couple of years trying to define resilience to disasters and how it operates. The fact that after more than 60 years of development and humanitarian practice, many communities in developing countries continue to struggle in front of recurrent natural, economical and social disturbances has put the focus on resilience as a possible solution. Policymakers and commentators see resilience as the way of reconciling relief and development initiatives (Barrett and Constasa 2013). Many of published papers try to de-code the complexity of resilience. Some of them are admitting that resilience cannot be conceived as only an outcome but also as the process that allows putting resilience in practice, which makes all the more difficult the necessary monitoring and assessment of resilience-oriented programmes.

Even if challenging, understanding how to improve resilience of communities while implementing development and humanitarian programmes is an urgent necessity. Building on its vision and on its experience, Plan International is promoting a way for *resilience-proofing* its activities with the objective of making resilience happen at community scale and to be able to demonstrate its contribution. The aim of the paper is to present the approach underway, which intends to connect the dots between academic knowledge on resilience and practitioners programming reality—as a reality check. It provides an innovative way of addressing both programming and accountability requirements regarding resilience, combining process-oriented (resilience enablers) and performance–oriented (characteristics of resilient systems) dimensions. Developed from a practitioner's perspective, the paper is designed to address specific needs for development programming and must be considered as such rather than as a purely academic paper.

Plan's Vision of Resilience

A strategic Vision

Children and their communities are the priority for Plan's intervention, considering children in the centre of a system where family, school and the education system, community and government converge. This interconnected system is what supports or undermines children's capacity to deal with disturbances. This specific perspective leads to a children-centered definition of resilience for Plan, i.e.: *resilience is defined as the ability of children and their communities to deal positively with disturbances that undermine the fulfilment of their rights* (Ocharan 2014).

Plan's resilience approach integrates the idea of 'multi-stress' into the understanding of potential disturbances, covering shocks (events suddenly breaking the functioning of a system) but also stresses (a steady but constant deterioration of the functioning of a system such as unrestrained urbanization, desertification or recurrent drought). While stresses don't necessarily create disasters, they erode the system over the long term thereby leaving communities with reduced capacity to respond to and recover from disturbances: "development resilience concerns the time path of well-being in the face of both stressors and shocks, and, especially, the likelihood that any adverse outcomes of a disturbance persist for an extended period of time" (Barrett and Constasa 2013). This is in contrast to traditional disaster risk reduction (DRR) work, which has the limitation of focusing more narrowly on natural disasters, and sometimes on conflicts or human-made disasters.

Being resilient to disturbances within this context highlights the reality that societies face different, often simultaneous shocks, such as conflicts, economic failures (financial crisis, food price spikes, unemployment), socio-political stresses (failed states and governance, uncontrolled migration, conflicts) or environmental stresses (depletion of natural resources, land degradation, pollution). In order to react to a wider and interacting range of disturbances children and their communities require a diversified range of capacities, improving their ability to absorb a shock, to adapt to it, and to transform if necessary. "Adaptive capacity" (and similar variants) is emerging as an integrating framework which emphasizes the ability to adjust to potentially radical changes in context, not just withstand shocks (Villanueva 2011).

A Programmatic Vision

As previously commented, there has been a growing preoccupation during the last 2–3 years—more particularly within the DME (Design, Monitoring, Evaluation) community—on how to define resilience, how to characterize resilient systems [considered as socio-ecological systems (Folke 2006)] and consequently measure a resilience performance; a consensus on how to measure resilience is still to build (Vaitla et al. 2012). Some recent publications are drawing transverse lessons on this issue with the objective of translating conceptual frameworks in operational ones (SRC 2010; ARUP 2014; IARWG 2012). If evidence base is still missing to help practitioners choose and implement available material in the best way (regarding their specific needs), some common messages can be highlighted when trying to characterize resilient systems.

Those messages have been used to support Plan's programmatic approach to resilience: as a support to translate the strategic approach into the programming cycle, a tentative framework of resilient communities has been defined as a set of conditions to be fulfilled, structured along six dimensions as described in Table 22.1.

Resilience building is highly contextual what requires to stay flexible in any assessment of a resilience situation: using this six-dimension framework, a project team (in collaboration with community's stakeholders) can provide an initial qualitative scoring for each of the dimensions and characterize potential pathways

Table 22.1 Key dimensions of resilient communities (adapted from key conceptual frameworks) (SRC 2010; ARUP 2014; IARWG 2012)

Key dimensions of resilient communities	Description
Informed	The community has access to information about potential disturbances that may affect them and are able to use information to advise their planning and decision making
Diverse	The community displays diversity in its range of functional groups, economic opportunities, reliance on natural resources and partnerships
Prepared	The community has plans and back-up systems to cope with disruptions. There is spare capacity in systems to accommodate severe disruptions
Flexible	Uncertainty and change are accepted by individuals and the community is able to accommodate and adjust to change. There is space for learning and innovation
Inclusive	Community decision making allows for meaningful participation of girls and boys and men and women, including people from commonly marginalised groups such as people with disabilities and ethnic minorities. There is a high degree of social and economic equity, including in the management and ownership of resources
Integrated	The community's planning is linked to and informed by processes and plans at other levels in different sectors

for improvement. Such characterization may be used as a support for further monitoring and assessment tasks.

In order to ensure that development intervention will support and enhance communities' resilience, both the content of programmes and the way they are implemented may be revised:

1. "What": *resilience enablers* have been designed as specific outputs or outcomes that may contribute to the community resilience, to be explicited in the context of pre-identified disturbances. Identifying these outputs and outcomes as resilience enablers allows to demonstrate, measure and document how programming is contributing to resilience. Resilience enablers are organized into two groups: (i) cross-cutting enablers, relevant to all projects (individual capacities, community cohesion, governance and participation and gender), (ii) sectorial enablers (related to specific programming areas).
2. "How": specific characteristics have been identified as pre-requisites that programmes should demonstrate to effectively contribute to resilience building.

Mainstreaming Resilience into Programming

Specific process and tools for mainstreaming resilience into Plan's programming are under construction with the objective of supporting teams to answer two types of questions:

- Does my programme/project—as it is currently designed—contribute to building resilience?
- How can I design my project so that it contributes to resilience building? What actions should my project include to help to build resilience?

Screening Existing Projects: A Resilience Lens

A *resilience lens* is designed to support a quick assessment of the degree of resilience building integration that any project, programme or planning document has. Table 22.2 describes this *resilience lens*:

- Four different levels of resilience building capacity are described on the left column.
- The central column offers criteria to gauge the level of capacity the programme or project may have.
- The links with the resilience dimensions (see Table 22.1) as previously described are commented in the right column.

Integrating Resilience in the Programming Cycle: Tools to Support the Programming Tasks

Taking into account specific requirements of project's cycle, the following tools are under development to support the programming teams:

- Resilience enablers, as entry points for the definition of resilience-impacting activities and their integration into the programme objectives and activities (i.e. to support the "what").
- Resilience building characteristics of the programme, as a checklist to identify if any specific programme has the qualities, totally or partially, to build resilience (i.e. to support the "how").

Table 22.2 Resilience lens (lessons from the ground)

Is the programme building resilience?	"Resilience lens" of an existing programme	Potential impact on the "resilience dimensions"
Resilience unaware It is not building resilience The programme does not recognise that the targeted population faces disturbances and does not enhance capacities to deal with them	Understanding of the 'resilience context' • Shows no understanding or analysis of disturbances that the community/programme may go through (such as floods, drought, earthquake, conflict and longer term environmental impacts) • There is no reference to or analysis of the exposure, vulnerabilities and capacities of children and their communities related to local hazards Contribution to building the resilience of communities • Objectives, outcomes and indicators do not explicitly target resilience building • Does not promote the participation of affected populations in risk identification and planning • Focuses on one time-scale of change only (short, medium or long term goals or changes) • Does not foster synergy between multiple levels of society (local, regional, national, global) • No specific human and financial resources are allocated to document how the programme builds resilience Resilience of the programme itself • Ignores usual or potential disturbances in the list of unforeseeable or 'killer assumptions' in the logical framework • Inflexible design—it is rigid in its implementation and it does not allow for variation in the programme influenced by the change of context	No positive impact

(continued)

Table 22.2 (continued)

Is the programme building resilience?	"Resilience lens" of an existing programme	Potential impact on the "resilience dimensions"
Resilience neutral It has some elements that might build resilience The programme recognises that the population may face disturbances but does not do anything about enhancing the capacity to deal with them	Understanding of the 'resilience context' • The participatory situation analysis includes an acknowledgment of disturbances that the target populations and the programme may experience, including conflict and longer term environmental impacts • There is some reference to, or identification of, the exposure, vulnerabilities and capacities of targeted populations in relation to the most common disturbances. However, these are not linked to the programme activities Contribution to building the resilience of communities • Includes some participation of target populations in risk identification and planning • It contemplates the possibility that changes in the community may be happening on different time-scales (short, medium or long term goals or changes) but does not include these in the programme management • Does not make a link between project/programme goals, outputs and outcomes and their potential as enablers for resilience Resilience of the programme itself • Focused on activities and outputs more than on outcomes, making it more difficult for project management to adapt to changes • The design allows the programme to make changes influenced by major changes of context (political, economic, etc.)	Occasional impact on the Informed, Flexible, Inclusive and the Integrated dimensions (awareness-raising on resilience, introduction lo long term and shifting context and integration of cross-scale perspective)

(continued)

Table 22.2 (continued)

Is the programme building resilience?	"Resilience lens" of an existing programme	Potential impact on the "resilience dimensions"
Resilience aware It is building resilience The programme seeks to improve the capacities of the targeted population to face disturbances. It acknowledges that by fostering some actions it will enhance resilience but is unable to document it	Understanding of the 'resilience context' • The participatory situation analysis includes an assessment of how target populations and programmes go through disturbances, including conflict and longer term environmental impacts • There is an identification of the exposure, vulnerabilities and capacities that target populations have in relation to the most common disturbances, many of which are linked to programme activities Contribution to building the resilience of communities • Target populations with specific vulnerabilities and capacities participate in risk identification and planning • Objectives and outcomes do target resilience building • Resilience issues and drivers are integrated throughout the programme • Promotes participation of affected populations on risk identification and risk planning and links these to the programme outcomes/goals • Likely to have no documentation of resilience building Resilience of the programme itself • The log frame is focused on outcomes, which allows for adaptive management of the project • Consideration has been given to how the project activities might be affected by potential disturbances • The project team understands the importance of dealing with uncertainty and adjusting to change	Significant impact on all the resilience dimensions of the communities, mainly the informed, flexible, inclusive and the Integrated dimensions (long term and shifting context as well as cross-scale dimensions are taken into account when designing the activities); prepared and diverse dimensions only partly addressed

(continued)

Table 22.2 (continued)

Is the programme building resilience?	"Resilience lens" of an existing programme	Potential impact on the "resilience dimensions"
Resilience championed It is championing resilience building The programme has an explicit intention to build resilience. It plans its activities against a pre-identified measurement of the capacities of the targeted population to deal with the disturbance and is able to demonstrate changes generated by our actions	Understanding of the 'resilience context' • An analysis of how targeted populations and programmes go through disturbances, including conflict and longer term environmental impacts is fully integrated in the participatory situation analysis • It includes scenario planning through new understandings of complexity of the social and economic systems, with diverse stakeholders. This can support understanding of potential changes in the economic, social, political and natural environment Contribution to building the resilience of communities • Target populations participate in the risk identification and planning and these inputs are key to the programme definition and management • Goals and outcomes of the programme target resilience building and associate indicators to measure it • It programmes in different time-scales to adapt to the changes that happen in the community • Specific human and financial resources are allocated to demonstrate how the programme builds resilience • Acknowledges that changes in the community happen on different time-scales (short, medium or long term goals or changes) and allows for programme implementation to adjust accordingly Resilience of the programme itself	Structural impact on all the resilience dimensions of the community

(continued)

Table 22.2 (continued)

Is the programme building resilience?	"Resilience lens" of an existing programme	Potential impact on the "resilience dimensions"
	• The log frame is focused on the outcomes including those related to building resilience, rather than activities and outputs, allowing for adaptive management and different potential pathways for implementation • It has crisis modifiers (identified in the scenario planning) that can be introduced so that alternative responses can be implemented in response to disturbances • Project management demonstrates high capacity to deal with uncertainty and to adjust to change	

Resilience Enablers

By tagging which specific factors may support resilience—in relation with the resilience dimensions we introduced, a more realistic planning and tracking of resilience building is promoted. Those enablers will entail activities very similar to activities yet usually included in programmes, but their design has to take into account key disturbances potentially impacting the community. As previously commented, resilience building is highly contextual: the importance of each factor and aspect will vary depending on the situation and location, which has to be characterized during preliminary assessments. Plan identifies two groups of enabling factors. Table 22.3 presents the cross-cutting enabling factors.

Characteristics of Programmes

Some specific characteristics of programmes are needed to demonstrate their potential contribution to resilience building; these characteristics overlap with Plan's internal standards. The Table 22.4 describes those characteristics.

Integrated Framework for Implementation

When applying resilience enablers, specific activities ("what") will be identified at a community and project scale to help increasing resilience. It may relate to usual

Table 22.3 Cross-cutting resilience enablers (lessons from the ground)

Category	Resilience enablers: a project may contribute to building resilience if
Individual capacities	• It promotes the capacity to deal with and accept uncertainty • It encourages flexibility in short and long term planning • It supports capacity to be open to and to adjust to change • It promotes the ability to work across different time-scales • It supports the ability to analyse and to understand the disturbances faced, and how these might change in the future
Community cohesion and empowerment	• The community is able to integrate diversity • It promotes inclusion, especially of traditionally marginalised 'groups' such as ethnic minorities and people with disabilities • It promotes the community's internal and external social networks and partnerships • It promotes the community's capacity to innovate, as well as use traditional knowledge and to learn from past experiences
Governance and participation	• It supports or promotes good governance systems and institutions • It promotes accountability of community leadership • It supports and strengthens local civil society • It builds the capacity of government institutions to promote and lead preparedness, planning and readiness to deal with disturbances • It promotes meaningful participation of all groups in the community, including boys and girls, in risk analysis, planning and action
Gender	• It strengthens the participation of women and girls in decision making and policy processes related to resilience building • It supports the mainstreaming of gender equality and women and girls empowerment in disaster and climate risk reduction policies and plans • It helps to secure women's rights over land and other resources and access to information and alternative livelihoods

development activities as implemented by Plan's impact programming areas but it may require new activities or activities calibrated in a different way.

Based on that framework, activities have to be translated in the overall programming cycle while checking their consistency with the required programme's qualities ("how").

As illustration, some specific activities applicable at different programming steps are listed in Table 22.5.

Table 22.4 Characteristics of programmes for resilience building (lessons from the ground)

Characteristics	Description
Accountable	Programmes must ensure adequate information is provided and transparent feedback mechanisms are available to all stakeholders, particularly the most vulnerable. Also they should promote and advocate for strengthened accountability across governance systems
Works across different levels	The analysis, decisions and actions taken at each programme level (local, sub-national, national, global) should be mutually informative and facilitate the development of a coherent and coordinated approach. Strategies to build resilience should engage all sectors of society and government. The goal of multi-sectorial and multi-stakeholder engagement should be to make building resilience central to development planning The project or programme should be able to address different timescales; the analysis and strategies and programmes should address current, identified risks and likely future scenarios
Inclusive	Inclusive project activities and outcomes are based on context-specific analysis of the differential needs, vulnerabilities, expectations and existing capacities of all population groups (particularly women, girls and boys, people with disability, ethnic minorities and others). Project activities should seek to complement local and traditional knowledge. Project management and implementation should guarantee meaningful participation from all population groups
Flexible	Analysis of disturbances should be responsive to a changing environment since there is a high degree of uncertainty about its presence and evolution. Similarly, strategies and programmes to build resilience should be flexible to accommodate new inputs. Project plans should be regularly revised and adapted to shifting risk to disturbances
Informed	Project design should be based on an understanding of past trends, present experiences and future projections of disturbance occurrence and the effects on the area and population. An assessment of the vulnerabilities and capacities of the population, systems and resources in relation to the most common disturbances should be the basis for decisions on the location, target populations, objectives and approach to building resilience through the enabling factors
Participatory	The project has strong participation of, and action by, the population at risk, including boys and girls Their first-hand knowledge of the issues affecting them is critical to ensuring that analysis and subsequent actions are based on empirical evidence
Promotes learning	The project has monitoring systems in place that allow learning for adaptive management. At the same time, it should promote replication of effective practices, encourage autonomous innovation and introduce, where appropriate, external technology to help address new or magnified challenges. Strategies and programmes should be monitored and evaluated to ensure that learning is captured and made available to other programmes
Prepared	Potential failures in the project implementation are anticipated and translated into information systems and action plans, to ensure failure is predictable, and not catastrophic. Critical functions of the project are identified and described in relation to potential disturbances. Back-up systems are in place to cope with short term disruptions of core functionality in event of disturbances. Spare capacity is purposely created within systems so that they can accommodate disruption

Table 22.5 Key actions for mainstreaming resilience at project's scale (lessons from the ground)

Stage of project cycle	Key actions to integrate resilience building at project's scale
Design	*Situation analysis/needs assessment* • Gather information about the 'resilience context' • Add a 'resilience perspective' when analysing the information collected • Make sure that the process of collecting the information about the resilience context is participatory and includes girls, boys, men and women • Information on the resilience context in the situation analysis is used to inform the *targeting* of the intervention; the choice of *resilience enablers*; and help to identify *how disturbances may affect the project's* outcomes/outputs or planned activities
	Project design There are two key questions to address when designing a project, in order to ensure that it contributes to the resilience of children and their communities • How can the **project activities** and interventions best contribute to resilience, by supporting enabling factors for resilience? • What **characteristics** should your project have to ensure that it best contributes to resilience? *Monitoring, evaluation and learning (MEL) framework* The MEL framework project should allow to capture changes in resilience and to support the process of building resilience by being flexible and promoting learning
Planning	When developing the implementation plan for the project, one should consider how the **project itself** can be made more resilient, by considering the impact of disturbances on project activities, outputs and outcomes and planning measures to mitigate risks

Conclusion

Resilience has marked a definitive break with the misconception that development is a linear process in which individuals, communities and countries can achieve step by step or through a 'building blocks' process. Considering communities as socio-ecological systems, resilience frameworks have to take into account their adaptive cycle, interacting across multi-scales [referred to as "panarchy" (Walker et al. 2004)].

Plan's approach and experimentation of resilience is based on this conceptual background, aiming at understand *how the fish swims and how we can contribute to allowing it swim better*. The toolbox under development manages to combine process-oriented and performance-oriented dimensions for resilience programming via the definition of (i) key dimensions of resilient systems/communities to help structure the objectives (short and long term) of a project and of (ii) resilience enablers to take into consideration in the programming cycle. It will help Plan International in addressing both programming and accountability requirements regarding resilience development. Based on the methodology introduced in this paper, a toolkit is under development to support Plan's staff to identify and implement changes to how and what to do across development programmes so that they better contribute to resilience building. An internal consultation is underway, whose results will help to fine-tune the methodology and the tools; further development steps of the methodology will include the selection and review of some typical in order to highlight specific opportunities for resilience integration.

References

ARUP (2014) City resilience framework. ARUP, London

Barrett CB, Constasa MA (2013) Toward a theory of resilience for international development applications – draft paper. Cornell University, Ithaca

Folke C (2006) Resilience: the emergence of a perspective for social–ecological systems analysed. Glob Environ Chang 16:253–267. http://www.sciencedirect.com/science/article/pii/S0959378006000379. Accessed 28 June 2015

Interagency Resilience Working Group (IARWG) (2012) The characteristics of resilience building – A discussion paper, London

Ocharan O (2014) Fishing a fish called "resilience". Plan International, London

Strengthening Climate Resilience (SRC) (2010) The resilience renaissance. Institute of Development Studies, Brighton

Vaitla B, Tesfay G, Rounseville M, Maxwell D (2012) Resilience and livelihoods change in Tigray Ethiopia. Feinstein International Center, Tufts University, Somerville

Villanueva PS (2011) Learning to ADAPT: monitoring and evaluation approaches in climate change adaptation and disaster risk reduction – challenges, gaps and ways forward, Strengthening Climate Resilience (SCR), Discussion Paper 9, Institute of Development Studies, Brighton

Walker B, Holling CS, Carpenter SR, Kinzig A (2004) Resilience, adaptability and transformability in social–ecological systems. Ecol Soc 9(2):5. http://www.ecologyandsociety.org/vol9/iss2/art5. Accessed 28 June 2015

Chapter 23
Wetlands Biodiversity, Livelihoods and Climate Change Implications in the Ruaha River Basin, Tanzania

Pantaleo K.T. Munishi, Halima Kilungu, Nice Wilfred, Bernadetha Munishi, and Stein R. Moe

Abstract Wetland ecosystems in Tanzania contribute significantly to livelihoods and biodiversity conservation. These ecosystems are vulnerable to both human and climate change induced impacts though the implication of climate change on these ecosystems is poorly known. We assessed a wide range of wetlands in the Ruaha River basin to quantify their biodiversity, livelihoods and potential climate change impacts. It was observed that biodiversity of wetlands is much higher than that in adjacent habitats. Wetlands are refuge for endangered plant species making them important in biodiversity conservation. Wetland cultivation and fisheries contributed over 40 % of the total household income and food. Over 90 % of the dry season agriculture is wetland dependent forming the major livelihood source for majority in the basin. The most visible climate change impacts on the wetlands of Ruaha are reduced water flows, drying of wetlands and degradation of their ecological functions. With climate change, water flows in the Ruaha have decreased by 10 % causing substantial reduction in extent of wetlands in the basin. Adapting the wetlands to climate change require evaluation of current and future climate vulnerability of

P.K.T. Munishi (✉)
Department of Forest Biology, Sokoine University of Agriculture, PO Box 3010, Morogoro, Tanzania
e-mail: pmunishi2001@yahoo.com

H. Kilungu
Department of Tourism and Hospitality, Open University of Tanzania, PO Box 23409, Dar es Salaam, Tanzania
e-mail: kilunguh@yahoo.com

N. Wilfred • B. Munishi
Ministry of Agriculture, Food Security and Cooperatives, PO Box 9192, Dar es Salaam, Tanzania
e-mail: niceafb@yahoo.co.uk; benney_munishi@yahoo.com

S.R. Moe
Department of Ecology and Natural Resource Management, Norwegian University of Life Sciences, Box 5003, 1432 Aås, Norway
e-mail: stein.moe@nmbu.no

W. Leal Filho (ed.), *Innovation in Climate Change Adaptation*, Climate Change Management, DOI 10.1007/978-3-319-25814-0_23

wetland ecosystems and related livelihoods. This information is important in informing policies, adaptation and mitigation strategies for wetland ecosystems.

Keywords Wetlands • Biodiversity • Livelihoods • Climate change

Introduction

Wetlands are among the world's most biologically rich ecosystems with high species diversity (Mvena 1999; Yanda et al. 2005; Munishi 2006). Wetlands ecosystems are second only to the rain forests in the number of wildlife and plant species that depend on them for feeding and habitat. Historically, wetlands have been regarded as wastelands but they can also be viewed as being among the last truly wild and untouched places in the world (Maltby 1986) making them of high repository of biological diversity with most wetlands offering important habitats to a variety of fauna and flora.

For most African countries, wetlands have a considerable importance in provision of innumerable benefits like drinking water, transport, harvestable plants and animals and food production (Munishi et al. 2010). Though not well quantified, Tanzania's wetlands contribute in diverse ways to livelihoods of many millions and are chiefly utilized for crop and livestock production (Kashaigili 2006; Munishi et al. 2010). Wetlands provide income in both dry and wet years for fairly large number of people engaged in agriculture because of their available water and high soil fertility (Mkavidanda and Kaswamila 2001; Kashaigili et al. 2005; Munishi et al. 2010; Kilungu and Munishi 2009).

The 2025 Tanzania's vision overall goal specifically includes references to 'sustainable development endeavours, on intergeneration equity basis, such that the present generation derives benefits from the rational use of natural resources of the country without compromising the needs of future generations'. Further more the National Strategy for Growth and Reduction of Poverty (NSGRP) or MKUKUTA recognize poverty as largely a rural phenomenon and that the rural poor depend solely or to a greater extent on natural resources (Bagachwa 1994; DPG 2005; URT 2006). Consequently the national environmental policy of 1997 and all the natural resource policies emphasize the clear cause-and-effect relationship between poverty and environmental degradation, and because of this they stress on the need for sectoral policies to address poverty issues by taking into account the need for wise use and sustainable resource exploitation (MNRT 2003, 2004, 2009).

Wetlands are amongst the most productive ecosystem in Tanzania and have significant economic, social, cultural and biological values. Apart from agricultural use wetlands are considered useful as sources of a variety of natural resources of significance to human welfare. Tanzania is endowed with exceptional wetland resources in which 10 % of the country is covered by wetland ecosystems. These ecosystems range from large lake systems to river floodplains, deltaic mangrove formations and associated catchments (Maltby 1986; Kamukara and Crafter 1993). Of the area covered by wetlands, 5.5 % is occupied by four Ramsar sites namely Malagarasi/Moyovosi (32,500 km^2), Lake Natron Basin (2250 km^2), Kilombero valley floodplains (7950 km^2) and Rufiji-Mafia-Kilwa (5969.7 km^2) (MNRT 2003). Broadly the country is divided into nine drainage (river) basins which include Lake Rukwa, Lake Victoria, Lake Tanganyika, Lake Nyasa, Rufiji River, Pangani River, Ruvuma River and Southern Coast, Wami-Ruvu River and Internal Drainage basin. Each of these basins includes a network of rivers and inland valley bottom wetlands scattered throughout the country. These vast biologically rich resources in Tanzania are unique in their biodiversity values as well as support to local livelihoods (MNRT 2003; Kilungu and Munishi 2009). The implications of climate change and human induced activities on the wetlands of the Great Ruaha River are diverse. Protection of the wetlands against biodiversity loss, while maintaining their socio-economic value under the wake of climate change does not however appear to be a serious alternative among majority of the society in Tanzania thus jeopardizing future sustainability of these wetlands.

This paper presents results of a study to assess the role of wetland ecosystems in conserving plant diversity, contribution to livelihood enhancement and implications of climate change in the Great Ruaha River basin Tanzania.

Methodology

The Great Ruaha River is a river in south-central Tanzania that flows through the Usangu wetlands and the Ruaha National Park east into the Rufiji River. Its basin catchment area is 83,970 km^{-2}. The population of the basin is mainly sustained by irrigation and water-related livelihoods such as fishing and livestock keeping. Great Ruaha is about 475 km long, its tributary basin has a catchment area of 68,000 km^2 and the mean annual discharge is 140 m^3/s. The Great Ruaha River supplies 22 % of the total flow of the Rufiji catchment system. Thirty-eight species of fish have been identified in the Great Ruaha River. The river's headwaters are in the Kipengere Range. From there the Great Ruaha River descends to the Usangu plains, an important region for irrigated agriculture and livestock in Tanzania. The river eventually reaches the Mtera Dam and then flows south to the Kidatu Dam which generate about 50 % of the Tanzania's electricity. The river continues southwards and flows across the Selous Game Reserve before reaching the Rufiji River. The major rivers contributing to the Great Ruaha River are Lukosi, Yovi, Kitete, Sanje, Little Ruaha, Kisigo, Mbarali, Kimani and Chimala whereas the small ones include Umrobo, Mkoji, Lunwa, Mlomboji, Mpombochi, Ipatagwa, Mambi, Isimani stream, Idete River, Mkungwe, Kinoga d Mswiswi Rivers and associated swamps and streams (Fig. 23.1). The wetlands of Little Ruaha River start from Bumilayinga, Nyololo, Mafinga and Ihalimba wards (Fig. 23.1). The River further extends to Kilolo and Iringa Districts crossing the Ipogoro Bridge along the Dar es Salaam—Mbeya road before joining the Great Ruaha River at Mawande village. The entire

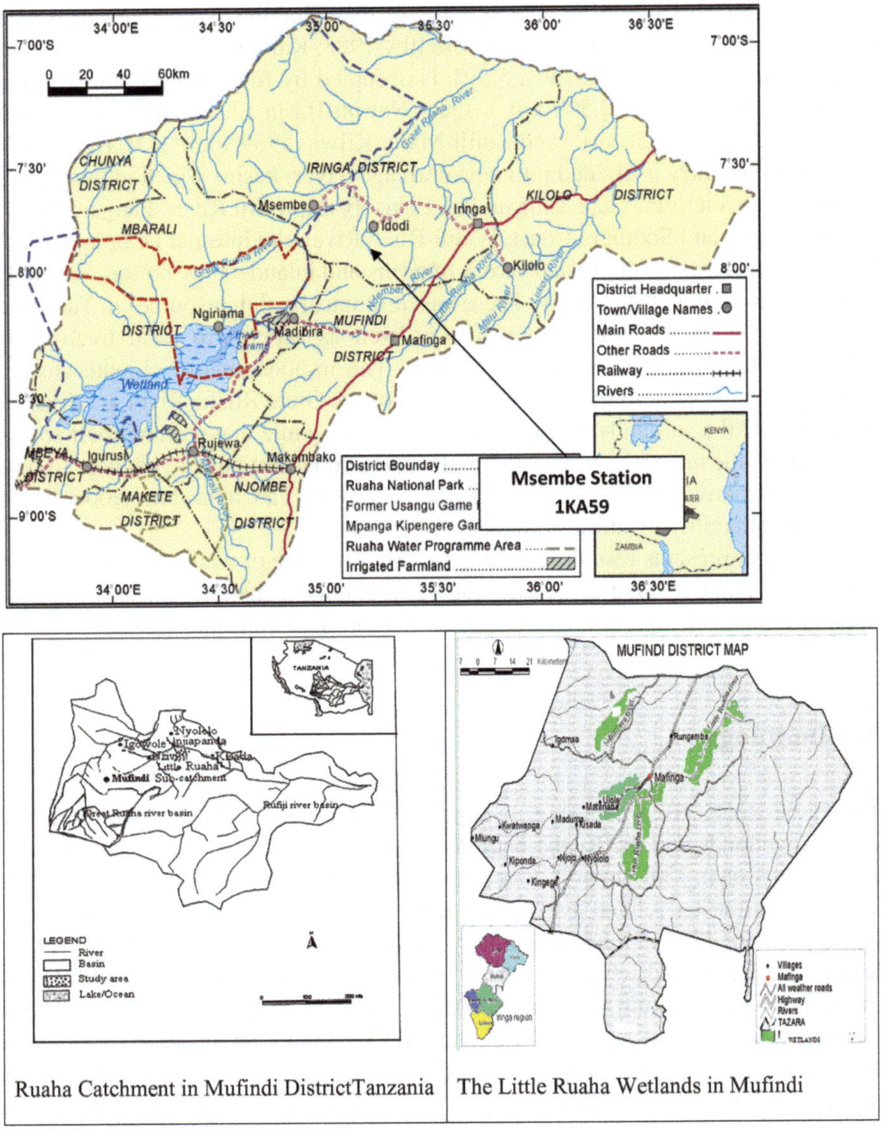

Fig. 23.1 Map of the Great Ruaha River Basin and Mufindi District showing the Study Sites

catchment, lies between latitudes 7° 15′ and 8° 35′ south and longitudes 35° 00′ East covering 4900 km^2 and altitude of 600 m to 2100 m above sea level.

Data Collection

Ecological Study

For ecological study, sites were selected based the presence and extent of valley bottom wetlands within the sites. Three sites were selected for the ecological studies which include Uchindile with Mpombochi River and Isimani stream plus associated swamps, Idete with Idete River and associated swamps/streams and Mapanda with Mkungwe and Kinoga Rivers and associated swamps/streams (Fig. 23.1). Prior to field surveys, the areas were stratified into broad vegetation types using existing topographic/land cover maps and reconnaissance survey. This stratification gave several major vegetation types including woodlands, grasslands, valley bottom wetlands/swamps undisturbed natural vegetation and plantations of different types. These broad types were further sub divided into sub categories depending on field conditions such as wooded grasslands, natural grasslands, pine plantations and eucalyptus plantations.

Temporary nested sample plots measuring 20×20 m were adopted and established systematically in clusters representing different vegetation types in each stratum on transect lines established along a predetermined compass direction. About 100 plots were established in each site based on area and an adopted minimum sampling intensity of 0.01 % ensuring a sufficient coverage and representation of each stratum. The location of each plot was recorded using Global Positioning System (GPS). At each quadrat all the plant species encountered were identified, recorded and percent cover determined. For species that were difficult to identify in the field voucher specimens were collected and identified at a herbarium.

Socio-economic Study

For the social economic studies villages were selected based on their proximity to wetland ecosystems. Multistage sampling was used where two divisions within the upper catchment of the Little Ruaha sub catchment were selected randomly. Two wards were then selected randomly from each Division. For each ward two villages were selected randomly for the study. The selection of villages was based on close proximity to the wetlands with the assumption that the respective communities were more involved with wetlands cultivation as compared to other villages. In this case, Matanana village (Bumilayinga Ward) being in the upper catchment and Luganga village (Ifwagi Ward) on the lower catchment were sampled. Other villages included in the study are Nyololo Njiapanda (Nyololo Ward), Kisada (Bumilayinga Ward) Igowole and Nzivi (Igowole Ward) (Fig. 23.1). In each village the village register was used as a sampling frame, households were then randomly selected and a questionnaire administered to heads of the selected households. Households were taken as sampling units in this study and total of 93 households were sampled

(44 and 49 households from Luganga and Matanana respectively). The questionnaire sought to get information on whether the household is involved in any kind of wetland utilization, socio-economic activities undertaken in the wetlands, costs and revenues from wetland utilization. Further to questionnaire survey PRA techniques including Focus Group Discussions (FGD) were used as well as participant field observation to supplement information from household surveys.

Hydrology and Climate

The data on climate impacts was based on the changes observed in the flow of the Ruaha River at Msembe station 1KA59 using historical flow data for 46 years—four decades (1958–2004) (Fig. 23.1).

Data Analysis

Ecological Data

All special features of interest in the area e.g. water sources, valley bottom wetlands, natural grassland ecosystem, miombo woodlands and wooded grasslands were identified and compiled into a list with a description of their species composition. The identified strata and field data were developed into tables showing a list of different vegetation types and their species composition, richness, abundance and dominance. The abundance and dominance of each species was determined from their percent cover estimates.

Socio-economic Data

Socio-economic data were analyzed and summarized into social economic activities undertaken by a household, agricultural utilization of wetlands and crops grown in dry/wet seasons and the proportional contribution of each wetland related socio-economic activities to household food security (food available for household consumption) and income. The economic benefits were assessed by using gross margin analysis, food available for consumption as indicator of food security was used to assess food security at household level; and the contingent valuation technique was applied to assess the value of wetlands services. The gross margin analysis was computed as:

$$GM = TR - TC$$

Where:

GM Average gross margin (Tshs/kg) or (Tshs/month)
TR Average total revenue (Tshs/kg) or (Tshs/month)
TC Average total variable cost (Tshs/kg) or (Tshs/month)

The food available for consumption at household level was determined to be 300 kg of cereal/person/year. This figure was used to offset post harvest losses (storage loss and handling loss) (FAO 1985; Ishengoma 1998). Food available for consumption was obtained by subtracting the amount of food crop that was sold from the total food produced per person per year. Standardization of food available for consumption was done using adult equivalent scale considering age category (Ishengoma 1998). In this case where the adult above 15 years old has unit equivalent of 1, ages 11–15 will have unit equivalent of 0.75 while children with age equal or less than 10 years will have unit equivalent of 0.36 (Ishengoma 1998).

Hydrology Data

The hydrologic data were analyzed for trend to give information on what have been the changes in the flow of the river over time that may have influenced the wetland status. Further, the data were analyzed for periods of extreme low flows that would be a determining factor on wetland conditions.

Results

Vegetation Types and Plant Species Composition in the Great Ruaha Basin

Several vegetation types were identified in the different sites. The major natural vegetation types identified included valley bottom wetlands, natural grasslands, wooded grasslands, and miombo woodlands. In addition planted exotic tree species formed specific vegetation types and included *Eucalyptus* and *Pine* plantations. There was a big variation in plant species composition and richness in all sites studied with the different vegetation types having different species composition and richness within sites. The variation in composition is a reflection of high diversity of plants in the areas. In all the three sites valley bottom wetlands ranked the highest in plant species richness when compared with other natural vegetation and combined natural vegetation and plantation of exotic species. This general trend shows that valley bottom wetlands in the Ruaha River Basin will likely be the major repositories of biodiversity of plant species. By the fact that vegetation composition may be a reflection of other taxa in an ecosystem there is a high possibility that fauna diversity will follow the same trend. Because plants respond to multiple environmental factors both biotic and abiotic, this richness and diversity of species is a reflection of the heterogeneity of the areas, diverse vegetation types, habitats and landscapes that allow co-existence of species in heterogeneous landscape (Carson and Root 2000; Franzén 2004; Tetsuya and Kuniyasu 2005; Munishi et al. 2007; Tomáš and František 2008).

Plant Species Composition by Vegetation Types in Uchindile Landscapes

Three natural vegetation types were identified in Uchindile including (i) Valley bottom Wetlands/Riverine/Riparian, (ii) Natural Grasslands and (iii) Wooded Grasslands. In addition to the natural vegetation there were two types of plantations of exotic tree species which are among the major vegetation found on the landscape. These are (i) Pine plantations and (ii) Eucalyptus plantations. The number of plant species varied between vegetation types. In regards to the natural vegetation Valley bottom Wetlands have the highest number of species compared to all other vegetation types (Fig. 23.2a). Further when we consider all the vegetation types Valley bottom Wetlands still rank the highest in plant species richness (Fig. 23.2b). Previous studies have also shown the valley bottom/riverine/riparian vegetation to have higher species richness than the other types of vegetation (Munishi 2006, 2007).

Plant Species Composition by Vegetation Types in Idete Landscapes

Four different natural vegetation types were identified in the Idete landscapes. These vegetation types include Valley bottom Wetlands/Riparian Areas, Natural Grasslands, Wooded Grasslands and Miombo Woodlands. Valley bottom Wetlands/riparian areas had the highest proportion of species compared to all other vegetation types followed by Wooded Grasslands. Miombo Woodlands and Natural Grasslands had almost the same species richness (Fig. 23.3). These findings indicate that the wetland ecosystems are the most diverse followed by wooded grasslands.

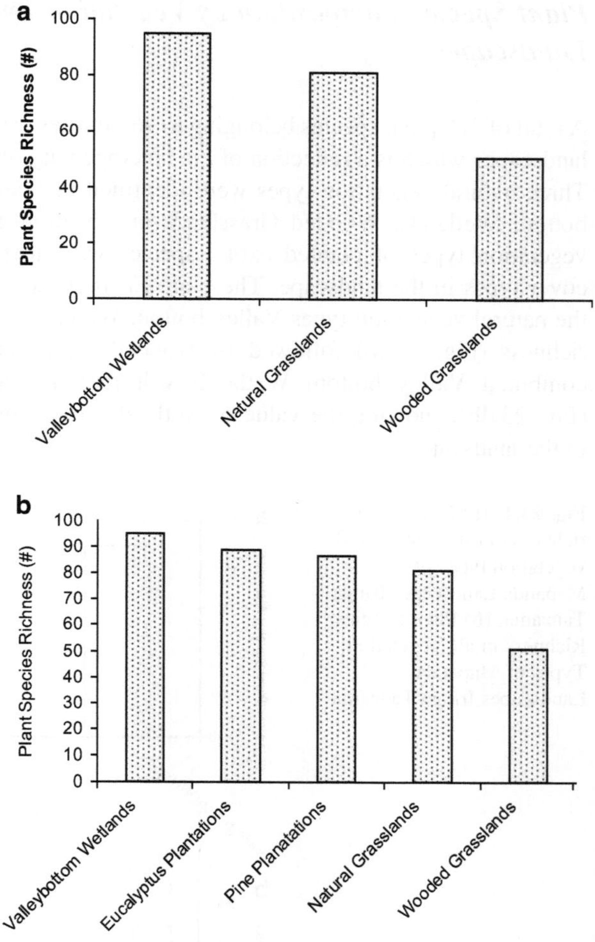

Fig. 23.2 (**a**) Average number of plant species by natural vegetation types in Uchindile. (**b**) Plant species richness in all vegetation types in Uchindile

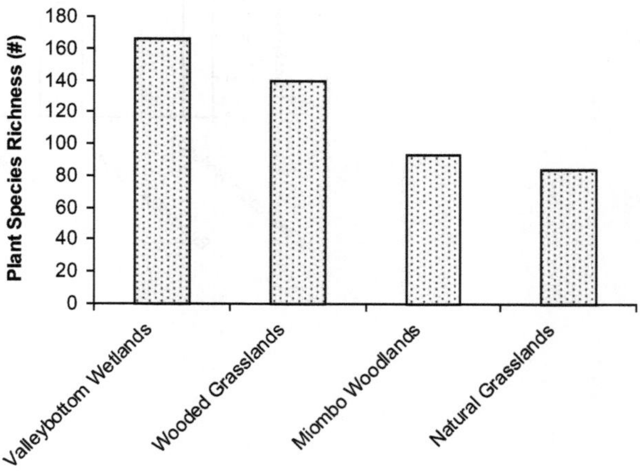

Fig. 23.3 Plant species richness in different natural vegetation types of Idete Landscapes Iringa Tanzania

Plant Species Composition by Vegetation Types in Mapanda Landscapes

A total of 320 plant species belonging to 96 families were identified in the Mapanda landscapes which is a reflection of the heterogeneity of the landscape and habitats. Three natural vegetation types were identified in Mapanda which include Valley bottom Wetlands, Wooded Grasslands and Natural grasslands. In addition two vegetation types of planted exotic species were identified as among the major cover types in the landscape. These are *Eucalyptus* and *Pine* plantations. Among the natural vegetation types Valley bottom Wetlands had the highest plant species richness (Fig. 23.4a) followed by wooded grasslands. Of all vegetation types combined Valley bottom Wetlands still rank the highest in species richness (Fig. 23.4b), showing the value of wetland ecosystems as repositories of species in the landscape.

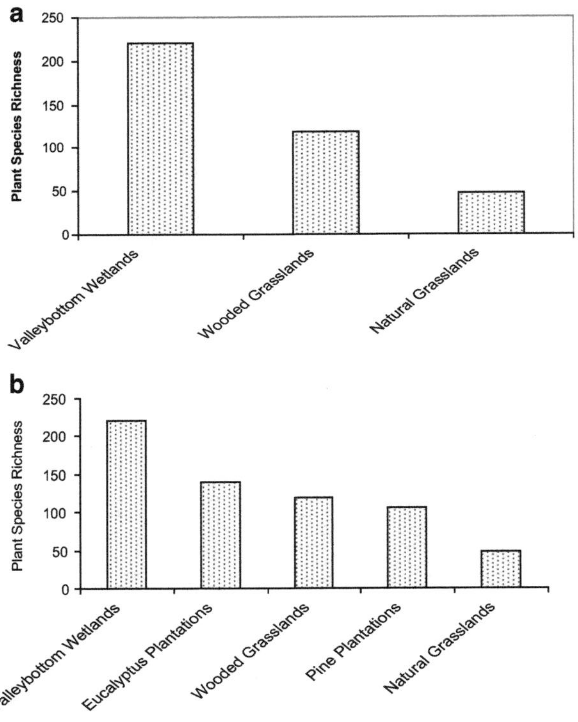

Fig. 23.4 (**a**) Plant species richness in different natural vegetation types of Mapanda Landscapes Iringa Tanzania. (**b**) Plant Species Richness in all Vegetation Types of Mapanda Landscapes Iringa Tanzania

Socio-economic Values of Valley Bottom Wetlands in the Little Ruaha River

Tanzania's wetlands contribute in diverse ways to livelihoods of many millions and wetlands are chiefly utilized for crop production and livestock. An assessment of wetland contribution to livelihoods in six villages of the Little Ruaha sub catchment of the Great Ruaha River showed that the total use value of productive activities carried out in upland and valley bottom wetlands was Tanzanian Shillings (Tshs) 3,415,458 (US$2732) per year per household in which 31 % of the total economic benefits accrued from utilization of Valley bottom Wetlands. Wetland based socio-economic activities included agricultural production practiced by over 98 % of the population followed by livestock grazing and fishing. Wetland based socio-economic activities carried out in valley bottoms commonly known by local people as *vinyungu* contribute about 15 of household food and 55–95 % of household income annually, equivalent to Tshs 3,234,721 (US$2588). In this respect valley bottom wetlands contribute significantly to household economy and food security. Planning for wetland friendly agricultural activities is pertinent in order to ensure wetlands conservation and sustainable contribution to household economy and food security without impairing the ecological integrity of the wetland ecosystems.

Climate Change Impacts on the Hydrology and Wetlands of the Great Ruaha River

There has been an appreciable decrease in water flows in the Ruaha River over time. An analysis of flows in the great Ruaha over 46 years period shows a decrease from an average of over 85 $m^3\ s^{-1}$ to below 70 $m^3\ s^{-1}$ (Fig. 23.5). This change is associated with rainfall variability in the basin, increased evapo-transpiration resulting from increasing temperatures as well as human induced impacts.

Extreme Low Flows

Zero flows were not experienced before 1990s in the great Ruaha River. Over time the flow has been decreasing and zero flows started to be experienced starting December 1990s. The low flows have been persistent since then and their duration have prolonged extending back to September. Occasionally zero flows are observed in January, a period that would be rain season in the region (Fig. 23.6). This situation marks extended periods of drying of the Ruaha River with consequences on resident biodiversity and socio-economic activities especially those which are water dependent.

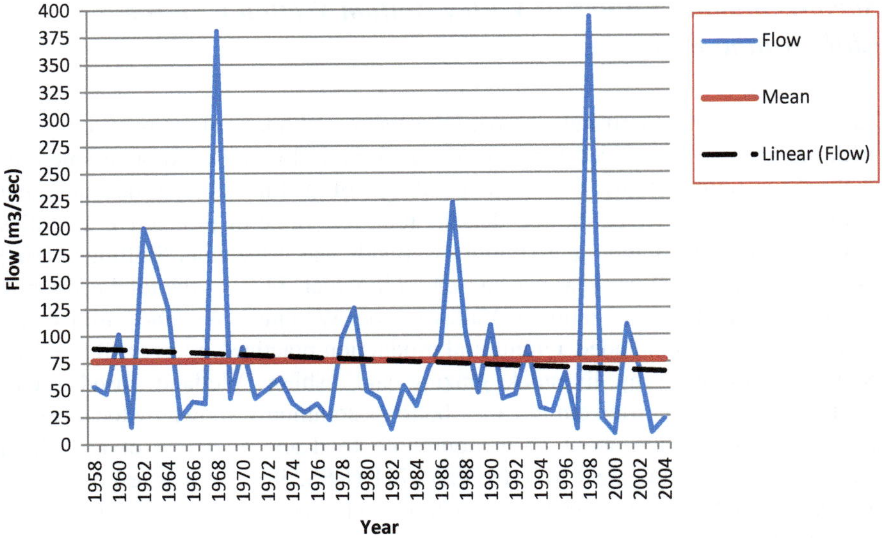

Fig. 23.5 Trends in the hydrology of the Great Ruaha River Tanzania 1953–2003 (40 years)

Discussion

The significance of valley bottom wetlands in acting as refuge for many species becomes more apparent during the peak of the dry season in which there would definitely be higher species richness for both flora and fauna in valley bottoms because of the wetter and favourable conditions compared to the other more terrestrial habitats. Most grassland ecosystems for example would have higher number of short-lived plant growth forms (annuals) during the rainy season which disappear during the dry season. However valley bottoms remain relatively wet and fertile during dry conditions making them potential sites for species assemblages and growth.

According to Dykema et al. (2014), with increase of temperature in the temperate regions, species diversity will decrease as cold water species are out competed by better adapted warm water species. Further, optimum habitat for most species will shift. This leads to disruption in food webs and upsets the balance of the ecosystem. Species with high water flow and flood tolerance may out-compete others that could previously exist prior to climate change. With a change in biodiversity, many rivers may lose some of their economic value. With climate change affecting temperatures all over the world, states could face less income generated by fewer cold water species.

Wetlands make appreciable contribution to rural livelihoods in terms of direct cash income and contribution to food security (Mkavidanda and Kaswamila 2001; Kilungu and Munishi 2009; Munishi et al. 2012), and many households that live close to wetland ecosystems in Tanzania and elsewhere utilize wetlands in

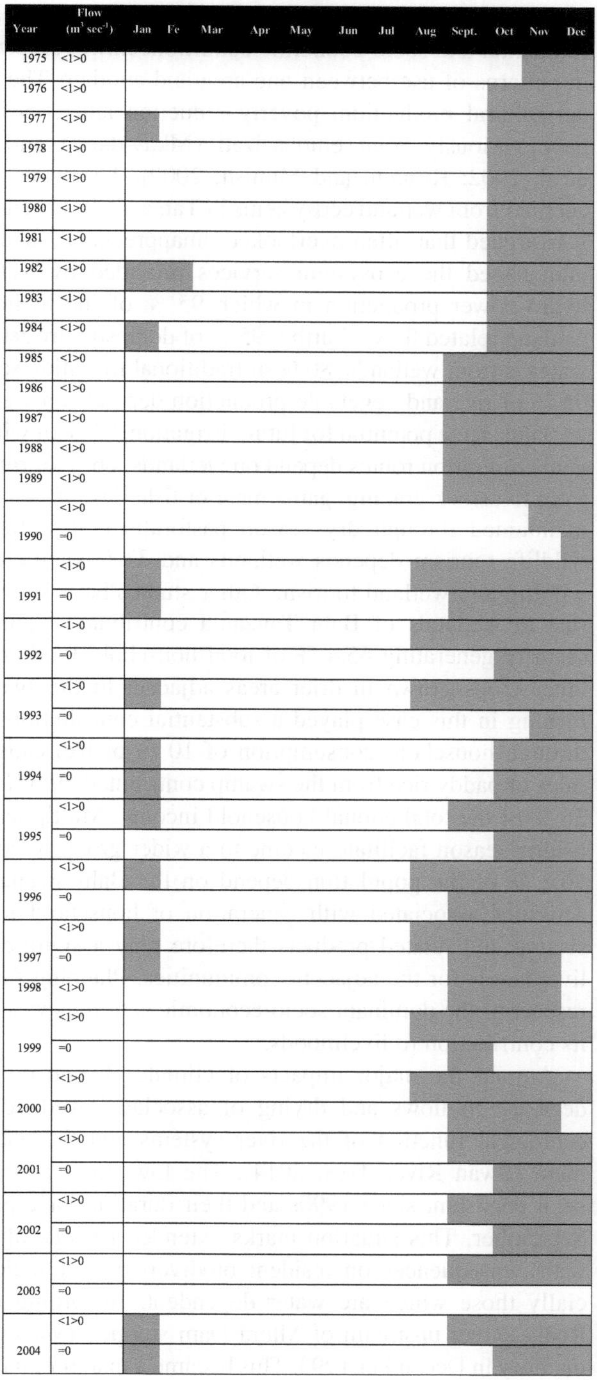

Fig. 23.6 Extreme dry season flows in the Great Ruaha River Tanzania

coping strategies during times of drought and food scarcity. Differences in environmental and socio-economic conditions however result into significant variation in patterns of use between one area and another. The significance of wetlands in agricultural production, poverty reduction and contribution to rural livelihoods have variously been emphasized (Mkavidanda and Kaswamila 2001; Ngailo et al. 2002; Kilungu and Munishi 2009). The wide range of economic benefits accrued from wetland ecosystems in Tanzania has been iterated (Yanda et al. 2008). It is argued that often overlooked, unappreciated, taken for granted and therefore unmanaged the ecosystem services provided by wetlands in Tanzania include hydro-power production in which 95 % of the hydropower production is from wetland related flows. Further 95 % of domestic, irrigation, industrial and livestock water is from wetlands, 80 % of traditional irrigation schemes depend on wetlands, 95 % of rice and vegetable production depends on wetlands, about 850,000 ha of wetlands have potential for future irrigation, 95 % of wildlife and wildlife corridors/ game migration routes depend on wetlands, 66 % of rural animal protein is derived from livestock grazing, game meat or fisheries, 95 % of the 25 million livestock is maintained through dry season pastoralism in wetlands, 95 % of coastal and wildlife tourism depends wetlands and 33 % of the country's GDP depends on wildlife and wetland tourism. Other studies have shown that cultivation of paddy rice in wetlands of Bahi Tanzania contributed significantly to household food security generating 65.4 % of total household food crop production compared to other crops grown in drier areas adjacent to the swamp (Munishi et al. 2010). Fishing in this case played a substantial contribution to household food security through household consumption of 10 % of fish caught. For household income, sales of paddy rice from the swamp contributed 59.6 % while fish sales contributed 36 % of the total annual household income. Multiplier activities emerging during fishing season facilitate income to a wider group of communities and on average, 56.2 % of the population depend on the Bahi swamp for daily socio-economic activities associated with generation of household food and income. The Bahi swamp and related products therefore play a significant role in enhancing local livelihoods for the adjacent communities. Planning for wise use of the swamp in respect of the dominant socio-economic activities was seen as a means to improve its contribution to livelihoods.

Among the major impacts of climate change on river systems globally are decrease in flows and drying of associated wetlands. Further it will alter the ecological function of the river systems and the way the society interact with them (Swan River Trust 2014). The low flows in the Great Ruaha River have been persistent since 1990s and their duration have prolonged extending back to September. This situation marks extended periods of drying of the Ruaha River with consequences on resident biodiversity and socio-economic activities especially those which are water dependent. According to Walsh (2012) the Great Ruaha River upstream of Mtera Dam stopped flowing for the first time in living memory in December 1993. This became a matter of national concern in 1995 when electricity shortages and rationing in the country were blamed by the national power supply company (TANESCO) on the continuing drying-up of the Great Ruaha. Since then different institutions and interest groups have sought to explain the river's increasing seasonality, focusing on resource use in and around its immediate source, the Usangu wetland, and laying the blame on different groups of resource users. Increasing government concern over power shortages culminated in the mass expulsion in 2006–2007 of livestock keepers from Usangu Wetlands a situation that has affected pastoralists' livelihood strategies in the Ruaha wetlands.

Munishi and Temu (1993) associated the drying of the Ruaha River with changing patterns of rainfall and anthropogenic activities in its upper catchment. Kashaigili (2006) observed that the frequency of occurrence of low flow events in the Great Ruaha River has increased over time. Between 1958 and 1973 there was not a single day with zero flow and the return period of a minimum 1-day duration flow of 0.84 $m^3\ s^{-1}$ was approximately 30 years. Between 1974 and 1985 short periods of zero flow occurred and a zero flow of 1-day duration had a return period of approximately 4 years. Post-1985, zero flows of 1-day duration occurred in all years and zero flow for durations of 60 days and greater were common. According to Mwakalila (2011a), the irrigated agriculture is most important as a user of water and impacts most heavily on wetlands. Abstraction of water for agriculture is leading to dried up rivers, falling ground water tables, increase in soil salinity and polluted waterways.

It has been shown that the rainfall in both the upper and lower catchments of the basin is quite variable. According to Mwakalila (2011b) rainfall in the basin varies from 250 to 2000 mm per annum and falls mainly from November to May. Rainfall amounts vary from year to year and in most of the basin the annual rainfall is much less compared to potential evapo-transpiration, indicating a negative balance (i.e. moisture deficit). Such situation may be worsened by climate change. Climate change and increasing water demand resulting from population growth has been earmarked as causing major degradation of water resources in the Sub-Saharan Africa (Ngigi 2009). The availability of water is affected in many ways by climate change. For example, changes in the quantity, timing, intensity and duration of rainfall as a result of climate change will contribute to greater water stress in the basin intensifying climate change impacts on wetlands. Decrease in water levels in the river will have diverse and negative impacts on wetland status especially biodiversity as well the potential of the wetlands to support socio-economic activities such as fishing, crop production among others. Flows in the river have constantly been below the long term average from the 1980s.

According to the Union of Concerned Scientists (2014) climate change is already beginning to affect plants and animals that live in freshwater lakes and rivers, altering their habitat and bringing life-threatening stress and disease. As air temperatures rise, water temperatures do also, particularly in shallow stretches of rivers and surface waters of lakes and streams and lakes may become unsuitable for cold-water fish but support species that thrive in warmer waters.

Adapting Wetlands to Climate Change

Among the major effective climate change response and adaptation to drying of wetlands is to sustain and restore wetlands and their functions. Other adaptation and mitigation measures that will especially adapt wetland livelihoods to climate change include irrigation development, strengthening integrated water resources management, development of both surface and subsurface water reservoirs, promotion of community based catchment conservation and management and promotion of new water serving technologies in irrigation. A more pragmatic approach to adapting the river basin and associated wetlands to climate change would be an evaluation of current and future climate vulnerability of wetland ecosystems and related livelihoods, formulation of scenarios for the future of wetlands under climate change impacts, identification of potential adaptation measures, capacity building for wetlands management, and implementation of adaptation measures on a pilot scale. This would then be followed by operationalization of the adaptations measures which consist of assessing the success of the measure, institutionalization, promotion and development of agreements and recommendations on how to sustain the promising adaptation measures. This will then be followed by programs and policy developments that ensure sustainable wetlands management at different levels of government. Furthermore, management practices in the river basin catchment will need to accommodate understanding of climate change impacts throughout the area. According to the Swan River Trust (2014) understanding the impacts of climate change on rivers is important information for mitigation measures. It is also important to improve local authority's understanding of the problems of degradation occurring as a result of anthropogenic activities in the river basin.

Conclusions

The most visible climate change impacts on the wetlands of Ruaha are reduced water flows, drying of wetlands and degradation of their ecological and socio-economic functions. With climate change, water flows in the Ruaha will decrease by 10 % causing substantial reduction in the extent of wetlands in the basin. Low flows have persisted since 1990s, with zero flows becoming more frequent and prolonged. A more pragmatic approach to adapting wetlands to climate change would be an evaluation of current and future climate vulnerability of wetland ecosystems and related livelihoods, formulation of scenarios for the future of wetlands under climate change impacts, identification of potential adaptation measures, capacity building for wetlands management, and implementation of adaptation measures on a pilot scale followed by operationalization of the adaptations measures. This information is important in informing policies, adaptation and mitigation strategies for wetlands.

Acknowledgements We acknowledge financial support from the Royal Norwegian Government through the Project NUFU-TZ 2007/10229 without which this study would have been impossible. Thanks to Sokoine University of Agriculture for institutional support in undertaking the NUFU Project and this study.

References

Bagachwa MSD (1994) Poverty alleviation in Tanzania: recent research issues. Dar es Salaam University Press, Dar es Salaam

Carson WP, Root RB (2000) Herbivory and plant species coexistence: community regulation by an out breaking phytophagous insect. Ecol Monogr 70:73–99

Development Partners Group (DPG) (2005) Brief on issues pertaining to tourist hunting. Development Partners Group, Dar es Salaam

Dykema J, Kriegel N, Jimenez A, Kahn C, Johnson J (2014) The effects of climate change on rivers. Macalester College, St. Paul, Macalester.edu/academics/environmentalstudies. (Last visited November 2015)

Food and Agriculture Organisation (FAO) (1985) Energy and protein requirements. Report of a joint FAO/WHO/UNU ad hoc Expert Consultation. Geneva, pp 724

Franzén D (2004) Plant species coexistence and dispersion of seed traits in a grassland. Ecography 27(2):218–224

Ishengoma C (1998) The role of women in household food security in Morogoro Rural and Kilosa District. Unpublished Ph.D. Thesis. Sokoine University of Agriculture, Morogoro, Tanzania, pp 337

Kamukara GL, Crafter SA (Eds) (1993) Wetlands of Tanzania. In: Proceedings of a seminar on wetlands of Tanzania, Morogoro, Tanzania 27–29 Nov 1991. IUCN Gland, Switzerland, p 179

Kashaigili JJ (2006) Effects of land use/cover changes on flow regime of the Usangu wetland and the Great Ruaha River in Tanzania. http://www.waternetonline.ihe.nl/downloads/uploads/symposium/zambia-2007

Kashaigili JJ, Kadigi RMJ, Lankford BA, Mahoo HF, Mashauri DA (2005) Environmental flows allocation in river basins: exploring allocation challenges and options in the Great Ruaha Basin, Tanzania. J Phys Chem Earth 30(11–16):689–697

Kilungu H, Munishi PKT (2009) Contribution of wetlands to household income and food security and livelihoods; a case of 'Nyumba ya Mungu' wetlands system in Northern Tanzania. Tanzania J Forest Nat Conservat 79(2):99–108

Maltby E (1986) Water logged wealth, why waste the world's wet places? International Institute for Environment and Development, London, p 19

Ministry of Natural Resources and Tourism (MNRT) (2003) Assessment needs for wetlands inventory and tools for assessing, mapping wetland types and their distribution. Report, Prepared by Institute of Resource Assessment, University of Dar es Salaam, p 41

Ministry of Natural Resources and Tourism (MNRT) (2004) An issue paper for the formulation of the national wetland strategy. Wildlife Division, Ministry of Natural Resources and Tourism, Dar es Salaam, p 98

Ministry of Natural Resources and Tourism (MNRT) (2009) Wildlife, wetlands and climate change in Tanzania. Wildlife Division, p 7

Mkavidanda TAJ, Kaswamila AL (2001) The role of traditional irrigation systems in alleviating poverty in Iringa Rural District, Tanzania. REPOA Research Paper No 01.2. Mkuki na Nyota Publishers, Dar es Salaam, p 36

Munishi PKT (2006) Ecological survey in the Kilombero forest project at Uchindile Kilombero District, Tanzania, p 87

Munishi PKT (2007) Ecological survey in the Uchindile and Mapanda forest projects of Green Resources (Ltd), Tanzania, p 48

Munishi PKT, Temu RPC (1993) The natural forests and environmental conservation in the Southern Highlands of Tanzania. In: Ekpere JA, Rees DJ, Mbwile KP, Lyimo NG (eds) Proceedings of an international workshop on agricultural research, training and technology transfer in the Southern Highlands of Tanzania: past achievements and future prospects. Uyole Agricultural Centre, Mbeya, pp 405–412

Munishi PKT, Shear TH, Wentworth T, Temu RPC (2007) Compositional gradients in plant communities in submontane rain forests of eastern Tanzania. J Trop For Sci 19(1):35–45

Munishi PKT, Shirima D, Jackson H, Kilungu H (2010) Analysis of climate change and its impacts on productive sectors, particularly agriculture in Tanzania. A Study for United Republic of Tanzania Ministry of Finance and Economic Affairs, p 105

Munishi PKT, Renatus R, Kideghesho JR, Kilungu H (2012) The role of Bahi swamp wetlands in enhancing household food security and income of adjacent communities. Tanzania J Forest Nat Conservat 82(2):120–137

Mvena ZSK (1999) The past, present and future of urban agriculture in Tanzania. J Agric Econ Dev 3:71–77

Mwakalila S (2011a) Assessing the hydrological conditions of the Usangu wetlands in Tanzania. J Water Resour Prot 3:876–882

Mwakalila S (2011b) Vulnerability of people's livelihoods to water resources availability in semi arid areas of Tanzania. J Water Resour Prot 3(9):8

Ngailo JA, Kaswamila AL, Senkoro CJ (2002) Rice farming system of the Wasukuma and its contribution to poverty alleviation. In: Presented at the 5th REPOA workshops, Dar es Salaam, 21–22 Mar 2002, p 23

Ngigi N (2009) Climate change adaptation strategies. Water resources management options for smallholder farming systems in Sub-Saharan Africa. Earth Institute, MDG Center, East and Southern Africa, New York, 189p. ISBN 9789290592648

Swan River Trust (2014) Annual report 2013–2014. http://www.swanrivertrust.wa.gov.au/home, p 107

Tetsuya S, Kuniyasu M (2005) Organic matter dynamics control plant species coexistence in a tropical peat swamp forest. Proc Biol Sci 272(1571):1503–1510

Tomáš H, František K (2008) Mechanisms of species coexistence in semi natural meadow communities; importance of species traits. http://botanika.bf.jcu.cz/suspa/ohrazeni/

Union of Concerned Scientists (2014) Climate hot map. Global warming effects around the world. http://www.climatehotmap.org/global-warmingeffects/lakes-and-rivers.html

United Republic of Tanzania (URT) (2006) Poverty reduction strategy paper (PRSP). IMF Country Report No. 06/142. International Monetary Fund, p 119

Walsh M (2012) The not-so-Great Ruaha and hidden histories of an environmental panic in Tanzania. J East Afr Stud 6(2):303–335

Yanda PZ, Majule AE, Mwakaje AG (2005) Wetland utilization, poverty alleviation and environmental conservation in semi-arid areas of Tanzania, the Case of Singida Region, Tanzania. In: Proceedings of East Africa integrated river basin management conference held on 7th–9th 2005 at ICE Hall, Sokoine University of Agriculture, Morogoro

Yanda PZ, Olson J, Moshy P (2008) Climate change vulnerability impacts and adaptation in Tanzania. CLIP Working Paper. Institute of Resource Assessment, University of Dar es Salaam, Dar es Salaam

Chapter 24
Addressing the Financing Gap for Adaptation in Fragile and Conflict-Affected Countries

Fiona Bayat-Renoux and Yannick Glemarec

Abstract This paper examines opportunities to address the financing gap for adaptation in fragile and conflict-affected countries through greater synergies between humanitarian, development and climate finance. Fragile states which have the least capacity to address the challenges of climate change, are often most at risk and vulnerable to its effects. However, the financing gap for adaptation in fragile countries coupled with the risk of renewed conflict linked to climate change is likely to accentuate the existing adaptation divide and create a vicious negative cycle. While creating synergies between different sources of finance is a potentially powerful way to address the adaptation financing gap and avoid mal-adaptation in fragile countries, joint action remains the exception. However, innovative practices to bridge efforts across different sources of finance are emerging in different country contexts, such as Jordan and Mali. As these innovative practices are very recent, it will be critical to leverage their learning potential to fully assess their effectiveness and replicability in similar development settings. Replicability will also be constrained by the differences in perspectives, objectives and financing between the humanitarian, development and climate adaptation communities. The resilience paradigm offers an opportunity to reconcile such differences and facilitate replicability through a shared theory of change across the different communities. The formulation process for the NAPs, which has recently begun in a number of fragile states, potentially provides a practical platform to bring different communities together in support of financing sustainable climate resilient development.

Fiona Bayat-Renoux is a Policy Advisor and Yannick Glemarec is the Deputy Executive Director at the United Nations Entity for Gender Equality and the Empowerment of Women (UN Women). They have worked extensively on climate change, natural disaster management, peace and security and sustainable development finance with both the United Nations Development Programme and UN Women.

F. Bayat-Renoux (✉) • Y. Glemarec
UN Women, 220 East 42nd Street, New York, NY 10017, USA
e-mail: Fiona.bayat-renoux@unwomen.org; Yannick.glemarec@unwomen.org

W. Leal Filho (ed.), *Innovation in Climate Change Adaptation*, Climate Change Management, DOI 10.1007/978-3-319-25814-0_24

Keywords Adaptation • Financing gap • Fragile • Resilience

The Financing Gap for Adaptation in Fragile and Conflict-Affected States

According to the OECD (2014a, b), 51 countries are considered 'fragile'.[1] Today, fragile and conflict-affected states are home to more than one-third of the world's population surviving on less than $1.25 a day. By 2018 this proportion will increase to more than half and by 2030 to nearly two-thirds (OECD 2014a, b).

Fragility, here defined as countries and economies characterized by weak capacity to carry out basic governance functions, limited ability to develop mutually constructive relations with society, and greater vulnerability to internal or external shocks covers a spectrum of contexts and affects a wide range of countries. Fragile states includes countries recovering from conflict (e.g. Liberia and Myanmar), as well those experiencing long-term insecurity or recurrent crises (e.g. the Central African Republic and the Democratic Republic of Congo), or those that are vulnerable to natural disasters (e.g. Kiribati and Tuvalu). While a number of fragile states are least developed countries (LDCs), close to half (23 of the 51 fragile states) are middle-income countries, many of which are rich in natural resources (OECD 2014a, b). Fragile and conflict-affected countries are also vulnerable to the 'conflict trap'. About 60 % of fragile and conflict-affected countries experience a relapse into additional episode of violence, often resulting in protracted crises (Collier and Sambanis 2002; Collier et al. 2003; Walter 2010).[2] For example, six of the top ten countries on the 2005 Fragile States Index remained on the top ten list in 2014[3] (Fund for Peace 2014). At the same time, long term trends show that fragility is not irreversible. Improvements have been particularly evident in the Balkans, and Liberia and Sierra Leone in Western Africa.

A key challenge in the twenty-first century is to support fragile and conflict-affected countries find sustainable solutions to peace, recovery and development. However, fragile states are often most at risk and vulnerable to the effects of climate change. The Climate Change Vulnerability Index, which measure the vulnerability of 233 countries against extreme weather change, sea level rise and reduced agricultural productivity, ranked Somalia, Burundi, Myanmar, Central African

[1] Annually, the OECD complies a list of countries and economies considered to be fragile. This list combines the latest harmonized list of fragile situation published by the World Bank, African Development Bank and Asian Development Bank, and the Fragility State list developed by the Fund for Peace (http://ffp.statesindex.org/), which scores countries based on 12 indicators of risk.

[2] The Walter study looked at 103 countries that experienced some form of civil war between 1945 and 2009 and found that only 44 avoided a subsequent return to civil war.

[3] These countries are: DRC, Sudan, Somalia, Chad, Yemen and Haiti. Fragile States Index Report (2014). http://library.fundforpeace.org/library/cfsir1423-fragilestatesindex2014-06d.pdf

Republic and Eritrea—all considered fragile countries—as the five countries with the highest overall vulnerability.[4]

Addressing the challenges of climate change requires governments to have the capacity to support public and private long-term planning based on climate scenarios; create an enabling policy environment for public and private investment in climate resilient infrastructure and supply chains as well as new economic opportunities; and facilitate adaptation at the individual, household, enterprise and society level through investments in essential social services and human capital. Regardless of income level, such capacities are extremely limited in fragile states, which face immediate challenges related to weak institutions, socio-political instability, and poor human development indicators. This, coupled with the high risk of relapse into conflict, creates short-term planning and investment horizons focused on responding to, and recovering from shocks, rather on than building long-term adaptive capacity to climate change.

There is also a growing body of research that suggests that climate change increases the chances of violence and conflict. The increase in frequency and impact of extreme hydro-metrological events and unusual and erratic weather patterns caused by climate change are likely to compound competition for natural resources, exacerbate traditional drivers of conflict, and reinforce the 'conflict trap'. Factors such as poverty and economic shocks which are associated with a higher risk of violent conflict are themselves sensitive to climate change. Evidence shows that between 1950 and 2010 "the frequency of violence between individuals rises 2.3 % and the frequency of intergroup conflict rises 13.2 % for each standard deviation change towards warmer temperatures" (Oppenheimer et al. 2014). With annual temperatures expected to rise by 2–4 standard deviations, there is potential for "large relative changes to global patterns of personal violence, group conflict and social instability in the future" (Oppenheimer et al. 2014).

Climate change has also been identified as key risk driver of population displacement. The Intergovernmental Panel on Climate Change (2014) has noted that "extreme weather events have in the past led to significant population displacement, and changes in the incidence of extreme events will amplify the challenges and risks of such displacements" (Adger et al. 2014). More than half of the 20 countries with the highest per capita displacement risk from 1993 have also experienced armed conflict during the same period (Ginnetti 2015). Population displacement can cause conflicts over access to resources among displaced communities and between displaced and host communities. It can also change the overall political economy of countries.

A recent study asserts that manmade climate change fueled a severe drought in 2006–2010 in Syria. The drought caused widespread crop failure and a mass

[4] The Climate change vulnerability index has been created by the Center for Global Development. Overall vulnerability is defined as physical impacts adjusted for coping ability. The index draws on a dataset described in Wheeler (2011). http://www.cgdev.org/page/mapping-impacts-climate-change

migration of farming families to overstretched urban centers. By 2010, internally displaced persons and Iraqi refugees made up roughly 20 % of Syria's urban population. The ineffective response of the Government to the crisis, when added to all other stressors, may have helped push Syria into open conflict (Kelley et al. 2015).

Conversely, violent conflict harms assets that facilitate adaptation. It will further inhibit the ability of governments to diversify the economy, mitigate conflicts among communities, plan urban development and reduce internal displacement risks. Paralysis of government in post-crisis situations will further limit national capacity to anticipate impacts of climate change, potentially generating a downward spiral of mal-adaptation.

Despite the threat posed by climate change most fragile countries are yet to take proactive action. One of the key reasons for inaction is the overall financing gap in fragile and conflict-affected countries. The IPCC 5th Assessment Report reported that the current global estimates of the cost of adaptation suggest a range of $70 billion to $100 billion per year by 2050 for developing countries, assuming that the world will embark on an emission pathway that will limit the increase in global average temperature to 2 °C.

However, it highlights that there is little confidence in these numbers. Findings from a review by UNEP (2014) of various studies and estimates of adaptation costs suggest that at a minimum the costs of adaptation are likely to be two to three times higher, while country studies indicate costs four to five times higher, particularly in LDCs and small island states (SIDs). Adaptation costs will also increase under higher emission scenarios. Indicative models indicate that costs could potentially double under a 4 °C pathway around 2050 compared to a 2 °C pathway (UNEP 2014).

In stark contrast, public adaptation-related finance, excluding domestic budgets, was estimated at $24.6 billion, off which about $22 billion was invested in developing countries in 2013.[5] Sub-Saharan Africa accounted for only 16 % of the total public adaptation finance, compared to 57 % of countries on the OECD (2014a, b) fragile states list which are from sub-Saharan Africa.

The UNDP Climate Public Expenditure and Institutional Reviews suggest that national governments are prioritizing adaptation in their total budgets allocated to climate change, including in fragile and conflict-affected countries. For example, annual expenditure on all climate change-related activities constitutes approximately 2 % of Gross Domestic Product (GDP) and around 6 % of Government Expenditure in Nepal (UNDP, 2011). Around three quarters of climate change expenditure relates to adaptation activities. However, total expenditure remains modest in absolute terms due to fiscal constraints in low income countries.

[5] UNEP (2014), highlights that it is unclear how much of adaptation finance reported is 'new and additional' given the lack of agreed definition of the terms 'new' and 'additional' and accounting methods.

There are no estimates of the current level of private sector adaptation finance (UNEP 2014). Tracking remains a challenge and no common methodology has been established. While private finance could play a major role is addressing the adaptation gap, it faces many investment barriers. In conflict-affected countries, insecurity, political instability and limited capacity create high levels of risks and costs of capital for investments by the private sector. Only 6 % of foreign direct investment to developing countries goes to fragile states (OECD 2015a, b).

In practice, the development and humanitarian communities have been at the forefront in supporting local communities in reducing their vulnerability to climate-related risks. However, development and humanitarian finances are coming under increasing pressure in fragile and conflict-affected states. According to the OECD (2015a, b), while official development assistance (ODA) reached a record high in 2013 ($134.8 billion), rising by 6.1 % compared to 2012, bilateral aid to sub-Saharan Africa fell by 4 % in real terms, and grants (as opposed to loans) increased by only 3.5 % (excluding debt relief). The rise in overall ODA in 2013 follows a fall by 4 % and 2 % in 2012 and 2011 respectively. According to the OECD (2014a, b) "(s)ince peaking in 2005, the volume of aid to fragile states has followed an erratic downward trend", falling by 2.4 % in 2011. The OECD (2014a, b) Global Outlook on Aid Report based on a survey of donors' forward spending plans, expects ODA growth to slow for fragile states. Two-thirds of countries in sub-Saharan African are projected to receive less aid in 2017 than 2014, while future increases are expected to benefit middle income countries in Asia, mostly as a result of loans from multilateral agencies. Similarly, despite a rise in humanitarian assistance in 2013 to $22 billion compared with $17.3 billion in 2012, over a third of estimated humanitarian needs were unmet. By the end of 2014, humanitarian needs reached an unprecedented level—$18.1 billion off which only 57 % were financed (UN OCHA 2015).

Potential of Synergies Between Different Sources of Finance to Address the Financing Gap for Adaptation in Fragile States

Humanitarian, development and climate action change assistance shares significant priorities, including the prioritization of life-and livelihood-saving activities; food security and agriculture; water availability, quality and accessibility; energy security and accessibility; women economic and political empowerment and essential infrastructure.[6] Figure 24.1 visualizes the linkages between humanitarian, development and climate action and possible synergies.

[6] The National Adaptation Programmes of Actions (NAPAs) supported by the United Nations Framework Convention on Climate Change (UNFCCC) address, inter alia, (i) disaster risk management; (ii) human health; (iii) food security and agriculture; (iv) water availability, quality

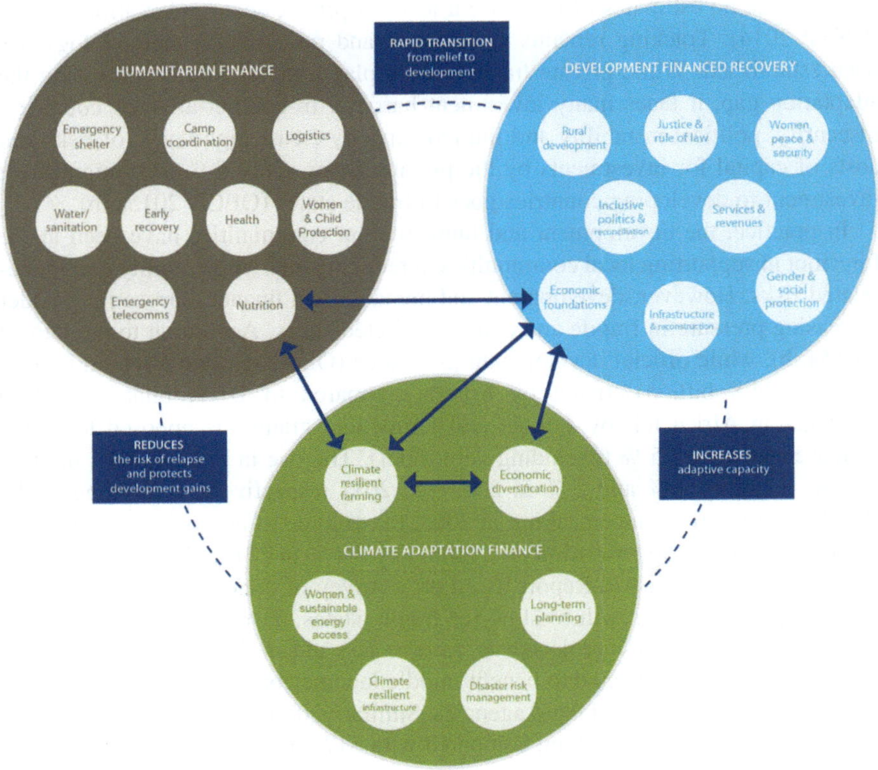

Fig. 24.1 Synergies between different sources of finance

Creating synergies between humanitarian, development and climate action could, potentially, be a powerful way to address the adaptation financing gap, strengthen national government adaptive capacities, and avoid mal-adaptation in fragile states. To test this hypothesis, the study undertook a global mapping and comparative analysis of pooled financing mechanisms in 52 fragile countries across humanitarian, development and climate adaptation finance.[7] See Annex 1 for a list of the pooled financing mechanisms reviewed in the global mapping.

---

and accessibility; (v) essential infrastructure; (vi) economic diversification; (vii) biodiversity and ecosystem management; (viii) land-use management and forestry; (ix) environmental management; (x) cultural heritage; (xi) coastal zones and associated loss of land and (xii) climate-resilient urban development.

[7] The global mapping exercise covers relevant pooled financing instruments administered by the UN system, multilateral development banks and multilateral funds with an independent legal status. It does not include pooled financing mechanisms administered by bilateral agencies or private entities. The mapping captured information about pooled funds across five major criteria: fund purpose, scope and size; establishment requirements and timelines; governance structure and decision making; fund allocation, implementation and reporting; and cost structure. Information

The mapping focused on pooled financing mechanisms, which provide a natural platform to bring different stakeholders together. The key objectives of pooled funds are enhancing aid coordination and coherence. They can be designed to partner with other funds in support of a common objective (Glemarec et al. 2015). As such, a global mapping of humanitarian, development and climate funds should reveal to which extent they reference proposed programmes to each other and partner to develop synergies. It should also enable an evaluation of the effects of such synergies where they exist.

The list of countries used for the purposes of the global mapping is based on the 47 fragile states and economies analysed in the OECD (2012) States of Fragility 2013 Report. In addition, Mali and Syria were included due to the establishment of the UN Peacekeeping Mission in Mali in 2013, and the ongoing conflict in Syria. Jordan, Lebanon, and Turkey, which were not on the list of fragile countries, were also included due to the impact of the Syrian conflict on these countries. For the purposes of this study, such countries are referred to as 'stable countries affected by crisis'.[8] To complement the global mapping exercise, detailed mappings of humanitarian, development-financed recovery and climate-adaptation financial flows (including of bi-lateral assistance) were conducted in eight fragile countries: Democratic Republic of Congo (DRC), Ethiopia, Haiti, Pakistan, Somalia, Sudan, South Sudan, and Yemen.[9] A qualitative case study and field visits were also conducted in Jordan, DRC, CAR and South Sudan.

Findings: Innovative Approaches and Emerging Practices

The global mapping and comparative analysis reveals that current financing practices reinforce rather than bridge a silo approach across sources of finance. With the exception of a few large humanitarian funds in protracted crisis settings,

on UN funds was drawn from the UNDP MPTF Office Gateway (2013) and information on the World Bank from the World Bank Concessional Finance and Global Partnership webpage as well as from its Financial Intermediary Funds: Meeting Global Development Challenges through International Partnerships Report. Information on climate adaptation finance was drawn from the joint UNDP/World Bank platform, which highlights 74 climate funding sources (Climate Finance Options: http://www.climatefinanceoptions.org/cfo/)

[8] While 'stable countries affected by crisis' are not in conflict, they face certain structural issues, including demographic pressure, group grievances and uneven development, which make them vulnerable to crisis and spillover effects from neighbouring countries. Such countries tend to score between 70.1 and 89.9 on the 2014 Fragile States Index (http://fsi.fundforpeace.org/rankings-2014).

[9] Bi-lateral related financial flows per country over the period 2008–2012 and for 2012 have been estimated based on data from the OECD Creditor Reporting System (CRS). Interviews and focus group discussions were used to gather complementary qualitative information on pooled operations in these eight countries. Field visits were carried out to two of the eight countries (DRC and South Sudan).

development and climate funds do not exist in the majority of countries or are established too late to promote synergies with humanitarian assistance. For example, of the top 20 recipients of humanitarian assistance over the period 2002–2011, UN country-level pass-through development financed funds have been established in only eight. Similarly, such pooled funds have only been established in 40 % of the 52 fragile states used for the analysis. When available, development and climate funds are too small and fragmented to act as gravity centres for aid coordination and alignment (Bayat-Renoux and Glemarec 2014).

However, a finer analysis of the variety of capitalization strategies and theories of change adopted by individual pooled financing instruments highlights some emerging practices to create synergies and address the financing gap for adaptation. These emerging practices are specific to country political and socio-economic contexts and take place at different stages in the prevention-crisis-response-recovery and sustainable development contiguum. Table 24.1 shows potential opportunities for synergies between different sources of finance within four country context categories.

In stable middle income countries with well-functioning governments which are affected by a crisis, opportunity for synergies between different sources are emerging in the response to the crisis. For example, several countries in the Middle East are facing a refugee crisis as a result of the Syrian conflict. In Jordan, this crisis is leading to increased energy demands on a country which has traditionally relied on imported fossil fuels for its electricity generation. Overall energy costs were likely to reach US$18 billion in 2013. This decreases fiscal space to respond to overall social and economic development goals. Rather than humanitarian assistance increasing its fossil fuel-fired power generation capacity to meet the increased energy demands from the Syrian crisis through diesel generators, the Jordan government envisages accelerating energy efficiency and renewable energy measures in buildings and residences across Jordan. These measures will benefit both Syrian refugees and host communities in the form of lower electricity bills, promoting private investment and job creation in clean energy, reducing fiscal expenditures and putting Jordan on a green development trajectory. It is estimated that replacing three million inefficient lights in residences and buildings with new compact fluorescent lamps alone would free up 144 MW of power in the years 2014 and 2015. At a cost of US$14.8 million, it would save US$50 million in energy subsidies alone and avoid about $300 million investment in additional fossil fuel infrastructure. In this case, limited public resources are used to leverage the private sector through donations (Ikea will contribute towards the bulb needed) and through investment (in bulbs and safe disposal). This approach requires a paradigm shift in addressing humanitarian crisis in middle-income stable countries and the development of a new range of humanitarian response instruments, fully aligned with national climate-compatible development systems. The Government of Jordan established a pooled financing mechanism to promote this integrated response to the Syrian refugee crisis in March 2015 (Government of Jordan 2015).

Fragility is not limited to low income countries. A number of fragile and conflict-affected countries have enormous natural resource endowments and

Table 24.1 Opportunities for synergies between different sources of finance

	Stable countries affected by a crisis	Fragile countries
Middle income	*Sources of finance* Domestic public and private resources; limited international public *Opportunities for synergies* Use scarce public resources to catalyse much larger levels of private capital to finance sustainable investment. Public resources should focus on removing key structural barriers to private investment, improving the policy environment, providing credit enhancement mechanisms, and reducing investment risk. Public resources can also be used to leverage international private resources through donations	*Sources of finance* Domestic public resources (in many cases, from extractive industries); high levels of remittances; limited international aid, limited domestic, foreign direct investment (concentrated and uneven)[a] *Opportunities for synergies* • Better management of domestic public revenues from extractive industries to finance climate resilient development, including through innovative financing instruments (e.g. levy on gas exports) • Use limited international public resources to strengthen national capacities and systems • Leverage remittances through public private partnerships (e.g. matching remittances with public finance to fund local sustainable development)
Low income	*Sources of finance* Limited domestic public and private finance; limited international aid[b] *Opportunities for synergies* • Create a political imperative for climate change by linking climate action to crisis management and development co-benefits • Increase the total amount of public domestic finance available by improving the revenue-collection capacity of the state and introducing innovative sources of finance through environmental taxes, levies and fines • Increase the amount of international finance by improving national capacity to access global climate finance	*Sources of finance* Limited domestic public and private finance; remittances; significant, but uneven levels of international aid; very limited foreign direct investment[c] *Opportunities for synergies* Better alignment between different sources of international public resources (humanitarian, development and climate finance) through greater reliance on national systems and a limited number of well capitalized country-level pooled funds

[a] According to the OECD (2014a, b), in non-LDC fragile states, remittances account for 66 % of inflows, followed by aid (18 %) and foreign direct investment (16 %). FDI is very heavily concentrated and concentrated in extractive industries

[b] Net disbursement of ODA to sub-Saharan Africa for non-fragile LDCs in 2013 amounts to \$16,113 million for 23 countries, compared to \$22,419 million to 22 fragile LDCs over the same period. OECD Statistics on resources to developing countries. http://www.oecd.org/dac/stats/statisticsonresourceflowstodevelopingcountries.htm

[c] In low income fragile states, aid in fragile LDCs account for 45 % of development finance, followed by remittances (42 %). FDI is limited at 13 % (OECD 2014a, b)

qualify as middle income countries. For example, 16 new countries—a number of them affected by or recovering from conflicts—are expected to join the ranks of oil and gas exporters in the next few years. Income generated from oil and gas revenues in these new oil-producing countries could largely exceed aid receipts in a relatively short time, as is already the case in Angola and the Republic of Congo (Bazilian et al. 2013). The exploitation of high-value natural resources has often been cited as a key factor in triggering, escalating or sustaining conflicts around the globe (UN Interagency framework for preventative action 2012). However, these revenue streams could finance climate resilient development and provide new sustainable livelihood opportunities to their populations if transparently and well managed. Channelling part of the revenues from extractive industries into a dedicated investment vehicle could fill part of the financing gap as well as stabilize and improve the predictability of funding. Furthermore, it could positively impact the governance of extractive industries, address a root cause of fragility and ensure that natural resource endowments are used strategically to finance economic diversification and development. Limited international public resources should be used to strengthen national capacities and systems. For example, in Angola, formulation efforts are underway to establish a national climate fund to pool and leverage funding from public, private, national, international and innovative sources in support of Angola's low carbon and climate-resilient economic development (UNDP and KPMG 2015). The fund will ensure coherence of climate investments by supporting under-funded priorities, and could support the design and implementation of innovative carbon-related financing mechanisms.

Low-income stable countries have a limited access to both domestic and international resources. They also tend to face structural credit scarcities. This makes prioritizing action on climate change over other development goals particularly challenging. In these contexts, it is critical to create a political imperative for climate change by linking climate action to crisis management and development co-benefits. For example, the productivity of the world's 450 million smallholder farms could be substantially improved through climate-smart agriculture practices (FAO 2013). Climate-smart agriculture could address the priorities of the three communities by improving food security in the short-term; by increasing agricultural productivity as a driver of economic development; and by encouraging adoption of climate compatible farming practices. It could create political support for robust climate action by aligning adaptation with national development goals. However, this would require removing a range of investment barriers in climate-smart smallholder agriculture. Smallholders, and among them particularly women because of discriminatory or uncertain land regimes,[10] typically lack access to resources to implement climate-smart practices, such as climate-resilient seeds,

[10] The negative consequences from the lack of land access and titling disproportionally affect women, who are estimated to own less than 2 % of land worldwide (UNEP et al. 2013; Lukatela, 2012). FAO (2011) estimated that if women had the same access to productive resources as men, this could raise total agricultural output in developing countries by 2.5–4 %, which could in turn reduce the number of hungry people in the world by 12–17 %.

fertilizers, machinery, links to local and global markets, technical support and finance. Access to long-term finance is notably critical as climate-smart agriculture usually requires larger upfront investments for greater future returns. The global demand for smallholder agriculture finance is estimated at $450 billion and is largely unmet (Dalberg 2014). International and domestic public finance will be essential to establish an enabling environment to remove investment barriers, including credit scarcity, and catalyze private investment in win-win options such as climate-smart agriculture and energy access in stable but low income countries. In turn, this will require developing the capacity of low income stable countries to increase fiscal revenues, notably through innovative sources from environmental taxes, levies and fees, and better access international climate finance.

In low income fragile countries, the bulk of the finance comes from the international public sector. Here, a smarter and better coordinated aid architecture is key to creating linkages between different sources of international finance. In some cases, this means that donors will have to be more flexible and risk tolerant to on-budget aid modalities that strengthen national institutions and enable long-term planning (OECD UNDP 2014). In others, specific financing vehicles might need to be established to manage risks and promote integrated approaches. Notably, well designed, capitalized and operationalized country-level pooled funds can mitigate risks and act as gravity centers to access, sequence and combine different sources of finance. For example the Government of Mali developed a five-year Green Economy and Climate Resilient Strategy to support the emergence of an economy that would be resilient to climate change, decrease the risk of conflicts over natural resources and reduce poverty. This strategy stems from this growing understanding that the issues of poverty, crisis, conflict and capacity to respond to climate change are intertwined. In January 2014, the Government of Mali stated the implementation of the strategy by combining domestic public resources with international public resources from two national funds (the National Fund for Economic and Social Stabilization and the National Climate Fund), two global vertical funds for adaptation (the Adaptation Fund and the Least Developed Country Fund) and one global bilateral fund (German International Climate Initiative). As the National Fund for Stabilization is phased out, the National Climate Fund is expected to grow and support coordinated, integrated implementation of the Green Economy and Climate Resilient Strategy. It should ensure that a common platform for national and international actors is available to share views on building national resilience. The overall strategy is expected to be implemented through a series of annual phases combining and sequencing different sources of development and climate finance (both public and private).

Barriers and Opportunity for Replicability of Emerging Practices

Despite the material benefits that better synergies could generate, joint action across the humanitarian, development and climate communities remains the exception. While there are important similarities between interventions financed by, humanitarian, development and climate adaptation assistance, substantial differences can exist in perspectives, objectives and financing. This limits interactions and presents a huge challenge for creating linkages and synergies across the three communities.

They operate over different spatial and temporal scales. Humanitarian assistance is focused on meeting short-term, urgent needs in target areas. It is based on annual humanitarian needs assessments, which rarely take into account long-term stressors. Implementation strategies are typically articulated in annual strategic response plans (IASC 2014a, b). Development assistance in fragile countries tends to focus on recovery and reconstruction articulated in 3–5 year national recovery or development plans. Climate assistance aims at anticipating, mitigating and responding to climate change and must take a medium to long-term perspective in the face of substantial uncertainties, in line with the economic life of housing, transport, water and energy infrastructure.

Humanitarian, development and climate assistance have different immediate aims, follow different principles and have evolved separately (Steets et al. 2011). Humanitarian aid aims to saves lives and is guided by the principles of independence, neutrality and impartiality (Good Humanitarian Donorship Initiative 2003). Development assistance, on the other hand, aims at contributing towards sustainable development, including pro-poor economic growth, poverty reduction and improvements in living standards, and follows principles of national ownership and alignment. Climate finance reflects the principle of Common but Differentiated Responsibility (CBDR), which was explicitly formulated in the context of the 1992 Rio Earth Summit and is a cornerstone of the United Nations Framework Convention on Climate Change (UNFCCC). Recognizing the differences in historical responsibilities and capacity to act, this international legal principle aims to operationalize the concept of burden sharing across countries to address climate change. Although its scope was more recently broadened to help populations and economies adapt to the unavoidable impacts of climate change, climate finance initially focused on the incremental costs of low greenhouse gas emission investment to mitigate the increase in global average temperatures.

Humanitarian, development and climate assistance have different financing instruments. Humanitarian finance is coordinated through a small number of well capitalized pooled funds. Development assistance on the other hand, tends to be fragmented in fragile and conflict-affected countries funded through a proliferation of bilateral and multilateral instruments. While there is international agreement[11] about the need to increase the proportion of international assistance through country systems in fragile states, currently the use of national public financial management and procurement systems still limited. Data from the latest Global Partnership for Effective Development Cooperation (2014) shows that only 31 % of disbursements for the government sector used national public financial management and procurement systems in fragile countries in 2013. The Ministry of Plan or line ministries tend to be the national counterparts for international development assistance in fragile countries. Public climate finance aims at catalyzing private investment. Climate adaptation finance is essentially channeled through development finance institutions (bilateral, multilateral and national) in the forms of low cost or market level project debt (82 %) and vertical funds such as the Least Developed Country Fund and the Adaptation Fund. These vertical funds operate through project modalities at the country level, with limited capacity to support greater aid alignment with other sources of finance at that level.

In order to facilitate and accelerate the replicability of emerging practices identified in the previous section, it is important to reconcile these differences in perspectives, objectives and financing. This could be achieved through a shared theory of change. The recent application of the concept of resilience to humanitarian action and development cooperation provides a unique opportunity to bridge efforts across the humanitarian, development and climate adaptation communities at the country level. The concept of resilience (Folke 2006) has the potential to capture the priorities of the different communities and potentially align spatial and temporal scales across the different sources of finance. It captures the humanitarian emphasis on ability of a system to resist, absorb and bounce back from the effects of a shock; the development community's focus preparedness and risk reduction; and the climate community's attention on the capacity of a system to anticipate, shape and adapt to change. It also provides a platform for different communities to develop a shared understanding of the risk landscape, a shared understanding of what makes a community or system resilient to such risks, and hence a shared vision of the resilience characteristics to strengthen (Mitchell 2013).

The shared theory of change for sustainable resilient development will highlight specific opportunities for cooperation between different funding sources. The formulation process for the national adaptation plans (NAPs) could provide a platform to

[11] The Global Partnership for Effective Development Cooperation was established during the Fourth High Level Forum on Aid Effectiveness in Busan in 2011 to ensure that development cooperation has the greatest impact on development results. It brings together a wired range of countries and organizations and its monitoring framework tracks progress on commitments made during the Fourth High Level Forum. The monitoring framework consists of a set of ten indicators, some of which are based on those contained in the Paris Declaration on Aid Effectiveness. Key indicators include the percentage of aid that is on budget and the use of country public financial management and procurement systems. Some fragile states through the g7+ echoed the importance of the use of country systems through the New Deal for Engagement in Fragile States (International Dialogue on Peacebuilding and Statebuilding 2011).

bring together different communities to develop this broad vision of sustainable climate resilient development and identify country specific opportunities for collaboration between different sources of finance. The Cancun Adaptation Framework that was adopted during the 16th UNFCCC Conference of the Parties in 2010 established a process for formulation and implementation of NAPs as a means of identifying medium- and long-term adaptation needs and developing and implementing strategies and programmes to address those needs (UNFCCC, 2011). Most countries are still at the initiation consultative stages of the NAP process. To date, 26 LDCs of which 13 are fragile states, have requested supported from the NAP Global Support Programme.[12] It is essential that this support includes expertise in identifying opportunities and developing financial mechanisms to develop these synergies and address the existing financing gap for adaptation in fragile and conflict-affected states.

Conclusion

Peace and security are recognized as pre-requisites for development, which is itself a pre-requisite for effective adaptation. Similarly, effective adaptation in turn is critical to ensure that in a changing climate, the development of today does not become the mal-adaptation of tomorrow. However, the financing gap for adaptation in fragile countries coupled with the risk of renewed conflict linked to climate change is likely to accentuate the existing adaptation divide and create a vicious negative cycle. One key opportunity to address the financing gap for adaptation is through greater synergies between different sources of finance.

The specific synergies will depend on the political and socio-economic country context. Innovative practices to bridge efforts leveraging a various sources of finance are emerging in different country contexts, such as Jordan and Mali. Similarly, national climate funds capitalized from innovative sources, including extractive industries could potentially narrow the financing gap for adaptation in such countries. However, most of these innovative practices are very recent (less than 2 years) and their effectiveness and replicability remain unproven. Moving forward, it will be critical to leverage the learning potential from these experiences to fully assess their effectiveness and replicability in similar development settings. Monitoring, assessing and evaluating both the processes and results will be important in this regard.

Replicability of emerging practices will also be constrained by the differences in perspectives, objectives, and financing that exist between humanitarian, development and climate adaptation communities. The resilience paradigm offers an opportunity to reconcile such differences and facilitate replicability through a

[12] The NAP Global Support Programme is a UNDP-UNEP programme financed by the Least Developed Country Fund (LDCF). http://www.undp-alm.org/projects/naps-ldcs. The Least Developed Countries Expert Group (2012) have developed technical guidelines for the NAP process.

shared theory of change across the different communities. The formulation process for the NAPs, which has recently begun in a number of fragile states, potentially provides a practical platform to bring different communities together in support of financing sustainable climate resilient development. Piloting this approach in a these countries may be valuable and could contribute towards strengthening the quality and level of financing of future NAPs.

Annex 1: List of Huamanitarian, Development and Climate Change Pooled Financing Mechanisms Reviewed in the Global Mapping (As of 31 December 2013)

	Global/regional	Country-based
Humanitarian financing mechanisms	Central Emergency Response Fund (CERF)	*Common Humanitarian Funds (CHFs)* in Central African Republic (CAR), Democratic Republic of Congo (DRC), Somalia, South Sudan, Sudan *Emergency Response Funds (ERFs)* in Afghanistan, Columbia, DRC, Ethiopia, Haiti, Indonesia, Kenya, Myanmar, Pakistan, State of Palestine, Syria, Yemen, Zimbabwe
Development financing mechanisms	UN Peacebuilding Fund (PBF) UN Trust Fund for Human Security (UNTFHS) UN Fund for Action Against Sexual Violence in Conflict (UN Action) FAO Technical Cooperation Programme (TCP) United Nations Voluntary Trust Fund for Victims of Trafficking in Persons Two United Nations Voluntary Trust Funds for (i) Victims of Torture and (ii) Contemporary Forms of Slavery World Bank Statebuilding and Peacebuilding Fund (SPF) African Development Bank (AfDB) Fragile States Facility European Union Instrument for Stability Global Agriculture and Food Security Program Global Fund to Fight AIDS, Tuberculosis and Malaria (Global Fund) Water and Sanitation Program	*UN development financed pooled funds* Comoros One UN Fund DRC Stabilization and Recovery Fund Ethiopia One Fund UNDG Haiti Reconstruction Fund UNDG Iraq Trust Fund Iraq UNDAF Trust Fund Kiribati One Fund Jordan Relief ad Resilience Fund Kyrgyzstan One Fund Lebanon Recovery Fund Libya Recovery Trust Fund Malawi One Fund Mali National Economic and Social Stabilization Fund UN Peace Fund for Nepal Occupied Palestinian Territory Trust Fund Pakistan One Fund Rwanda One UN Fund Sierra Leone Multi-Donor Trust Fund

(continued)

	Global/regional	Country-based
		Somalia UN Multi-Partner Trust Fund South Sudan Recovery Fund Sudan: Darfur Community Peace and Stability Fund Sudan: UN Fund for Recovery, Reconstruction and Development in Darfur Syria Transition and Recovery Trust Fund Yemen National Dialogue and Constitutional Reform Trust Fund *World Bank Trust Fund Programs* Afghanistan Reconstruction Trust Fund National MDTF Sudan MDTF South Sudan
Climate change financing mechanisms	*Adaptation* The Global Environment Trust Fund (GEF) Least Developed Countries Fund for Climate Change (LDCF) Special Climate Change Fund (SCCF) UNFCCC Adaptation Fund Adaptation for Smallholder Agriculture Programme (ASAP) Pilot Program for Climate Resilience (PPCR) Asian Development Bank (ADB) Climate Change Fund (CCF) Global Facility for Disaster Reduction and Recovery (GFDRR) MDG Achievement Fund *Energy Access Focused Funds* ClimDev-Africa Special Fund (CDSF) End User Finance for Access to Clean Energy Technologies in South and South-East Asia (FACET) Global Energy Efficiency and Renewable Energy Fund (GEEREF) Strategic Climate Fund Program on Scaling-Up Renewable Energy in Low Income Countries (SREP) Pilot Program for Climate Resilience (PPCR) Sustainable Energy Fund for Africa (SEFA) The Seed Capital Assistance Facility (SCAF)	*National Climate Funds* Mali Climate Fund Ethiopia Climate Resilient Green Economy Facility Micronesia Conservation Trust Bangladesh Climate Change Trust Fund Bangladesh Climate Change Resilience Fund Rwanda National Climate and Environment Fund (FONERWA)

References

Adger WN, Pulhin JM, Barnett J, Dabelko GD, Hovelsrud GK, Levy M, Oswald Spring Ú, Vogel CH (2014) Chapter 12: Human security. In: Climate change 2014: impacts, adaptation, and vulnerability. Part A: Global and sectoral aspects. Contribution of working group II to the 5th assessment report of the intergovernmental panel on climate change. Cambridge University Press, Cambridge, p 758

Bayat-Renoux F, Glemarec Y (2014) Financing recovery for resilience. United Nations Development Programme, New York

Bazilian M et al (2013) Oil, energy poverty and resource dependence in West Africa. J Energy Nat Resour Law 33:33–53

Climate Finance Options. http://www.climatefinanceoptions.org/cfo/

Collier P, Sambanis N (2002) Understanding civil war: a new agenda. J Confl Resolut 46(1):3–12

Collier P et al (2003) Breaking the conflict trap: civil war and development policy. Oxford University Press, Oxford

Dalberg (2014) Catalyzing smallholder agricultural finance. Dalberg Global Development Advisors

FAO (2011) The state of food and agriculture 2010–2011: women in agriculture – closing the gender gap for development, Rome. http://www.fao.org/docrep/013/i2050e/i2050e.pdf

FAO (2013) Climate-smart agriculture sourcebook, Rome. http://www.fao.org/docrep/018/i3325e/i3325e.pdf

Folke C (2006) Resilience: the emergence of a perspective for social-ecological systems analyses. Global Environ Change 15:253–267

Fund for Peace (2014). Fragile States Index 2014. Washington DC, USA. http://library.fundforpeace.org/library/cfsir1423-fragilestatesindex2014-06d.pdf

Ginnetti J (2015) Disaster-related displacement risk: measuring the risk and addressing its drivers. International Displacement Monitoring Centre, Norwegian Refugee Council, Geneva. http://www.internal-displacement.org/assets/publications/2015/NRC-Displacement-Risk-Analysis-EFA-FINAL.pdf

Glemarec Y, Bayat-Renoux F, Bardoux P, Saenz E (2015) Designing funds for performance. UNDP MPTF Office, New York

Global Partnership for Effective Development Cooperation (2014) Global monitoring framework. http://effectivecooperation.org/files/about-trackingprogress/INDICATORS.pdf

Good Humanitarian Donorship Initiative (2003) Principles and good practice of humanitarian donorship. http://goodhumanitariandonorship.org/Libraries/Ireland_Doc_Manager/EN-23-Principles-and-Good-Practice-of-Humanitarian-Donorship.sflb.ashx

Government of Jordan (2015) Press release: Jordan, UN, sign agreement to launch Jordan Resilience Fund. http://www.jrpsc.org/press-releases/2015/3/30/jordan-un-sign-agreement-to-launch-jordan-resilience-fund

Haken N et al (2014) Fragile states index 2014 report. The Fund for Peace Publication, Washington, DC. http://www.library.fundforpeace.org/library/cfsir1423-fragilestatesindex2014-06d.pdf

IASC (2014a) Humanitarian needs overview: guidance. https://www.humanitarianresponse.info/en/system/files/documents/files/HNO%20Guidance%202015_Final.pdf

IASC (2014b) Strategic response plan: guidance. http://www.humanitarianresponse.info/en/system/files/documents/files/SRP%20Guidance%202015%20final_2.pdf

Intergovernmental Panel on Climate Change (IPCC) (2014) Climate change 2014: impacts, adaptation, and vulnerability. Working group II contribution to the 5th assessment report of the IPCC. IPCC, Geneva

International Dialogue on Peacebuilding and Statebuilding (2011) A new deal for engagement in fragile states. http://www.pbsbdialogue.org/documentupload/49151944.pdf

Kelley C, Cane M, Kushnir Y (2015) Climate change in the fertile crescent and the implications of the recent Syrian drought. PNAS 112(11):3241–3246

Least Developed Countries Expert Group (2012) National adaptation plans. Technical guidelines for the national adaptation plan process. UNFCCC. http://unfccc.int/files/adaptation/cancun_adaptation_framework/application/pdf/naptechguidelines_eng_high__res.pdf
Lukatela A (2012) Gender and post-conflict governance: understanding the challenges. In: UN women sourcebook on women, peace and security. UN Women, New York
Mitchell A (2013) Risk and resilience: from good idea to good practice. OECD Development Co-operation Working Papers, No 13. OECD Publishing. http://www.oecd-ilibrary.org/docserver/download/5k3ttg4cxcbp.pdf?expires=1432251354&id=id&accname=guest&checksum=E2D53A6A37BA2579D4BDE852E4EF877C
OECD (2012) States of fragility 2013: resource flows and trends in a shifting world. OECD, Paris
OECD (2014a) States of fragility 2014: domestic revenue mobilization in fragile states. OECD, Paris
OECD (2014b) Global outlook on aid. OECD, Paris
OECD, UNDP (2014) Making development co-operation more effective: 2014 progress report. OECD. http://www.oecd-ilibrary.org/docserver/download/4314021e.pdf?expires=1432250169&id=id&accname=guest&checksum=C2FDE5DAFB193CB5589EBA26BF14E443
OECD (2015a) States of fragility 2015: meeting post-2015 ambitions. OECD, Paris
OECD (2015b). http://www.oecd.org/development/aid-to-developing-countries-rebounds-in-2013-to-reach-an-all-time-high.htm. Accessed 26 Jan 2015
Oppenheimer M, Campos M, Warren R, Birkmann J, Luber G, O'Neill B, Takahashi K (2014) Emergent risks and key vulnerabilities. In: Climate change 2014: impacts, adaptation, and vulnerability. Part A: Global and sectoral aspects. Contribution of working group II to the 5th assessment report of the intergovernmental panel on climate change
Organisation for Economic Co-operation and Development (OECD). Creditor Reporting System (CRS) on OECD.Stat Extracts page. OECD website. http://stats.oecd.org/Index.aspx?datasetcode=CRS1
Steets J et al (2011) Donor strategies for addressing the transition gap and linking humanitarian and development assistance: a contribution to the international debate. Global Public Policy Institute (GPPI), Berlin
UN OCHA (2015) Financing tracking service website: tracking global humanitarian aid flows. http://fts.unocha.org. Accessed 26 Jan 2015
UNDP, KPMG (2015) Feasibility assessment for the Angola National Climate Fund. Angola
UNDP Multi-Partner Trust Fund Office Gateway (2013). http://www.mptf.undp.org
UNEP, UN Women, PBSO, UNDP (2013) Women and natural resources. Unlocking the peacebuilding potential. Nairobi, New York. http://www.google.com/url?sa=t&rct=j&q=&esrc=s&source=web&cd=1&cad=rja&uact=8&ved=0CB8QFjAA&url=http%3A%2F%2Fpostconflict.unep.ch%2Fpublications%2FUNEP_UN-Women_PBSO_UNDP_gender_NRM_peacebuilding_report.pdf&ei=4uVwVZ_UGu2KsQTX1oHgCA&usg=AFQjCNFnJhQqTdG5RRJ2LJ0sMP-DnPYkjg&sig2=ADCRF0bZPxigGTU4t1PZrg&bvm=bv.95039771,d.cWc
United Nations (2010) Report of the secretary-general's high-level advisory group on climate change financing. http://www.un.org/wcm/webdav/site/climatechange/shared/Documents/AGF_reports/AGF_Final_Report.pdf
United Nations Development Programme (2011) Nepal-climate public expenditure and institutional review. Katmandu. http://www.climatefinance-developmenteffectiveness.org/archive/images/stories/Nepal_CPEIR_Report_2011.pdf
United Nations Environmental Programme (2014) The adaptation gap report 2014. United Nations Environment Programme (UNEP), Nairobi
United Nations Framework Convention on Climate Change (2011) Report of the conference of the parties on its sixteenth session, held at Cancun from 29 November to 10 December 2010, Addendum, Part Two: Action taken by the Conference of the Parties
United Nations Interagency Framework for Preventative Action (2012) Land and conflict. http://www.un.org/en/events/environmentconflictday/pdf/GN_Land_Consultation.pdf

Walter BF (2010) Conflict relapse and the sustainability of post-conflict peace. A background paper for the 2011 world development report. University of California, San Diego. http://www.google.com/url?sa=t&rct=j&q=&esrc=s&source=web&cd=1&cad=rja&uact=8&ved=0CB8QFjAA&url=http%3A%2F%2Fweb.worldbank.org%2Farchive%2Fwebsite01306%2Fweb%2Fpdf%2Fwdr%2520background%2520paper_walter_0.pdf&ei=9OhwVbG3OMjIsQSUlIOgBg&usg=AFQjCNFslmNNbH10vgaQHFWHwnkeM6Ye9g&sig2=mzBaFJwvRX0KWP2c-G_ACw&bvm=bv.95039771,d.cWc

Wheeler D (2011) Quantifying vulnerability to climate change: implications for adaptation assistance. Working Paper 204. Center for Global Development, Washington, DC. http://www.cgdev.org/files/1424759_file_Wheeler_Quantifying_Vulnerability_FINAL.pdf

World Bank website. Concessional finance and global partnerships page: all trust fund programs. http://web.worldbank.org/WBSITE/EXTERNAL/EXTABOUTUS/ORGANIZATION/CFPEXT/0,,contentMDK:20143620~menuPK:64060225~pagePK:64060249~piPK:64060294~theSitePK:299948~isCURL:Y,00.html

World Bank. Financial intermediary funds: meeting global development challenges through international partnerships. World Bank, Washington, DC

Chapter 25
Evaluating Differences in Barriers to Climate Change Adaptation Between the Poor and Nonpoor in Coastal Tanzania

Frederick Ato Armah, Isaac Luginaah, Herbert Hambati, Ratana Chuenpagdee, and Gwyn Campbell

Abstract Understanding context-specific barriers to effective climate change adaptation encountered by individuals is necessary to elicit a nuanced understanding of adaptation processes in order to support decision-making. Recent scholarship highlights poverty as a critical element in barriers to climate change adaptation in developing countries. However, even within developing countries marked heterogeneities in poverty and barriers to climate change adaptation exist. As part of a larger study in countries along the Indian Ocean coastline, this paper examines the gap in barriers to climate change adaptation gap between the poor and nonpoor.

A nationally-representative cross sectional survey of 1253 individuals (606 males and 647 females) was carried out in Coastal Tanzania and four counterfactual decomposition techniques were used to analyse the primary data. Differentials in climate change adaptation barriers are predominantly due to group differences in the *magnitudes* of the determinants (differences in group character-

F.A. Armah (✉)
Environmental Health and Hazards Laboratory, Department of Geography, University of Western Ontario, 1151 Richmond Street, London, ON, Canada N6A 5C2
e-mail: farmah@uwo.ca

I. Luginaah
Department of Geography, University of Western Ontario, 1151 Richmond Street, London, ON, Canada N6A 5C2
e-mail: iluginaa@uwo.ca

H. Hambati
Department of Geography, University of Dar es Salaam, Dar es Salaam, Tanzania
e-mail: hhambati@udsm.ac.tz

R. Chuenpagdee
Department of Geography, Memorial University of Newfoundland, St. John's, NL Canada A1B 3X9
e-mail: ratanac@mun.ca

G. Campbell
Indian Ocean World Centre (IOWC), Peterson Hall, 3460 McTavish Street, Room 100, Montréal, QC H3A 1X9, Canada
e-mail: Gwyn.Campbell@mcgill.ca

W. Leal Filho (ed.), *Innovation in Climate Change Adaptation*, Climate Change Management, DOI 10.1007/978-3-319-25814-0_25

istics) rather than differences in the *effects* of the determinants (estimated coefficients). Self-rated ability to handle personal pressure and unexpected difficulties accounted for the largest share of contribution to the overall explained gap in the barrier to climate change adaptation between the poor and non-poor, suggesting that climate change adaptation differentials between the poor and non-poor in coastal Tanzania are likely due to psychosocial factors.

Keywords Barriers • Climate change • Poverty • Adaptation • Differentials, coast • Tanzania

Introduction

Understanding the relationship between climate change, the human responses it necessitates, and how cognition and emotion (as potential personal barriers) shape such responses, is an increasingly urgent research and policy need. Not only have existing belief and knowledge systems, values, and norms affected how residents responded to environmental challenges in the past (Agrawal 2010), they are also the fundamental mediating mechanisms that will translate the impact of external interventions to facilitate adaptation to climate change (Adger et al. 2009; Agrawal 2010). Historical experience and knowledge about adaptation possibilities is critical to future policy formulations regarding adaptation (Füssel and Klein 2006; Tompkins and Adger 2005).

This paper directs attention towards a subset of such relationships, focusing on personal barriers and poor populations in the context of climate variability and change. The term *barriers* refer to the obstacles that hinder the planning and implementation of climate change adaptation (Biesbroek et al. 2013; Eisenack et al. 2014). Adaptation means anticipating the adverse effects of climate change and taking appropriate action to prevent or minimize the damage they can cause, or taking advantage of opportunities that may arise.

It is critically important to understand the role of personal barriers in shaping adaptation, especially the role of poverty, in order to address the challenges of the most vulnerable. Adaptation to climate change is highly local, and its effectiveness depends on local and extra-local initiatives through which incentives for individual and collective action are structured (Agrawal 2010; Burton et al. 2005; Rojas Blanco 2006). Future efforts to address climate change and design strategic initiatives to enhance the poor's adaptive capacity can, therefore, profitably examine personal adaptive responses, their socio-cultural context and correlates, and the role of poverty in facilitating or encumbering adaptation (Agrawal 2010).

Although the relationship between poverty and adaptation to climate change (and its associated barriers) is rather complex (Naser 2011; Thornton and Herrero 2008), it is frequently suggested that poverty translates into vulnerability, and by extension into weak adaptive capacity. For instance, it is widely accepted that wealthy nations are better prepared to bear the costs of adaptation to climate change impacts and risks than poorer nations (Parry 2007). It is also recognized that poverty is directly related to vulnerability (Fothergill and Peek 2004; Lwasa 2010). Poverty should not be considered synonymous with vulnerability; it is a surrogate of the ability to cope (Dow 1992). There appears to be sufficient evidence that poorer nations and disadvantaged groups within nations are especially vulnerable to disasters (Munasinghe 2000; Parry 2007). However, this view which is rather widespread in the burgeoning literature on climate change adaptation has been critically interrogated. Magnan et al. (2009), for example, identify two underlying biases of this notion. In the first place, by considering the poor as being intrinsically unable to cope with climate variability and change (higher levels of barriers to adaptation) induces the risk of obscuring true, specific, and potentially replicable adaptive capacities. Also, equating poverty with low adaptive capacity leads to the conviction that the rich are presupposed to have high adaptive capacities. Other research scholars have critiqued this presupposition by noting the relationship is non-linear and a complex nexus (see Alwang et al. 2001; Cafiero and Vakis 2006; Naser 2011; Teller and Hailemariam 2011).

This dichotomy in the literature reflects partial understanding of the nexus between poverty and adaptation to climate change, which is a rather complex relationship. For this reason, disparities in climate change adaptation between the poor and nonpoor will continue to engage the attention of both research scholars and policy makers. Within such research milieu, it is imperative to decompose disparity in adaptation to climate change into contributing factors with a view to explaining its distribution by a set of factors that vary systematically with socioeconomic status given that even within poor populations heterogeneities exist. This is a fundamental motivation for this study. In particular, variations in barriers to climate change adaptation may be explained by variations in education, income, insurance coverage, distance to health facilities, and quality of care at local facilities. Even if policy makers have managed to mitigate inequalities in some of these dimensions, inequalities between the poor and nonpoor may remain in others (O'Donnell et al. 2008). The decomposition methods used in this study will potentially reveal how far inequalities in barriers to adaptation to climate change can be explained by inequalities in, say, education rather than inequalities in, say, distance to health facilities.

We disaggregated existing disparities with the aim of eliciting a deeper understanding of the specific factors that account for the climate change adaptation outcomes gap between the poor and nonpoor in coastal Tanzania. This approach is novel for two reasons. First, it focuses predominantly on personal barriers, which have, until now, received negligible research and policy attention. Secondly, this study is one of the first to apply counterfactual decomposition techniques to barriers to climate change adaptation.

Theoretical Context

The capacity to adapt is dynamic and is influenced by a society's productive base, including natural and man-made capital assets, social networks and entitlements, human capital and institutions, governance, national income, health, and technology (Biesbroek et al. 2013; Moser and Ekstrom 2010; Parry 2007). It is also affected by multiple climate and non-climate stresses, as well as development policy. Many adaptation actions have multiple drivers, such as economic development and poverty alleviation and are embedded within broader developmental, sectoral, regional and local planning initiatives, such as water resources planning, coastal defense, and disaster risk reduction strategies (Dovers and Hezri 2010; Moser 2012; Parry 2007).

Adaptation to climate change together with its associated barriers is meaningless unless it is contextualized. In particular, clarity is required to identify whether it is individuals, households, communities, community sector organizations and/or local, state and federal governments that serve them that face barriers to effective adaptation (Adger et al. 2009; Dovers and Hezri 2010). This is important due to the significant heterogeneity in adaptive capacities of individuals even within the same community. Adaptation takes place in a social, political, and institutional context (Adger et al. 2009). It is not enough to consider the adaptation measures of individuals, households, and communities. It is imperative to take into consideration the broader social and political contexts in which local people strive to adapt to changing circumstances and to address barriers to adaptation (Adger et al. 2009; Eisenack et al. 2014). This will have significant implications for the way adaptation responses are framed and enacted (Biesbroek et al. 2013; Eisenack et al. 2014; Moser and Ekstrom 2012). Another important consideration is the distinction between adaptation as a climate change *response* and adaptation as climate change *readiness*.

Effective adaptation can be impeded by one type of barrier or as a result of multiple barriers interacting (Biesbroek et al. 2013; Moser and Ekstrom 2011). Potential barriers to effective adaptation take many forms, including market failures, policy and regulatory barriers, governance and institutional barriers, and behavioural barriers (Moser and Ekstrom 2012). These are the dominant categorizations of barriers to adaptation (see Biesbroek et al. 2013; Eisenack et al. 2014; Ford et al. 2011; Leary et al. 2008; Jantarasami et al. 2010; Moser and Ekstrom 2010; Naess et al. 2005). It is of concern that the narrow categorization of barriers as 'market failures', 'regulatory barriers', 'behavioural and cultural barriers' and 'organizational barriers' does not give sufficient prominence to the structural barriers to adaptation facing many disadvantaged individuals in sub-Saharan Africa. Poverty and constrained choices due to the lack of resources and appropriate information are masked when barriers are articulated in the terms used in the foregoing literature on institutional barriers.

A greater focus on socioeconomic disadvantage and social exclusion as barriers to effective adaptation is needed (Hedger et al. 2008). Also, attention to personal (psychosocial, cognitive and emotive) barriers is required. Studies have shown that the behaviour and attitudes of family members and friends can have a strong impact on the decisions and actions of individuals (Gifford 2011; Patt and Schröter 2008). For example, Ajzen and Fishbein (2005) found that individuals have difficulty maintaining an attitude that differs from that of those around them. Further, the way in which people process information is strongly influenced by existing attitudes (Gardner et al. 2009). People tend to ignore or not seek out information that is inconsistent with their current views, and additional information often tends to reinforce their pre-existing views (Kahneman 2011). The preceding theoretical constructs were used to explain the adaptive capacities of poor versus nonpoor in coastal Tanzania.

Materials and Method

Study Area

Tanzania is a coastal country lying between longitude 29° and 49° East and latitude 1° and 12° south of the Equator (Francis and Bryceson 2001). The marine waters comprise 64,000 km^2 as territorial waters and 223,000 km^2 as offshore waters (EEZ) (Mngulwi 2003). Tanzania's coastline stretches for 800 km. It has five coastal regions-Tanga, Pwani, Dar-es-Salaam, Lindi and Mtwara. The five coastal regions cover about 15 % of the country's total land area and are home to approximately 25 % of the country's population. According to the 2012 Population and Housing census, the total population was 44,928,923 compared to 12,313,469 in 1967 (National Bureau of Statistics 2013), reflecting an annual growth rate of 2.9 %. The under 15 age group represented 44.1 % of the population, with 35.5 % being in the 15–35 age group, 52.2 % being in the 15–64 age group, and 3.8 % being older than 64 (National Bureau of Statistics 2013).

Overall Tanzania on average is sparsely populated with population density of 51 persons per km^2, lower significant variation exists across regions. The population density varies from 1 person per km^2 in arid regions to 51 per km^2 in the mainland's well-watered highlands to 134 per km^2 in Zanzibar (United Republic of Tanzania 2013). The population density for the Dar es Salaam region is 3133 persons per km^2 (the most densely populated) and that of Lindi is only 13.1 persons per km^2 (National Bureau of Statistics 2013). This suggests wide disparities in population density across regions. This study specifically focuses on Dar-es-Salaam, Pwani and Tanga. The three coastal regions selected for analysis were chosen for two main reasons. First, the three regions are of historical significance to the Indian Ocean World project. Second, these regions were selected because of the five regions in the coastal zone, they are the most ethnically diverse (that is, representative of the different geographical locations) and thus, had better prospects of providing heterogeneous survey responses. Dar es Salaam is the capital of the

Dar es Salaam Region, which is one of Tanzania's 26 administrative regions. The Dar es Salaam Region consists of three local government areas or administrative districts: Kinondoni to the north, Ilala in the center of the region, and Temeke to the south. Pwani (coast) is the 21st most densely populated region. It is bordered to the north by the Tanga Region, to the east by the Dar es Salaam Region and the Indian Ocean, to the south by the Lindi Region, and to the west by the Morogoro Region. Tanga region has a population of 2,045,205 (United Republic of Tanzania 2013). It is bordered by Kenya and Kilimanjaro Region to the north; Manyara Region to the west; and Morogoro and Pwani regions to the south. Its eastern border is formed by the Indian Ocean.

Data Collection

The study design was approved by the Committee of Research Ethics of the University of Western Ontario, Canada. Research approval was also granted by the Commission on Science and Technology (COSTECH) in Tanzania. A cross-sectional survey was conducted with 1253 individuals in three regions (Dar es Salaam, Tanga and Pwani) along the coastline of Tanzania. The data were collected between March and September 2013. The study population included male (606) and female (647) participants between the ages of 18 and 70 years. The study used multistage sampling to obtain representative estimates of the population of residents of the three regions (Armah et al. 2015a). Within each region, a list of villages based on the 2012 Population and Housing Census was divided further into households. Figure 25.1 presents a schematic overview of the site.

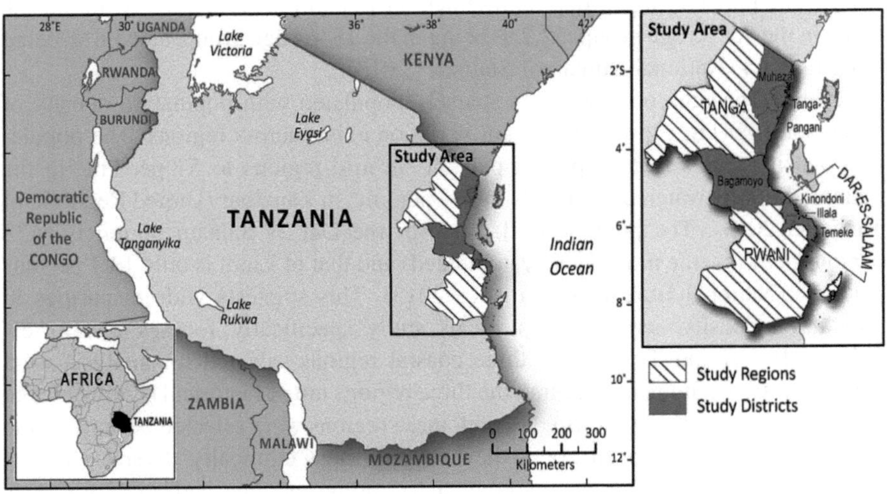

Fig. 25.1 Schematic overview of the study site (*Source*: Armah et al. 2015b)

According to Armah et al. (2015a), the list of villages was divided into clusters ensuring that each cluster would provide adequate numbers of eligible respondents to be included in the survey. This approach both corrects for sampling bias and weights the cases to match census percentages of males and females of various age groups and by ethnicity. The enumeration areas (EAs) and their total number of households were listed geographically by urban and rural areas. Where EAs did not include the minimum number of households, then geographically adjacent EAs were amalgamated to yield sufficient households. This provided the frame for selecting the clusters to be included in the survey according to a stratified systematic sampling technique in which the probability for the selection of any cluster was proportional to its size. A sampling interval was calculated by dividing the total number households by the number of clusters. A random number between one and the sampling interval was computer generated. The EA in which the random number fell was identified as the first selected cluster. The sampling interval was applied to that number and then progressively until the 20 (urban) and 15 (rural) clusters were identified. These clusters made up the sample for the survey. Households were randomly selected from these clusters for interview.

Measures

Outcome Variable

The literature indicates that complex approaches, such as factor analysis or latent variable analysis are very useful in providing deeper understanding of multi-dimensional constructs. Initially, all respondents were asked whether they experienced a barrier to adaptation to climate change or not. Out of 1253, 1130 responded in the affirmative and were further asked to identify specific barriers to adaptation to climate change they had previously experienced. From exploratory analyses of the questions capturing barriers to adaptation to climate change, we retained nine questions, all of which were ordered and were recoded such that higher values indicate a specific barrier. The questions are on a scale of 1–10 (lowest to highest) please indicate your level of agreement with the following: In order to adapt to climate change I don't know what steps to take (knowledge), I lack the skills needed (knowledge), I lack personal energy or motivation (cognitive), I lack the time (personal resources), I lack money or the resources needed (financial resources), I lack help from others (cultural), I feel I don't make a difference (cognitive, emotion), I don't believe in climate change (cognitive, personal values, cultural), and I believe government will protect me (cognitive, institutional). We derived a composite index of barriers to adaptation to climate change through principal component and factor analysis. All factors loaded on a single construct. Cronbach's alpha for the index was 0.789.

The Independent Variables

Previous research have established links between health (both perceived and observed) and adaptation to climate change (see Haines et al. 2006; Kinney 2008; McCarthy 2001; Wolf et al. 2010). Perceived (self-rated or self-reported), which has both emotive and cognitive dimensions, mediates adaptive actions (Costello et al. 2009). Respondents were asked to evaluate their health status, ability to handle work pressure and responsibilities, and ability to handle personal crisis and unexpected responsibilities. Each of these three variables were coded as 1 = poor, 2 = fair, 3 = good, 4 = very good and 5 = excellent. Poverty, a binary variable coded as 1 when the household is below the poverty line, and otherwise as 0, was used to stratify the data. Socio-demographic variables (including age, sex, and marital status, level of education, income, occupation, and ethnicity) that have frequently been shown to be associated with barriers to adaptation to climate change were included as predictors. On the whole, educated individuals are less likely to experience deleterious consequences of climate change and to encounter maladaptation because they supposedly have a better understanding and appreciation for effective adaptation-related matters. Socio-culturally, educated individuals are also less subservient to norms and practices that adversely affect their adaptation choices and adaptive capacity. Residential locality (rural, urban) was also included in the model since the common presumption in the literature is that rural-urban residence distinguishes clearly between poor and good sanitation, housing structure and availability of disaster relief and adaptation resources. In Tanzania, not only are rural populations disadvantaged socio-economically, but they are historically under served in disaster infrastructure and emergency relief personnel. Besides the availability of climate change adaptation infrastructure, urban residents are also more likely than their rural counterparts to flout customs and taboos that could negatively affect adaptation to climate change. Again, Tanzania displays a distinctive regional disparity in development with roots in colonial development policy.

Counterfactual Decomposition Techniques

Although this method has been most widely used in economics to study gender- and race-based discrimination in the labor market, it can be applied to explain differences in any continuous outcome across any two groups. The counterfactual decomposition method used in this study, known as the Blinder-Oaxaca decompositions (Oaxaca 1973; Jann 2008), explains the gap in the means of climate change adaptation barrier scores between two groups (in this instance, between the poor and the nonpoor/better-off) in coastal Tanzania. O'Donnell et al. (2008) gives a comprehensive account on the technique. The gap is decomposed into that part that is due to group differences in the magnitudes of the determinants of barrier to climate change adaptation scores, on the one hand, and group differences in the effects of these determinants, on the other hand. For example, residents in coastal Tanzania may be less adaptive not only because they have less access to piped water but also because they are less knowledgeable about how to obtain the maximum health benefits from piped water (see Jalan and Ravallion 2003; Wagstaff and Nguyen 2003).

Barrier to climate change adaptation scores (y_i) is our outcome variable of interest. We have two groups, which we shall call the poor and the nonpoor. We assume climate change adaptation barrier score is explained by a vector of determinants, x, according to a regression model:

$$y_i = \begin{cases} \beta^{poor} x_i + \varepsilon_i^{poor} \, if \; poor \\ \beta^{nonpoor} x_i + \varepsilon_i^{nonpoor} \, if \; nonpoor \end{cases} \tag{25.1}$$

Where the vectors of β parameters include intercepts. The nonpoor are assumed to have a more advantageous regression line (lower scores on barrier to climate change adaptation) than the poor. Also, the nonpoor are assumed to have a higher mean of x. We assume exogeneity, thus the conditional expectations of the error terms in Eq. (25.1) are zero. The gap in mean barrier to adaptation scores between the poor (y^{poor}) and nonpoor ($y^{nonpoor}$) is given by:

$$y^{poor} - y^{nonpoor} = \beta^{poor} x^{poor} - \beta^{nonpoor} x^{nonpoor} \tag{25.2}$$

Where x^{poor} and $x^{nonpoor}$ are vectors of the independent variables evaluated at the means for the poor and nonpoor, respectively. For our set of independent variables, we write the following:

$$\begin{aligned} y^{poor} - y^{nonpoor} &= (\beta_0{}^{poor} - \beta_0{}^{nonpoor}) + (\beta_1{}^{poor} x_1{}^{poor} - \beta_1{}^{nonpoor} x_1{}^{nonpoor}) \\ &\quad + (\beta_2{}^{poor} x_2{}^{poor} - \beta_2{}^{nonpoor} x_2{}^{nonpoor}) \ldots \\ &\quad + \ldots (\beta_n{}^{poor} x_n{}^{poor} - \beta_n{}^{nonpoor} x_n{}^{nonpoor}) \\ &= G_0 + G_1 + G_2 \ldots + \ldots G_n \end{aligned} \tag{25.3}$$

so that the gap in adaptation barrier scores between the poor and the nonpoor can be thought of as being due in part to (i) differences in the intercepts (G_0), (ii) differences in x_1 and β_1 (G_1), and (iii) differences in x_2 and β_2 (G_2). For example, G_1 might measure the part of the gap in mean score of barrier to climate change adaptation (y) due to differences in educational attainment (x_1) and the effects of educational attainment (β_1), and G_2 might measure the part of the gap due to the gap in age of respondents (x_2) and differences in the effects of age of respondents (β_2). Estimates of the difference in the gap in mean adaptation score can be obtained by substituting sample means of the x's and estimates of the parameters β's into Eq. (25.2).

We further estimated how much of the overall gap or the gap specific to any one of the x's (e.g., G_1 or G_2) is attributable to (i) differences in the x's (sometimes called the explained component) rather than (ii) differences in the β's (sometimes called the unexplained component). In doing so, two options were considered. In the first, the differences in the x's were weighted by the coefficients of the poor group and the differences in the coefficients were weighted by the x's of the nonpoor group, whereas in the second the differences in the x's were weighted by the coefficients of the nonpoor group and the differences in the coefficients were weighted by the x's of the poor group. Either way, we had a way of partitioning the gap in outcomes between the poor and nonpoor into a part attributable to the fact that the poor have worse x's than the nonpoor, and a part attributable to the fact that *ex hypothesi* they have worse β's than the nonpoor. These formulations are expressed as follows:

$$y^{poor} - y^{nonpoor} = \Delta x \beta^{\mathrm{poor}} + \Delta \beta x^{\mathrm{poor}} + \Delta x \Delta \beta = \mathrm{E} + \mathrm{C} + \mathrm{CE} \tag{25.4}$$

From Eq. (25.4), the gap in mean score of barrier to climate change adaptation can be thought of as deriving from a gap in x's or endowments (E), a gap in β's or coefficients (C), and a gap arising from the interaction of endowments and coefficients (CE). So, in effect, Eq. (25.5) places the interaction in the unexplained part, whereas the Eq. (25.6) places it in the explained part.

$$y^{poor} - y^{nonpoor} = \Delta x \beta^{\mathrm{poor}} + \Delta \beta x^{\mathrm{nonpoor}} = \mathrm{E} + (\mathrm{CE} + \mathrm{C}) \tag{25.5}$$

$$y^{poor} - y^{nonpoor} = \Delta x \beta^{\mathrm{nonpoor}} + \Delta \beta x^{\mathrm{poor}} = (\mathrm{E} + \mathrm{CE}) + \mathrm{C} \tag{25.6}$$

We also write Oaxaca's decomposition as a unique case of another equation

$$\begin{aligned} y^{poor} - y^{nonpoor} = {} & \Delta x [D \beta^{poor} + (I - D) \beta^{nonpoor}] \\ & + \Delta \beta [x^{poor} (I - D) + x^{nonpoor} D] \end{aligned} \tag{25.7}$$

Where, I is the identity matrix and D a matrix of weights. In the simple case, where x is a scalar rather than a vector, I, is equal to one and D is a weight. In this case, $\mathrm{D} = 0$ in Eq. (25.5), and $\mathrm{D} = 1$ in Eq. (25.6).

In addition to the above formulations, we consider three more formulations. Cotton (1988) suggested weighting the differences in the x's by the mean of the coefficient vectors, which yields

$$\mathrm{diag(D)} = 0.5(\mathrm{Cotton}) \tag{25.8}$$

Where diag (D) is the diagonal of D. Reimers (1983) suggested weighting the coefficient vectors by the proportions in the two groups, so that if f_{NP} is the sample fraction in the nonpoor group, we obtain

$$\mathrm{diag(D)} = f_{\mathrm{NP}}(\mathrm{Reimers}) \tag{25.9}$$

Finally, we include the decomposition proposed by Neumark (1988), which makes use of the coefficients obtained from the pooled data regression, β^{P}:

$$y^{poor} - y^{nonpoor} = \Delta x\beta^{P} + [x^{poor}(\beta^{poor} - \beta^{p}) + x^{nonpoor}(\beta^{p} - \beta^{nonpoor})](\text{Neumark}) \quad (25.10)$$

The foregoing equations were implemented in STATA 13SE software. The detailed Blinder-Oaxaca decomposition of wage differentials is not invariant to the choice of reference group when a set of dummy variables is used. If we use dummy variable(s) as predictors, as in this study, then the detailed coefficients effect attributed to individual variables is not invariant to the choice of left-out group (s). This invariance or identification problem is well documented in the literature. The "normalized" regression equation where the estimate is simply the average of three sets of estimates with varying reference groups has been proposed to address this problem. The oaxaca.ado and mvdcmp.ado file in STATA 13 (StataCorp, College Station, TX, USA) SE was operationalized to address this issue.

Results

In this section, we present results on the characteristics of the participants and the findings based on the four counterfactual decomposition techniques. In particular, we show that while studies after studies on adaptation to climate change suggest large and statistically significant returns to education, only a small fraction of differences in barriers to adaptation can be accounted for by changes or differences in educational achievement between the poor and the nonpoor.

Sample Characteristics

Non parametric Pearson's chi-square test for independence of the two categorical distributions (poor versus nonpoor) was calculated, using the observed frequencies of the background characteristics of the respondents as the expected frequencies against which to compare the frequencies of income poverty. The chi-square statistic reported for variables firmly rejects the hypothesis that respondents' background characteristics and income poverty categories are independent (Table 25.1). The total number of respondents in Table 25.1 is 1253. Although, the chi-square statistic shows a significant relationship, Cramer's V statistic values are less than 0.3, indicating that the association between the background characteristics of respondents and poverty is not strong. The exceptions are poverty and occupation (0.44), poverty and education (0.34), and poverty and residential locality (0.47). Only 2 % of respondents in the poor category rated their health status as excellent. None of the respondents in the poor category rated their ability to handle personal pressure and unexpected difficulties as excellent. Interestingly, not more than 2 %

Table 25.1 Background characteristics of respondents by income poverty status

Variables	Nonpoor (%)	Poor (%)	Pearson's χ^2 (*df*)
Sex			χ^2 (1) = 33.1199 Pr = 0.000
Male	82.5	17.5	Cramer's V = 0.16
Female	68.5	31.5	
Self-rated health status			χ^2(1) = 11.6577 Pr = 0.001
Poor	65.5	34.5	Cramer's V = −0.09
Good	77.0	23.0	
Self-rated ability to handle work pressure and responsibilities			χ^2(1) = 6.4817 Pr = 0.011 Cramer's V = −0.07
Poor	71.5	28.5	
Good	77.8	22.2	
Self-rated ability to handle personal pressure and unexpected difficulties			χ^2(1) = 5.8781 Pr = 0.015 Cramer's V = −0.07
Poor	59.3	40.7	
Good	76.8	23.2	
Age			χ^2(3) = 48.5027 Pr = 0.000
18–35	80.4	19.6	Cramer's V = 0.20
36–50	78.1	21.9	
51–65	72.0	28.0	
More than 65	45.8	54.2	
Marital status			χ^2(1) = 3.6375 Pr = 0.056
Unmarried	72.1	27.9	Cramer's V = −0.05
Married	77.0	23.0	
Ethnicity			χ^2(2) = 15.3462 Pr = 0.000
Zaramo	66.9	33.1	Cramer's V = 0.11
Sambaa	70.2	29.8	
Others	78.3	21.7	
Religion			χ^2(2) = 71.5999 Pr = 0.000
Christian	89.3	10.7	Cramer's V = 0.24
Muslim	67.6	32.4	
Traditional religion	75.0	25.0	
Employment status			χ^2(1) = 57.6081 Pr = 0.000
Unemployed	42.4	57.6	Cramer's V = −0.21
Employed	77.8	22.2	
Educational attainment			χ^2(3) = 279.4208 Pr = 0.000
No Education	27.1	72.9	Cramer's V = 0.34
Primary	65.4	34.6	
Secondary	91.8	8.2	
Tertiary	98.6	1.4	
Residential locality			χ^2(1) = 146.4910 Pr = 0.000
Rural	75.2	25.8	Cramer's V = 0.47
Urban	87.5	12.5	

of public servants (government worker) and civil servants (NGO staff) were in the poor category.

As observed in Table 25.2, several significant relationships (both direct and inverse) exist between the explanatory variables. However, most of them are weak except between self-rated ability to handle personal pressure and unexpected difficulties and self-rated ability to handle work pressure and responsibilities ($r = 0.62$ $p < 0.001$).

Table 25.3 reports the mean values of y (barriers to climate change adaptation scores) for the two groups, and the difference between them. It then shows the contribution attributable to the gaps in endowments (E), the coefficients (C), and the interaction (CE). Table 25.3 is the threefold Blinder-Oaxaca decomposition of the mean outcome difference. The endowments term represents the contribution of differences in explanatory variables across groups (poor and nonpoor), and the coefficients term is the part that is due to group differences in the estimated coefficients. The interaction term accounts for the fact that cross-group differences in explanatory variables and coefficients can occur at the same time. In this study, the gap in endowments accounts for the great bulk of the gap in outcomes (barriers to climate change adaptation scores).

Table 25.4 shows how the explained and unexplained portions of the gap in climate change adaptation vary depending on the decomposition used. The first and second columns correspond to the Oaxaca decomposition in Eqs. (25.5) and (25.6), where $D = 0$ and $D = I$, respectively (supplementary material). The third and fourth columns correspond to Cotton's and Reimers' decompositions, where the diagonal of D equals 0.5 and $f_{NP} = 0.749$ (in our case), respectively. The final column labeled "*" is Neumark's decomposition. Several variations of computing counterfactual do not change the main results qualitatively. Whatever decomposition is used, it is obviously the difference in the mean values of the x's (explained component) that accounts for the vast majority of the difference in climate change adaptation between poor and nonpoor residents in coastal Tanzania. The only exception is the Oaxaca decomposition where $D = I$ in which case the differences in the effects of the determinants (coefficients or unexplained component) rather accounts for the main difference in climate change adaptation scores between poor and nonpoor residents in coastal Tanzania. By and large, however, differences in the effects of the determinants play a tiny part in explaining inequalities in climate change adaptation between the two groups.

Based on Oaxaca's decomposition $D = 0$, differences in the mean values of x's (gaps in endowments) account for about 127 % of the differentials in barriers to climate change adaptation between the poor and nonpoor. Based on Cotton's decomposition, differences in the mean values of x's (gaps in endowments) explain about 82 % of the differentials in barriers to climate change adaptation between the poor and nonpoor. About 59 and 63 % of the differentials in barriers to climate change adaptation between the poor and nonpoor in coastal Tanzania is explained by the mean values of x's (gaps in endowments) using the Reimer's and Neumark's decompositions, respectively. Only, about 36 % of the differentials in barriers to climate change adaptation between the poor and nonpoor are explained by the

Table 25.2 Correlation coefficients of explanatory variables and poverty

	SRHS	SRWP	SRPD	Age	Marital	Ethnic	Religion	Occup	Locality	Educ	Poverty
SRHS	1										
SRWP	0.38^{***}	1									
SRPD	0.35^{***}	0.62^{***}	1								
Age	-0.28^{***}	-0.12^{***}	-0.07^{*}	1							
Marital	-0.23^{***}	-0.07^{*}	-0.11^{**}	0.47^{***}	1						
Ethnic	0.13^{***}	0.09^{**}	0.13^{***}	-0.11^{**}	-0.1^{**}	1					
Religion	-0.16^{***}	-0.14^{***}	-0.16^{***}	0.10^{**}	0.12^{***}	-0.36^{***}	1				
Occup	0.06^{*}	0.06^{*}	0.05^{*}	-0.10^{**}	-0.07^{*}	0.06^{*}	−0.04	1			
Locality	-0.18^{***}	-0.19^{***}	-0.19^{***}	0.14^{***}	0.08^{**}	-0.24^{***}	0.33^{***}	-0.14^{***}	1		
Educ	0.32^{***}	0.25^{***}	0.21^{***}	-0.18^{***}	-0.22^{***}	0.20^{***}	-0.37^{***}	0.20^{***}	−0.43	1	
Poverty	-0.18^{***}	-0.09^{**}	-0.14^{***}	0.14^{***}	0.17^{***}	-0.13^{***}	0.24^{***}	−0.04	0.34	−0.45	1

Key: $^{*}p < 0.05$, $^{**}p < 0.01$, $^{***}p < 0.001$; SRHS = Self-rated health status, SRWP = Self-rated ability to handle work pressure and responsibilities, SRPD = Self-rated ability to handle personal pressure and unexpected difficulties, Occup = occupation, locality = residential locality, Educ = educational attainment

Table 25.3 Summary of decomposition results

Summary of decomposition results	
High: poverty = 0.0000	
Low: poverty = 1.0000	
Mean prediction high (H)	0.047
Mean prediction low (L)	−0.141
Raw differentials (R) {H − L}	0.188
Due to endowments (E)	0.239
Due to coefficients (C)	0.121
Due to interaction (CE)	−0.172

Table 25.4 Proportion of explained and unexplained components

D	0	1	0.5	0.749	*
Unexplained (U) {C + (1 − D) CE}	−0.051	0.121	0.035	0.078	0.069
Explained (V) {E + D * CE}	0.239	0.067	0.153	0.111	0.119
% Unexplained {U/R}	−27.3	64.3	18.5	41.3	36.7
% Explained {V/R}	127.3	35.7	81.5	58.7	63.3

Note: D in the 4th column = relative frequency of high group
*Reference = pooled model over both categories

differences in the mean values of x's (gaps in endowments) when Oaxaca's decomposition D = 1 is used. This implies that, when Oaxaca's decomposition D = 1 is used, differences in the effects of the determinants (coefficients or unexplained component) rather accounts for about 64 % of the differentials in barriers to climate change adaptation between the poor and nonpoor in coastal Tanzania.

Table 25.5 affords us the opportunity to observe how far gaps in individual x's contribute to the overall explained gap. For example, focusing on the final column corresponding to Neumark's decomposition, we realize that the gaps in the three demographic variables (i.e., self-rated ability to handle work pressure and responsibilities, age, and ethnicity) actually favor the poor whereas the gaps in the remaining variables all disfavor the poor. Of the latter, it is the gap in Self-rated ability to handle personal pressure and unexpected difficulties that accounts for the bulk of the explained gap. It is not so much the correlates of poverty (poor water and sanitation, low educational levels) that account for climate change adaptation inequalities between poor and nonpoor residents in coastal Tanzania; it is rather a psychosocial problem, in the form of lack of ability to handle stress (personal pressure) and unexpected difficulties.

Table 25.6 provides the coefficient estimates, means, and predictions for each x for each group, the "high group" in this case being the nonpoor and the "low group" being the poor. For the first Oaxaca decomposition (Eq. 25.5), columns 2 and 3 of Table 25.5 allow us to identify how the gap in each of the β's contributes to the overall unexplained gap. For the other decompositions, the contributions of the individual β's can be found by subtracting the explained part given in Table 25.5 from the group difference in the variable specific predictions given in Table 25.6.

Table 25.5 Decomposition results for variables

				D			
Variables	E(D = 0)	C	CE	1	0.5	0.749	*
SRHS	0.083	−0.681	−0.066	0.018	0.051	0.034	0.030
SRWP	−0.094	0.150	0.006	−0.088	−0.091	−0.089	−0.090
SRPD	0.131	−0.576	−0.056	0.076	0.103	0.090	0.086
Age	−0.051	−0.137	0.019	−0.032	−0.042	−0.037	−0.034
Marital	0.060	0.002	−0.000	0.060	0.060	0.060	0.064
Ethnicity	−0.018	0.063	0.013	−0.006	−0.012	−0.009	−0.009
Religion	−0.028	−0.293	0.040	0.012	−0.008	0.002	0.003
Occup	−0.008	0.045	0.011	0.004	−0.002	0.001	0.001
Locality	0.018	0.076	−0.017	0.001	0.009	0.005	0.007
Educ	0.147	−0.119	−0.124	0.023	0.085	0.054	0.060
_cons	0.000	1.591	0.000	0.000	0.000	0.000	0.000
Total	0.239	0.121	−0.172	0.067	0.153	0.111	0.119

Key: SRHS = Self-rated health status, SRWP = Self-rated ability to handle work pressure and responsibilities, SRPD = Self-rated ability to handle personal pressure and unexpected difficulties, Occup = occupation, locality = residential locality, Educ = educational attainment
*Neumark's decomposition

Table 25.6 Coefficients, means and predictions of the models

	High model			Low model			Pooled
Variables	Coefficient	Mean	Predicted	Coefficient	Mean	Predicted	Coefficient
SRHS	0.065	3.108	0.203	0.305	2.835	0.866	0.110
SRWP	−0.774	2.786	−2.155	−0.830	2.673	−2.218	−0.791
SRPD	0.361	2.371	0.856	0.628	2.162	1.357	0.412
Age	0.041	4.779	0.194	0.065	5.567	0.363	0.043
Marital	−0.144	2.117	−0.306	−0.145	2.532	−0.368	−0.154
Ethnicity	−0.006	5.708	−0.035	−0.019	4.754	−0.092	−0.010
Religion	−0.046	1.608	−0.075	0.111	1.863	0.206	−0.012
Occup	0.003	7.498	0.019	−0.005	5.982	−0.030	0.000
Locality	−0.003	1.314	−0.003	−0.047	1.690	−0.080	−0.020
Educ	0.026	1.771	0.045	0.163	0.870	0.142	0.067
_cons	1.305	1.000	1.305	−0.287	1.000	−0.287	1.015
Total			0.047			−0.141	

Key: SRHS = Self-rated health status, SRWP = Self-rated ability to handle work pressure and responsibilities, SRPD = Self-rated ability to handle personal pressure and unexpected difficulties, Occup = occupation, locality = residential locality, Educ = educational attainment

We emphasize that the unimportance overall of the unexplained portion is due to offsetting effects from different β's. The poor have a higher intercept in the decomposition equation, but this is largely offset by the fact that the ability to handle stress and unexpected difficulties is weaker for the poor.

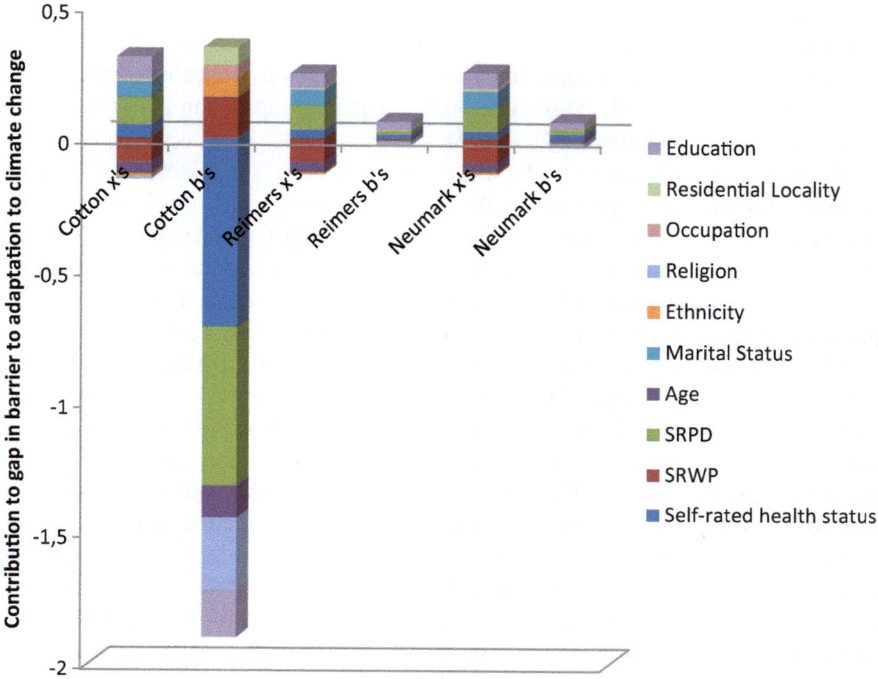

Fig. 25.2 Contributions of differences in means and in coefficients to the gap in barriers scores between the two groups

Figure 25.2 indicates the contribution of the difference in the means of each x and the difference in coefficients on each x. For the Cotton, Reimers and Neumark decompositions, the contributions of the individual x's was obtained by taking the group difference in the variable specific predictions given in Table 25.6 and subtracting the explained part given in Table 25.5 from this. Regarding the means of the x's, Fig. 25.1 suggests that most of the explained part of the climate change adaptation gap is attributable to the gaps in self-rated ability to handle personal pressure and unexpected difficulties and self-rated ability to handle work pressure and responsibilities.

Discussion

This paper set out to decompose the gap in climate change adaptation outcomes into the part that is due to group differences in the magnitudes of the determinants (i.e., the explained component) of barrier to climate change adaptation scores, on the one hand, and group differences in the effects of these determinants (i.e., the unexplained component), on the other hand. In other words, we disaggregated the characteristics effects (explained variation) and coefficients effects (unexplained variation) for the two mutually exclusive groups (poor and nonpoor). This technique is especially useful for identifying and quantifying the separate contributions of group differences in measurable characteristics, such as education, age, marital status, and geographical location, to ethnic and gender gaps (Jann 2008) in climate change adaptation outcomes.

We found that difference in the mean values of the *x*'s (explained component) accounts for the vast majority of the difference in barriers to climate change adaptation between poor and nonpoor residents in coastal Tanzania. To the best of our knowledge, no previous research on climate change adaptation has attempted to decompose disparities in adaptation barriers by poverty status. Thus, it is difficult to compare our findings with other results in the existing literature.

The contribution of each of the variables to the overall explained gap using the various decompositions provides interesting insight on the relative importance of each of the variables. Using the Neumark decomposition, Self-rated ability to handle personal pressure and unexpected difficulties alone accounted for 75 % of the overall explained gap in barriers to climate change adaptation between the poor and nonpoor. This was followed by marital status (53 %), education (50 %) and self-rated health status (25 %) in that order. Occupation, religion and residential locality (rural-urban) jointly accounted for less than 10 % of the overall explained gap. Similar results were obtained using the Reimer's decomposition where self-rated ability to handle personal pressure and unexpected difficulties alone accounted for 81 % of the overall explained gap in barriers to climate change adaptation between the poor and nonpoor. This was followed by marital status (54 %), education (49 %) and self-rated health status (30 %) in that order. These trends change slightly when Cotton's decomposition is used. Although self-rated ability to handle personal pressure and unexpected difficulties still accounts for the largest proportion (67 %) in the overall explained gap, education overtakes marital status as the second largest contributor (56 %) to the explained gap. Marital status (39 %) and self-rated health status (33 %) then follow. In the Cotton's and Reimer's decompositions, occupation, religion and residential locality (rural-urban) still cumulatively accounted for less than 10 % of the overall explained gap. This implies that, although the magnitudes of contribution of each variable differ across decomposition techniques, the trends and order of contribution remains almost the same.

The fact that self-rated ability to handle personal pressure and unexpected difficulties accounted for the largest share of contribution to the overall explained gap in barriers to climate change adaptation between the poor and nonpoor in coastal Tanzania regardless of decomposition technique used suggests that climate adaptation differentials may be due to psychosocial issues rather than poverty *per se*. Across decomposition, the magnitude of self-rated ability to handle personal pressure and unexpected difficulties varies but is not surpassed by other biosocial or socio-cultural variables. This does not mean that biosocial or socio-cultural variables are not important. It simply implies that the issue is multi-faceted (Swim et al. 2009), and though important, the individual contributions of biosocial or socio-cultural variables is lower in magnitude than the psychosocial factor, that is, self-rated ability to handle personal pressure and unexpected difficulties.

This finding highlights the importance of perception and cognition in stimulating or inhibiting adaptive actions of individuals. Human perceptions and judgments about climate change are important because they affect levels of concern and, in turn, the motivation to act (Swim et al. 2009). Adaptation includes a range of coping actions that individuals and communities can take, as well as psychological processes (e.g., appraisals and affective responses) that precede and follow behavioural responses (Swim et al. 2009). Available research suggests that the psychosocial impacts of climate change are likely to be moderated by a number of individual and contextual factors that increase or decrease the severity of the impact, as well as the perception of the impact (Leiserowitz 2007).

In general, cognitive adaptation approaches (Taylor and Stanton 2007) and protection motivation approaches (Weinstein et al. 2000) are premised on the kinds of cognitive and emotional appraisal processes and coping processes, which are elicited in the context of climate change and other risks that contain implicit or explicit threats and induce fear (Fiske and Taylor 2008; Swim et al. 2009). An individual's perceptions of climate change impacts can be moderated by social norms (Leiserowitz 2005) and by their environmental identity (Clayton and Opotow 2003). The impacts of climate change, and by extension adaptive actions are also likely to be mediated by various types of cognitive appraisals, such as estimates of personal risk and attributions of responsibility (Leiserowitz 2007), and media representations of climate change adaptation impacts (Reser and Swim 2011). Emotional reactions are critical components of information processing and also have a direct relation to physical and psychological health (Groopman 2004; Moser 2007). It is also hypothesized that certain strong emotional responses such as fear, despair, or a sense of being overwhelmed or powerless can inhibit thought and action (Moser 2007; Nicholsen 2002), which in turn may either constrain or serve as a barrier to effective adaptation to climate change.

Limitations of the Study

A limitation of this study is that, while decompositions are useful for quantifying the contribution of various factors (psychosocial, biosocial, sociocultural) to a difference or change in barriers to climate change adaptation outcomes, they may not necessarily deepen our understanding of the mechanisms underlying the relationship between these factors and climate change adaptation outcomes. In that sense, decomposition methods do not seek to recover behavioural relationships or deep structural parameters. By indicating which factors are quantitatively important and which are not, however, decompositions provide useful indications of particular hypotheses or explanations to be explored in more detail. For example, if decomposition indicates that differences in educational attainment account for a large fraction of the poverty-climate change adaptation gap, then exploring in more detail how the poor and nonpoor choose their adaptive behaviours is imperative.

Policy Implications

Climate change adaptation is a multi-faceted and complex phenomenon, rooted in an extensive body of interdisciplinary science and with deeply challenging policy implications (e.g., Prins et al. 2010). Given that the empirical evidence presented in this paper indicates that self-rated ability to handle personal pressure and unexpected difficulties, educational attainment and self-rated health status accounts for a large portion of the overall explained gap in barriers to climate change adaptation between the poor and nonpoor, there is need for policy that systematically addresses these gaps in endowments. In developing countries such as Tanzania, government can stimulate policy action to address the gaps in outcomes in two fundamental ways: information through extension services (e.g. community radios), and provision of social support services. In so far as self-rated ability to handle personal pressure and unexpected difficulties was the foremost factor explaining gaps in barrier to climate change adaptation between the poor and nonpoor, it may be that improved psychosocial health would improve climate change adaptation, although the precise mechanism underlying this is unclear. There are many area-specific differences in the propensity of coastal residents to adapt to climate change and further analysis would be required to understand the underlying factors. Adaptation, however, is undertaken only by those who perceive climate change. The perception of climate change appears to hinge on residents experiences and the availability of free advice on social support and services specifically related to climate change adaptation. However, while the policy options for promoting an increased adaptation to climate change are rather limited the perception of climate change is already high in coastal Tanzania. The opinions of residents of coastal Tanzania who perceive climate change as a risk should be taken into consideration with respect to the type, scale and form of adaptation strategies to be initiated across spatio-temporal scales. This is critical to the widespread acceptance or rejection of proposed climate adaptation strategies of individuals.

Conclusion

This study aimed to disaggregate disparities in climate change adaptation outcomes between two mutually exclusive groups (poor and nonpoor) in coastal Tanzania based on characteristics effects (explained variation) and coefficients effects (unexplained variation). Self-rated ability to handle personal pressure and unexpected difficulties accounted for the largest share of contribution to the overall explained gap in barriers to climate change adaptation between the poor and nonpoor in coastal Tanzania regardless of the decomposition technique used. This indicates that climate adaptation differentials between the poor and nonpoor in coastal Tanzania are likely due to psychosocial issues rather than other biosocial and socio-cultural correlates of poverty *per se*. This paper is unique in two critical ways. First, it focused on personal barriers rather than the institutional barriers to climate change adaptation that has received much attention in the extensive body of literature on climate change adaptation. Secondly, it used decomposition techniques hitherto not considered in the climate adaptation research domain. Adaptation to climate change together with its associated barriers is meaningless unless it is contextualized. Specifically, clarity is required to identify whether it is individuals, households, communities, community sector organizations and/or local, state and federal governments that serve those who face barriers to effective climate change adaptation. This is imperative considering the significant heterogeneity in adaptive capacities of individuals even within the same community.

Acknowledgements Many thanks to Karen Van Kerkoerle, of the Cartographic Unit, Department of Geography, University of Western Ontario, Canada for drawing the map of the study areas. We acknowledge research funding from 'the Indian Ocean World: The Making of the First Global Economy in the Context of Human Environment Interaction" project within the framework of Major Collaborative Research Initiative (MCRI). The funders had no role in study design, data collection and analysis, decision to publish, or preparation of the manuscript.

References

Adger WN, Dessai S, Goulden M, Hulme M, Lorenzoni I, Nelson DR, Naess LO, Wolf J, Wreford A (2009) Are there social limits to adaptation to climate change? Clim Change 93 (3–4):335–354

Agrawal A (2010) Local institutions and adaptation to climate change. In: Mearns R, Norton A (eds) Social dimensions of climate change: equity and vulnerability in a warming world. World Bank, Washington, DC, pp 173–198

Ajzen I, Fishbein M (2005) The influence of attitudes on behavior. In: Albarracín D, Johnson BT, Zanna MP (eds) The handbook of attitudes. Erlbaum, Mahwah, NJ, pp 173–221

Alwang J, Siegel PB, Jorgensen SL (2001) Vulnerability: a view from different disciplines, vol 115. Social protection discussion paper series

Armah FA, Luginaah I, Yengoh GT, Hambati H, Chuenpagdee R, Campbell G (2015a) Analyzing the relationship between objective-subjective health status and public perception of climate change as a human health risk in coastal Tanzania. Hum Ecol Risk Assess 21(7):1936–1959. doi:10.1080/10807039.2014.1003203

Armah FA, Quansah R, Luginaah I, Chuenpagdee R, Hambati H, Campbell G (2015b) Historical perspective and risk of multiple neglected tropical diseases in Coastal Tanzania: compositional and contextual determinants of disease risk. PLoS Negl Trop Dis 9(8):e0003939. doi:10.1371/journal.pntd.0003939

Biesbroek GR, Klostermann JE, Termeer CJ, Kabat P (2013) On the nature of barriers to climate change adaptation. Reg Environ Chang 13(5):1119–1129

Burton I, Malone E, Huq S (2005) Scoping and designing an adaptation project. In: Lim B, Spanger-Siegfried E (eds) Adaptation policy frameworks for climate change: developing strategies, policies, and measures. Cambridge University Press, Cambridge, p 258

Cafiero C, Vakis R (2006) Risk and vulnerability considerations in poverty analysis: recent advances and future directions. Social Protection Discussion Paper No. 610. World Bank, Washington, DC. http://documents.worldbank.org/curated/en/2006/10/7141525/risk-vulnerability-considerations-poverty-analysis-recent-advances-future-directions

Clayton S, Opotow S (eds) (2003) Identity and the natural environment. MIT Press, Cambridge, MA

Costello A, Abbas M, Allen A, Ball S, Bell S, Bellamy R, Friel S, Groce N, Johnson A, Kett M, Lee M, Levy C, Maslin M, McCoy D, McGuire B, Montgomery H, Napier D, Pagel C, Patel J, de Oliveira JA, Redclift N, Rees H, Rogger D, Scott J, Stephenson J, Twigg J, Wolff J, Patterson C (2009) Managing the health effects of climate change: lancet and University College London Institute for Global Health Commission. Lancet 373(9676):1693–1733

Cotton J (1988) On the decomposition of wage differentials. Rev Econ Stat 70(2):236–243

Dow K (1992) Exploring differences in our common future (s): the meaning of vulnerability to global environmental change. Geoforum 23(3):417–436

Dovers SR, Hezri AA (2010) Institutions and policy processes: the means to the ends of adaptation. Wiley Interdiscip Rev Clim Chang 1(2):212–231

Eisenack K, Moser SC, Hoffmann E, Klein RJ, Oberlack C, Pechan A, Rotter M, Termeer CJ (2014) Explaining and overcoming barriers to climate change adaptation. Nat Clim Chang 4 (10):867–872

Fiske ST, Taylor SE (2008) Social cognition: from brains to culture. McGraw-Hill, New York

Ford JD, Berrang-Ford L, Paterson J (2011) A systematic review of observed climate change adaptation in developed nations. Clim Change 106(2):327–336

Fothergill A, Peek LA (2004) Poverty and disasters in the United States: a review of recent sociological findings. Nat Hazards 32(1):89–110

Francis J, Bryceson I (2001) Tanzanian coastal and marine resources: some examples illustrating questions of sustainable use. Chapter 4:76–102

Füssel HM, Klein RJ (2006) Climate change vulnerability assessments: an evolution of conceptual thinking. Clim Change 75(3):301–329

Gardner J, Dowd A-M, Mason C, Ashworth, P (2009) A framework for stakeholder engagement on climate adaptation. CSIRO climate adaptation flagship working paper no. 3. http://www.csiro.au/resources/CAFworking-papers.html

Gifford R (2011) The dragons of inaction: psychological barriers that limit climate change mitigation and adaptation. Am Psychol 66(4):290

Groopman J (2004) The anatomy of hope. Random House, New York

Haines A, Kovats RS, Campbell-Lendrum D, Corvalán C (2006) Climate change and human health: impacts, vulnerability and public health. Public Health 120(7):585–596

Hedger M, Greeley M, Leavy J (2008) Evaluating climate change: pro-poor perspectives. IDS Bull 39(4):75–80

Jalan J, Ravallion M (2003) Does piped water reduce diarrhea for children in rural India? J Econ 112(1):153–173

Jann B (2008) A Stata implementation of the Blinder-Oaxaca decomposition. Stata J 8(4):453–479

Jantarasami LC, Lawler JJ, Thomas CW (2010) Institutional barriers to climate change adaptation in US national parks and forests. Ecol Soc 15(4):33

Kahneman D (2011) Thinking, fast and slow. Macmillan, New York

Kinney PL (2008) Climate change, air quality, and human health. Am J Prev Med 35(5):459–467

Leary N, Adejuwon J, Barros V, Burton I, Kulkarni J, Lasco R (eds) (2008) Climate change and adaptation. Earthscan, London

Leiserowitz A (2005) American risk perceptions: is climate change dangerous? Risk Anal 25:1433–1442

Leiserowitz A (2007) Communicating the risks of global warming: American risk perceptions, affective images, and interpretive communities. In: Moser SC, Dilling L (eds) Creating a climate for change. Cambridge University Press, New York, pp 44–63

Lwasa S (2010) Adapting urban areas in Africa to climate change: the case of Kampala. Curr Opin Environ Sustain 2(3):166–171

Magnan A, Garnaud B, Billé R (2009) Adaptive capacity of the poor. In: IOP conference series: earth and environmental science, vol 6, no. 36. IOP Publishing, p 362017

McCarthy JJ (ed) (2001) Climate change 2001: impacts, adaptation, and vulnerability: contribution of working group II to the 3rd assessment report of the intergovernmental panel on climate change. Cambridge University Press, Cambridge

Moser SC (2007) More bad news: the risk of neglecting emotional responses to climate change information. In: Moser SC, Dilling L (eds) Creating a climate for change. Cambridge University Press, New York

Moser SC (2012) Adaptation, mitigation, and their disharmonious discontents: an essay. Clim Change 111(2):165–175

Moser SC, Ekstrom JA (2010) A framework to diagnose barriers to climate change adaptation. Proc Natl Acad Sci 107(51):22026–22031

Moser SC, Ekstrom JA, Susanne Moser Research and Consulting, Stanford University, University of California, Berkeley (2012) Identifying and overcoming barriers to climate change adaptation in San Francisco Bay: results from case studies. California Energy Commission, Sacramento

Mngulwi BS (2003) Country review: United Republic of Tanzania. Review of the state of world marine capture fisheries management: Indian Ocean, 447. http://www.fao.org/3/a-a0477e/a0477e13.htm

Munasinghe M (2000) Development, equity and sustainability in the context of climate change. In: Munasingha M, Swart R (eds) Climate change and its linkages with development, equity and sustainability: Proceedings of the IPCC expert meeting on development, equity and sustainability. Colombo, 27–29 Apr 2000. IPCC and World Meteorological Organization, Geneva, pp 13–66

Naess LO, Bang G, Eriksen S, Vevatne J (2005) Institutional adaptation to climate change: flood responses at the municipal level in Norway. Glob Environ Chang 15(2):125–138

Naser MM (2011) Climate change, environmental degradation, and migration: a complex nexus. William Mary Environ Law Policy Rev 36:713

National Bureau of Statistics (2013) Tanzania in figures 2012. Ministry of Finance, June 2013, p 23. Retrieved July 2014 from http://www.nbs.go.tz/takwimu/references/Tanzania_in_figures2012.pdf

Neumark D (1988) Employers' discriminatory behaviour and the estimation of wage discrimination. J Hum Resour 23:279–295

Nicholsen SW (2002) The love of nature and the end of the world. MIT Press, Cambridge

O'Donnell O, Van Doorslaer E, Wagstaff A, Lindelow M (2008) Analyzing health equity using household survey data. A guide to techniques and their implementation. The International Bank for Reconstruction and Development/The World Bank, Washington, DC

Oaxaca R (1973) Male-female wage differentials in urban labor markets. Int Econ Rev 14 (3):693–709

Parry ML (ed) (2007) Climate change 2007: impacts, adaptation and vulnerability: contribution of working group II to the fourth assessment report of the Intergovernmental Panel on climate change (Vol 4). Cambridge University Press

Patt AG, Schröter D (2008) Perceptions of climate risk in Mozambique: implications for the success of adaptation strategies. Glob Environ Chang 18(3):458–467

Prins G, Galiana I, Green C, Grundmann R, Hulme M, Korhola A, Laird F, Nordhaus T, Rayner S, Sarewitz D, Schellenberger M, Stehr N, Tezuka H (2010) The Hartwell Paper: a new direction for climate policy after the crash of 2009. Institute for Science, Innovation & Society, Oxford

Reimers CW (1983) Labor market discrimination against Hispanic and black men. Rev Econ Stat 65(4):570–579

Reser JP, Swim JK (2011) Adapting to and coping with the threat and impacts of climate change. Am Psychol 66(4):277–289

Rojas Blanco AV (2006) Local initiatives and adaptation to climate change. Disasters 30 (1):140–147

Swim J, Clayton S, Doherty T, Gifford R, Howard G, Reser J, Stern P, Weber E (2009) Psychology and global climate change: addressing a multi-faceted phenomenon and set of challenges. A report by the American Psychological Association's task force on the interface between psychology and global climate change. Retrieved 31 Jan 2014

Taylor SE, Stanton AL (2007) Coping resources, coping processes, and mental health. Annu Rev Clin Psychol 3:377–401

Teller C, Hailemariam A (2011) The complex nexus between population dynamics and development in Sub-Saharan Africa: a new conceptual framework of demographic response and human adaptation to societal and environmental hazards. In: Teller C, Hailemariam A (eds) The demographic transition and development in Africa. Springer, Dordrecht, pp 3–16

Thornton PK, Herrero M (2008) Climate change, vulnerability and livestock keepers: challenges for poverty alleviation. In: Rowlinson P, Steele M, Nefaoui A (eds) Livestock and Global Climate Change: Proceedings of the Livestock and Global Climate Change Conference, Hammamet, 17–20 May 2008. Cambridge, CUP: 21–24

Tompkins EL, Adger WN (2005) Defining response capacity to enhance climate change policy. Environ Sci Policy 8(6):562–571

United Republic of Tanzania (2013) Population distribution by administrative units, p 1. http://ihi.eprints.org/2169/1/Age_Sex_Distribution.pdf. Retrieved 15 June 2014

Wagstaff A, Nguyen NN (2003) Poverty and survival prospects of Vietnamese children under Doi Moi. In: Glewwe P, Agrawal N, Dollar D (eds) Economic growth, poverty and household welfare: policy lessons from Vietnam. World Bank, Washington, DC

Weinstein ND, Lyon JE, Rothman AJ, Cuite CL (2000) Preoccupation and affect as predictors of protection action following natural disaster. Br J Health Psychol 5:351–363

Wolf J, Adger WN, Lorenzoni I, Abrahamson V, Raine R (2010) Social capital, individual responses to heat waves and climate change adaptation: an empirical study of two UK cities. Glob Environ Chang 20(1):44–52

MIX
Papier aus verantwortungsvollen Quellen
Paper from responsible sources
FSC® C105338

If you have any concerns about our products,
you can contact us on
ProductSafety@springernature.com

In case Publisher is established outside the EU,
the EU authorized representative is:
Springer Nature Customer Service Center GmbH
Europaplatz 3, 69115 Heidelberg, Germany

Printed by Libri Plureos GmbH
in Hamburg, Germany